INTERNATIONAL GCSE (9–1)

ALAN SMITH
SOPHIE GOLDIE

Mathematics
for Edexcel Specification A

THIRD EDITION

HODDER EDUCATION
AN HACHETTE UK COMPANY

The Publishers would like to thank the following for permission to reproduce copyright material:

Photo credits: p. 97 © Art of Travel/Alamy Stock Photo

Orders: please contact Bookpoint Ltd, 130 Milton Park, Abingdon, Oxon OX14 4SB. Telephone: (44) 01235 827720. Fax: (44) 01235 400454. Lines are open 9 a.m to 5 p.m., Monday to Saturday, with a 24-hour message answering service. Visit our website at www. hoddereducation.co.uk

Personal Tutor CD: © Sophie Goldie, Alan Smith, 2007, with contributions from Andy Sturman; developed by Infuze Limited; cast: Nicolette Landau; recorded at Alchemy Soho.

Contents

Contents

Contents

Introduction

This book has been written to provide complete coverage of the Higher Tier Pearson Edexcel International GCSE Mathematics Specification A (9–1) for first teaching from September 2016.

The International GCSE written examination consists of two papers, each worth 50% of the overall grade. Candidates may earn a wide range of grades (from 4 through to 9 in the new numbered grading scale), and the papers reflect this with approximately 40% of the marks spread across grades 4 and 5, and the remaining 60% spread evenly across grades 6, 7, 8 and 9.

Students following the Higher International GCSE course may have differing individual target grades, and this is reflected in the ordering of the content. There are two chapters that reach well back into earlier work, forming a bridge with Key Stage 3 material; these are:

>   Chapter 4 (Working with algebra)
>   Chapter 11 (Working with shape and space)

Some of the early work covered in these chapters may be skipped, or used as revision material, as appropriate.

Many of the worked examples in the text are available as multimedia presentations at www.hoddereducation.co.uk/EdexcelIGCSEMaths.

Examination candidates will require a good quality scientific calculator for both papers, but are reminded to show all relevant working too: the examiners may penalise candidates who omit working even though the final answer is correct.

There are many differences between calculators, so make sure that you are fully familiar with your own particular model – do not buy a new calculator (or borrow one) just before the examination is due!

Each chapter begins with a Starter; this is an exercise, activity or puzzle designed to stimulate thinking and discussion about some of the ideas that underpin the content of the chapter. The main chapter contains explanations of each topic, with numerous worked examples, followed by a corresponding exercise of questions. At the end of each chapter there is a Review Exercise, containing questions on the content for the whole chapter, followed by a set of Key Points. Many of the Review questions are from past Edexcel GCSE or International GCSE examination papers; this is indicated in the margin.

Each chapter concludes with an Internet Challenge, intended to be done (either at school or at home) in conjunction with an internet search engine. The Internet Challenge material frequently goes beyond the strict boundaries of the International GCSE specification, providing enrichment and leading to a deeper understanding of mainstream topics. The Challenges may look at the history of mathematics and mathematicians, or the role of mathematics in the real world. When doing these, it is hoped that you will not just answer the written questions, but also take the time to explore the subject a little deeper – the internet contains a vast reservoir of very well-written information about mathematics. The reliability of internet information can be variable, however, so it is best to check your answers by referring to more than one site if possible.

The exercises in this book are colour coded to indicate their approximate level of difficulty.

The colour coding used is:

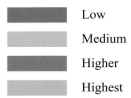 Low

Medium

Higher

Highest

All of the content has been checked very carefully against the new International GCSE specification, to ensure that all examination topics are suitably covered. If you have mastered all the relevant topics covered in this book then you should be able to approach the examinations confident in the knowledge that you are fully prepared.

Finally, it is hoped that you may consider studying mathematics as a sixth form subject. Many of the topics encountered in the early months of a post-16 Mathematics course are natural developments of the content of the later chapters in this book, and if you are able to handle these confidently then you should feel well-prepared for A-level or IB should you choose to study it.

Good luck on exam day!

Alan Smith and Sophie Goldie

Go to **www.hoddereducation.co.uk/EdexcelIGCSEMaths** to access the multimedia presentations and the answers.

CHAPTER 1

Fractions, decimals and rounding

In this chapter you will revise earlier work on:

- equivalent fractions
- adding and subtracting fractions and decimals
- changing fractions to decimals and vice versa
- rounding and approximation.

You will learn how to:

- multiply and divide using fractions
- order a list of fractions and decimals
- change recurring decimals into exact fractions
- use your calculator efficiently
- calculate upper and lower bounds.

You will also be challenged to:

- investigate Egyptian fractions.

Starter: Half and half

Here are some puzzles on the theme of a half.

1 A climbing plant grows 50 cm in week 1. Each following week it grows half as much as it grew the week before. How tall will the plant become if it grows for ever?

2 What fraction is a half of a half of a half?

3 Shade half the squares to divide the shape into two symmetrical halves.

4 Two people wish to share a cake equally. They have a knife with which to cut it. How can they divide the cake so that each person is happy that he has at least half?

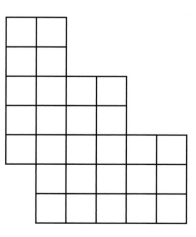

1.1 Equivalent fractions

You should already have met fraction diagrams like these:

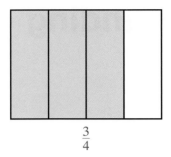

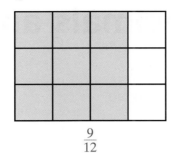

$$\frac{3}{4} \qquad\qquad \frac{9}{12}$$

The diagram on the left shows 3 shaded parts out of 4, and the one on the right shows 9 shaded parts out of 12. Clearly, both diagrams indicate the same shaded proportion of the large rectangle. Thus you can conclude that $\frac{3}{4}$ and $\frac{9}{12}$ are identical in value.

Fractions with identical values are called **equivalent fractions**.
It is easy to create a pair of equivalent fractions – you simply multiply (or divide) the top (numerator) and bottom (denominator) of a given fraction by the same number:

It is often helpful to apply this procedure in reverse:

This process is known as **cancelling down**. When a fraction has been cancelled as fully as possible, it is said to be written in its **simplest terms**.

EXAMPLE

Cancel down the following fractions into their simplest terms.

a) $\dfrac{6}{10}$ **b)** $\dfrac{18}{54}$ **c)** $\dfrac{144}{360}$

SOLUTION

a) $\dfrac{6}{10} = \dfrac{3}{5}$

6 and 10 are both even, so divide each of them by 2.

b) $\dfrac{18}{54} = \dfrac{\cancel{18}^{9}}{\cancel{54}^{27}}$

You can cancel 18 and 54 by dividing by 2…

$= \dfrac{\cancel{9}^{1}}{\cancel{27}^{3}}$

…but 9 and 27 are still both divisible by 9…

$= \dfrac{1}{3}$

…so cancel again.

c) $\dfrac{144}{360} = \dfrac{\cancel{144}^{12}}{\cancel{360}^{30}}$

$= \dfrac{\cancel{12}^{6}}{\cancel{30}^{15}}$

$= \dfrac{\cancel{6}^{2}}{\cancel{12}^{5}}$

$= \dfrac{2}{5}$

Always continue the cancelling procedure until there are no common factors remaining.

Equivalent fractions are a helpful tool when wanting to compare the sizes of fractions written with different denominators, as in the next example.

EXAMPLE

Arrange these fractions in order of size, smallest first.

$$\dfrac{3}{5}, \dfrac{1}{4}, \dfrac{7}{10}, \dfrac{1}{2}$$

SOLUTION

Rewriting the fractions with denominator 20, we have:

$$\dfrac{12}{20}, \dfrac{5}{20}, \dfrac{14}{20}, \dfrac{10}{20}$$

Examine the denominators. 5, 4, 10 and 2 all divide into 20.

So, in increasing order, the fractions are:

$$\dfrac{5}{20}, \dfrac{10}{20}, \dfrac{12}{20}, \dfrac{14}{20}$$

Restoring them to fractions in their lowest terms, the required order is:

$$\dfrac{1}{4}, \dfrac{1}{2}, \dfrac{3}{5}, \dfrac{7}{10}$$

Equivalent fractions are also used when adding (or subtracting) fractions with different denominators. The fractions must be rewritten to have the same denominator before starting the addition.

EXAMPLE

Work out $\dfrac{2}{7} + \dfrac{1}{4}$.

SOLUTION

7 and 4 both divide into 28, so:

$$\frac{2}{7} + \frac{1}{4} = \frac{2 \times 4}{7 \times 4} + \frac{1 \times 7}{4 \times 7}$$

$$= \frac{8}{28} + \frac{7}{28}$$

$$= \frac{8 + 7}{28}$$

$$= \frac{15}{28}$$

Add 8 and 7 to give 15. *Do not* try to add 28 and 28!

The same principle is used for subtraction.

EXAMPLE

Work out $\dfrac{7}{10} - \dfrac{1}{4}$.

SOLUTION

10 and 4 both divide into 20, so:

$$\frac{7}{10} - \frac{1}{4} = \frac{7 \times 2}{10 \times 2} - \frac{1 \times 5}{4 \times 5}$$

$$= \frac{14}{20} - \frac{5}{20}$$

$$= \frac{14 - 5}{20}$$

$$= \frac{9}{20}$$

Some fraction problems may require the use of mixed numbers, made up of part whole number and part fraction. For addition and subtraction you can often process the whole number parts separately from the fraction parts, and then combine everything at the end.

EXAMPLE

Work out $1\frac{3}{5} + 2\frac{3}{4}$.

SOLUTION

Add the whole numbers first: $1 + 2 = 3$
Now add the fractions:

$$\frac{3}{5} + \frac{3}{4} = \frac{12}{20} + \frac{15}{20}$$
$$= \frac{27}{20}$$
$$= 1\frac{7}{20}$$

$$\frac{27}{20} = \frac{20}{20} + \frac{7}{20} \text{ to give } 1\frac{7}{20}$$

So the total is $3 + 1\frac{7}{20} = 4\frac{7}{20}$

EXAMPLE

Work out $7\frac{3}{10} - 3\frac{4}{5}$.

SOLUTION

$$7\frac{3}{10} - 3\frac{4}{5} = 4 + \frac{3}{10} - \frac{4}{5}$$
$$= 3 + 1\frac{3}{10} - \frac{4}{5}$$
$$= 3 + \frac{13}{10} - \frac{4}{5}$$
$$= 3 + \frac{13}{10} - \frac{8}{10}$$
$$= 3 + \frac{13 - 8}{10}$$
$$= 3 + \frac{5}{10}$$
$$= 3\frac{1}{2}$$

Here $\frac{3}{10}$ is not large enough to allow $\frac{4}{5}$ to be subtracted…

…so you exchange 1 unit from the whole number part to make $\frac{13}{10}$ as shown.

Always check for cancelling opportunities at the end. Here $\frac{5}{10}$ cancels down to $\frac{1}{2}$.

Your IGCSE examination will test your ability to manipulate fractions *without* a calculator, so it is important to learn the methods thoroughly. (Even though a calculator is allowed in the exam, marks may be withheld unless you show all the relevant manual steps in full.)

Your calculator should be equipped with a fraction key, probably labelled $\boxed{a^{\,b}/c}$. You should practise entering simple fractions like $\frac{2}{3}$ and mixed fractions like $4\frac{2}{3}$ using this key. Many calculators will allow you to enter a fraction, and then, by pressing the fraction key agian, will show you the decimal equivalent. Most calculators will also automatically convert mixed fractions into top-heavy (improper) fractions, and back again. Try your calculator to see how to do this.

You can also use the fraction key to cancel a fraction into its simplest form. For example, $\frac{144}{180}$ would be entered as 144 $\boxed{a^{b/c}}$ 180 $\boxed{=}$ to obtain $\frac{4}{5}$.

EXAMPLE

Work out $1\frac{3}{4} + 2\frac{5}{6}$, giving your answer as

a) a mixed fraction and

b) a top-heavy (improper) fraction.

SOLUTION

a) Using the fraction key, $1\frac{3}{4} + 2\frac{5}{6} = 4\frac{7}{12}$

> On some calculators, you press $\boxed{\text{SHIFT}}$ and $\boxed{a^{b/c}}$ to change a mixed fraction to a top-heavy fraction.

b) Using the fraction key, $4\frac{7}{12} = \frac{55}{12}$

EXERCISE 1.1

Write these fractions in their simplest terms.

1 $\frac{9}{36}$ **2** $\frac{15}{35}$ **3** $\frac{8}{82}$ **4** $\frac{13}{52}$ **5** $\frac{11}{55}$ **6** $\frac{42}{56}$

7 $\frac{24}{40}$ **8** $\frac{12}{20}$ **9** $\frac{8}{14}$ **10** $\frac{9}{15}$ **11** $\frac{49}{91}$ **12** $\frac{35}{45}$

Arrange these fractions in order of size, smallest first.

13 $\frac{11}{20}, \frac{1}{2}, \frac{3}{5}, \frac{5}{8}$ **14** $\frac{2}{3}, \frac{7}{8}, \frac{5}{6}, \frac{3}{4}$ **15** $\frac{13}{15}, \frac{5}{6}, \frac{2}{3}, \frac{4}{5}$

Work out the answers to these without using a calculator.

16 $\frac{4}{9} + \frac{1}{3}$ **17** $\frac{5}{16} + \frac{1}{4}$ **18** $\frac{1}{3} + \frac{1}{2}$ **19** $\frac{3}{8} + \frac{1}{6}$ **20** $\frac{5}{8} - \frac{1}{4}$

21 Show that:

a) $\frac{3}{4} - \frac{1}{10} = \frac{13}{20}$ **b**) $\frac{4}{7} - \frac{1}{4} = \frac{9}{28}$ **c**) $3\frac{2}{3} + 2\frac{1}{2} = 6\frac{1}{6}$ **d**) $5\frac{7}{9} - 2\frac{1}{3} = 3\frac{4}{9}$

Use pencil and paper to work out the answer to each of these. Then use your calculator fraction key to check your answers.

22 $\frac{7}{12} + \frac{3}{16}$ **23** $\frac{11}{36} + \frac{13}{42}$ **24** $4\frac{19}{20} + \frac{3}{8}$ **25** $10\frac{5}{9} + \frac{3}{8}$

26 $5\frac{1}{8} - 2\frac{6}{7}$ **27** $5\frac{3}{4} - 2\frac{1}{8}$ **28** $8\frac{1}{2} - 5\frac{9}{10}$ **29** $6\frac{1}{4} - 3\frac{1}{2}$

30 $4\frac{2}{3} + 3\frac{5}{6}$ **31** $8\frac{1}{4} - 5\frac{1}{10}$

1.2 Multiplying and dividing with fractions

It is often easier to multiply two fractions than to add them! There is no need to worry about establishing the same denominator before you start. Simply multiply the two top numbers (numerators), and multiply the two bottom ones (denominators).

Note: Multiplying by $\frac{1}{n}$ is the same as dividing by n.

EXAMPLE

Work out **a)** $18 \times \dfrac{1}{6}$ **b)** $\dfrac{3}{5} \times \dfrac{2}{7}$

SOLUTION

a) $18 \times \dfrac{1}{6} = 18 \div 6$

$\phantom{18 \times \frac{1}{6}} = \underline{3}$

b) $\dfrac{3}{5} \times \dfrac{2}{7} = \dfrac{3 \times 2}{5 \times 7}$

$\phantom{\frac{3}{5} \times \frac{2}{7}} = \dfrac{6}{35}$

In many cases there will be an opportunity for cancelling. The cancelling may be done either before or after the multiplication – but it is much more efficient to do this beforehand, since the numbers you work with are then much smaller. The next example shows how this is done.

EXAMPLE

Work out $\dfrac{4}{9} \times \dfrac{15}{22}$

SOLUTION

$$\frac{4}{9} \times \frac{15}{22} = \frac{\cancel{4}^{\,2}}{9} \times \frac{15}{\cancel{22}^{\,11}}$$

$$= \frac{\cancel{4}^{\,2}}{\cancel{9}^{\,3}} \times \frac{\cancel{15}^{\,5}}{\cancel{22}^{\,11}}$$

$$= \frac{2}{3} \times \frac{5}{11} = \frac{2 \times 5}{3 \times 11}$$

$$= \frac{10}{33}$$

> 4 on the top cancels with 22 on the bottom.
> They do not have to be *directly* above/below each other.
> Similarly for 15 on the top and 9 on the bottom.

Division works in a very similar way, with just one extra stage: to divide one fraction by another, turn the second fraction upside down and then multiply them.

EXAMPLE

Work out $\dfrac{5}{8} \div \dfrac{10}{13}$

SOLUTION

$$\dfrac{5}{8} \div \dfrac{10}{13} = \dfrac{5}{8} \times \dfrac{13}{10}$$

$$= \dfrac{\cancel{5}^{1}}{8} \times \dfrac{13}{\cancel{10}^{2}}$$

$$= \dfrac{13}{16}$$

> Cancelling may take place only once the division has been converted into a multiplication.

> Do not try to cancel the 8 with the 2 – they are both on the bottom (denominator) of the fraction.

To multiply or divide by a whole number, just use the result that the integer n may be written as the fraction $\frac{n}{1}$. Note: Dividing by $\frac{1}{n}$ is the same as multiplying by n.

EXAMPLE

Work out $\frac{5}{24} \times 4$.

> This is the same as finding $\frac{5}{24}$ of 4.

SOLUTION

$$\dfrac{5}{24} \times 4 = \dfrac{5}{24} \times \dfrac{4}{1}$$

> Write the 4 as $\frac{4}{1}$ and then proceed as before.

$$= \dfrac{5}{{}_{6}\cancel{24}} \times \dfrac{\cancel{4}^{1}}{1} = \dfrac{5 \times 1}{6 \times 1} = \underline{\dfrac{5}{6}}$$

You need to be careful when multiplying or dividing mixed numbers. Unlike addition or subtraction, you cannot process the whole numbers separately from the fractional parts. Instead, you need to use 'top-heavy' or **improper** fractions.

EXAMPLE

Work out $1\frac{3}{4} \times 2\frac{2}{3}$.

SOLUTION

$$1\tfrac{3}{4} = \dfrac{(1 \times 4) + 3}{4} = \dfrac{7}{4}$$

and $2\tfrac{2}{3} = \dfrac{(2 \times 3) + 2}{3} = \dfrac{8}{3}$

> First, convert the mixed fractions to top-heavy fractions.

Then:

$$1\tfrac{3}{4} \times 2\tfrac{2}{3} = \dfrac{7}{4} \times \dfrac{8}{3}$$

$$= \dfrac{7}{{}_{1}\cancel{4}} \times \dfrac{\cancel{8}^{2}}{3}$$

$$= \dfrac{7 \times 2}{1 \times 3}$$

$$= \dfrac{14}{3}$$

$$= \underline{4\tfrac{2}{3}}$$

> Here the answer is a top-heavy fraction, so you convert it back into mixed fraction form:
> $\dfrac{14}{3} = \dfrac{12 + 2}{3} = \dfrac{12}{3} + \dfrac{2}{3} = 4\tfrac{2}{3}$

EXERCISE 1.2

Work out these multiplications and divisions. Show all your working clearly.

1 $\frac{1}{2} \times \frac{4}{5}$ **2** $\frac{5}{6} \times \frac{2}{3}$ **3** $\frac{3}{4} \times \frac{11}{15}$ **4** $\frac{10}{21} \times \frac{14}{15}$ **5** $\frac{3}{4}$ of 20

6 $\frac{1}{3} \times \frac{1}{5}$ **7** $\frac{2}{7} \times \frac{5}{8}$ **8** $\frac{2}{3}$ of 180 **9** $\frac{4}{9} \div \frac{5}{6}$ **10** $\frac{3}{8} \div \frac{3}{4}$

11 $\frac{5}{6} \div \frac{8}{9}$ **12** $\frac{2}{3} \div 12$ **13** $\frac{3}{8} \div \frac{6}{7}$ **14** $\frac{4}{9} \div 8$ **15** $\frac{1}{12} \div \frac{1}{9}$

16 $\frac{3}{5} \div \frac{3}{5}$ **17** $\frac{15}{16} \times \frac{1}{3}$ **18** $\frac{1}{10} \div \frac{1}{20}$ **19** $\frac{7}{12} \times \frac{27}{28}$ **20** $\frac{5}{7} \div \frac{10}{21}$

21 Show that:

    a) $1\frac{1}{3} \times \frac{2}{7} = \frac{8}{21}$ **b)** $3\frac{1}{3} \times 1\frac{2}{5} = 4\frac{2}{3}$ **c)** $3\frac{1}{4} \times \frac{2}{7} = \frac{13}{14}$ **d)** $1\frac{1}{5} \times 2\frac{1}{2} = 3$

    e) $\frac{4}{5} \div 1\frac{3}{5} = \frac{1}{2}$ **f)** $4\frac{1}{2} \div \frac{7}{8} = 5\frac{1}{7}$ **g)** $1\frac{1}{5} \div \frac{2}{5} = 3$ **h)** $2\frac{1}{4} \div 2\frac{7}{10} = \frac{5}{6}$

Use pencil and paper to work out the answer to each of these. Then use your calculator fraction key to check your answers.

22 $3\frac{3}{4} \times 5\frac{1}{2}$ **23** $5\frac{3}{7} \times 1\frac{2}{19}$ **24** $4\frac{3}{8} \times 7\frac{1}{5}$ **25** $18 \div 1\frac{1}{2}$ **26** $66\frac{2}{3} \times 1\frac{1}{2}$ **27** $3\frac{5}{14} \div 6\frac{5}{7}$

1.3 Decimals and fractions

You need to know how to use the following types of decimal numbers.

Terminating decimals stop after a finite number of decimal places. 0.32 is an example of a terminating decimal. Terminating decimals can be written as exact fractions.

Recurring decimals do not stop after a finite number of decimal places, but they do settle into a pattern of digits that repeats indefinitely.

0.316 316 316 316 316 316… is an example of a recurring decimal. It would normally be written as $0.\dot{3}1\dot{6}$, with a dot over the start and finish of the repeating pattern. Recurring decimals can be written as exact fractions.

EXAMPLE

Express the following as exact fractions. Give your answers as a fraction in its lowest terms.

a) 0.4 **b)** 0.32 **c)** 0.875

SOLUTION

a) 0.4 means $\frac{4}{10} = \frac{2}{5}$

b) 0.32 means $\frac{32}{100} = \frac{16}{50} = \frac{8}{25}$

c) 0.875 means $\frac{875}{1000} = \frac{175}{200} = \frac{35}{40} = \frac{7}{8}$

EXAMPLE

Express $0.\dot{4}$ as an exact fraction.

SOLUTION

Let $x = 0.\dot{4}$

Multiply both sides by 10:

$$10x = 4.\dot{4}$$

Write these results one below the other, and subtract:

$$\begin{array}{r} 10x = 4.\dot{4} \\ x = 0.\dot{4} \\ \hline 9x = 4 \end{array}$$

> Make sure you line up the decimal points. When you do the subtraction, the recurring decimals should disappear.

Divide both sides by 9,

$$x = \frac{4}{9}$$

You may need to multiply both sides by 100, or 1000, if the pattern of repeating digits is of length 2, or 3. Choose the right multiplier so that the digits move left by one full pattern.

EXAMPLE

Express $2.\dot{3}9\dot{6}$ as an exact mixed number. Give your answer in its lowest terms.

SOLUTION

> Since there are three figures in the recurring pattern, use a multiplier of $10 \times 10 \times 10 = 1000$

First, detach the whole number, 2, and consider the decimal part, 396

Let $x = 0.\dot{3}9\dot{6}$

Multiply by a power of 10 large enough to move all the digits along by one pattern, which in this case would be $\times 1000$.

Then $1000\,x = \dot{3}9\dot{6}$

Writing these results one below the other, and subtracting:

$$\begin{array}{r} 1000x = 396.\dot{3}9\dot{6} \\ x = 0.\dot{3}9\dot{6} \\ \hline 999x = 396 \end{array}$$

Thus

$$x = \frac{396}{999}$$

$$= \frac{132}{333}$$

$$= \frac{44}{111}$$

Finally, restore the whole number part, of 2, so $2.\dot{3}9\dot{6} = \underline{2\frac{44}{111}}$

When you convert a fraction into a decimal, the answer will be either a terminating decimal or a recurring decimal.

EXAMPLE

Write $\frac{1}{8}$ as an exact decimal.

SOLUTION

$\frac{1}{8} = 1 \div 8 = \underline{0.125}$

EXAMPLE

Write $\frac{1}{7}$ as a recurring decimal.

SOLUTION

$\frac{1}{7} = 1 \div 7 = \underline{0.\dot{1}4285\dot{7}}$

EXERCISE 1.3

Write these terminating decimals as exact fractions. Give each answer in its lowest terms.

1 0.24	**2** 0.72	**3** 0.3	**4** 0.625
5 0.91	**6** 0.025	**7** 1.94	**8** 0.38
9 2.125	**10** 0.303		

Write these recurring decimals as exact fractions, in their lowest terms. Show your method clearly.

11 $0.\dot{7}$	**12** $0.\dot{2}\dot{9}$	**13** $1.\dot{3}$	**14** $0.5\dot{2}0\dot{4}$
15 $0.4\dot{3}$	**16** $0.5\dot{4}$	**17** $0.\dot{3}2\dot{1}$	**18** $1.3\dot{4}\dot{2}$

Write these fractions as decimals.

19 $\frac{5}{8}$	**20** $\frac{3}{7}$	**21** $\frac{4}{9}$	**22** $\frac{9}{20}$

23 Which is larger: $0.\dot{2}\dot{7}$ or 0.28?

24 Andy says, '$0.\dot{7}$ is exactly twice as big as $0.3\dot{5}$.' Is Andy right or is he wrong? Explain your answer carefully.

1.4 Rounding and approximation

When quantities are written using decimals, it is often sensible to round them to a certain number of figures. Rounding may be described using either **decimal places** (d.p.) or **significant figures** (s.f.).

EXAMPLE

Round these numbers to 3 decimal places:
a) 14.2573   b) 0.0258   c) 0.14962

SOLUTION

a) $14.2573 = 14.257|3 = \underline{14.257}$ (3 d.p.)

> Make the cut after three decimal places. The first digit after the cut, 3, is *less than 5*, so round *down*.

b) $0.0258 = 0.025|8 = \underline{0.026}$ (3 d.p.)

> When the first digit after the cut is *5 or more*, then round *up*. Here it is 8, so the 0.025 rounds up to 0.026

c) $0.149\,62 = 0.149|62 = \underline{0.150}$ (3 d.p.)

Rounding to a certain number of significant figures can be confusing. The best way is to look at the number from the left-hand direction, and pick out the first non-zero digit – this is the first significant figure. Then count the significant figures across to the right.

The confusing thing is whether 0s should be counted as 'significant'. These illustrations show some different cases you might encounter:

Number	Comment	Significant figures
24 000	The 2 and the 4 are significant figures, telling you that the number contains 2 lots of 10 000 and 4 lots of 1000. The three zeros are not significant figures; they merely act as placeholders. This number has 2 significant figures.	24 000
305 000	The first zero is a significant figure, because it is between the 3 and 5. The other zeroes are placeholders only. This number has 3 significant figures.	305 000
0.000 27	The first significant figure is the 2. The previous zeroes merely show you which decimal column the 2 goes in to (they are placeholders). So this number has 2 significant figures.	0.000 27
0.014 03	Plainly the 1, 4 and 3 are significant figures. The 0 between them also is, but the other zeroes are not. This number contains 4 significant figures.	0.014 03
0.250	Watch out! You might think only the 2 and the 5 are significant. The final zero, however, has no role as a placeholder – instead, it tells you that this number contains 0 thousandths. This means it is a significant figure, so this one has 3 significant figures.	0.250

EXAMPLE

Round these numbers to the indicated number of significant figures.
a) 156 230 (2 s.f.) **b)** 0.0896 (2 s.f.) **c)** 10.09 (3 s.f.)

SOLUTION

a) $156\,230 = 15\,|\,6230 = \underline{160\,000}$ (2 s.f.)
b) $0.0896 = 0.089\,|\,6 = \underline{0.090}$ (2 s.f.)
c) $10.09 = 10.0\,|\,9 = \underline{10.1}$ (3 s.f.)

You can use rounding to 1 significant figure to get an estimate to a calculation without having to find an exact answer.

EXAMPLE

Estimate the value of $\dfrac{54 \times (7.6 + 2.3)}{63.1 - 37.98}$.

SOLUTION

$$\frac{54 \times (7.6 + 2.3)}{63.1 - 37.98} \approx \frac{50 \times (8 + 2)}{60 - 40}$$

$$\frac{50 \times (8 + 2)}{60 - 40} = \frac{50 \times 10}{20}$$

$$= \frac{500}{20}$$

$$= 25$$

So $\dfrac{54 \times (7.6 + 2.3)}{63.1 - 37.98} \approx \underline{25}$

Rounding is also used in a slightly different way to establish **upper** and **lower** **bounds** for calculations. The next example illustrates a straightforward case.

EXAMPLE

A rectangular sports field measures 21 metres wide by 83 metres long; each dimension is measured correct to the nearest metre.

a) Write down the smallest value that the width of the sports field could be.
b) Work out the smallest possible perimeter for the field.
c) Calculate the maximum possible value for the area of the field.
    State your answer correct to 3 significant figures.

SOLUTION

a) 21 metres to the nearest metre means the width lies between 20.5 and
    21.5 metres.
    Thus the smallest possible width is $\underline{20.5 \text{ metres}}$.
b) Smallest perimeter $= 20.5 + 82.5 + 20.5 + 82.5$
                     $= \underline{206 \text{ metres}}$

c) The maximum possible area is computed using a rectangle 21.5 metres by 83.5 metres.

$$\text{Area} = 21.5 \times 83.5$$
$$= 1795.25 \text{ square metres}$$
$$= \underline{1800 \text{ square metres}} \text{ (3 s.f.)}$$

Note that it would not be right to use, say, 21.49×83.49, because these numbers are too low, so you have no choice but to use 21.5×83.5, even though these numbers would not round to 21 and 83. In effect you are establishing a kind of upper limit, called an upper bound, which the true area of the field cannot actually quite attain. This is quite an advanced concept, and is addressed again on page 16.

EXERCISE 1.4

Round the following decimal numbers to the indicated number of decimal places.

1 3.141 59 (3 d.p.)　　　　**2** 3.141 59 (4 d.p.)　　　　**3** 16.237 (1 d.p.)

4 0.2357 (2 d.p.)　　　　**5** 14.08 (1 d.p.)　　　　**6** 14.80 (1 d.p.)

7 6.224 02 (4 d.p.)　　　　**8** 1.895 (2 d.p.)

Round the following to the specified number of significant figures.

9 15.42 (3 s.f.)　　　　**10** 15.42 (1 s.f.)　　　　**11** 14.257 (3 s.f.)

12 359 262 (4 s.f.)　　　　**13** 365.249 (2 s.f.)　　　　**14** 9.8 (1 s.f.)

15 0.002 07 (2 s.f.)　　　　**16** 10.99 (3 s.f.)

17 By rounding to 1 significant figure, work out an approximate answer to:

a) 129.1×9.99　　　　**b)** $128.72 + 879$　　　　**c)** $\frac{802.31}{38.9}$

d) $\frac{31.9 \times 17}{95}$　　　　**e)** $\frac{38.9 \times 107}{816 - 298}$　　　　**f)** $\frac{28 \times 176}{873 - 588}$

g) $\frac{1.8 \times 6.1}{3.14}$　　　　**h)** $\frac{3.8 \times (2.93 + 7.1)}{8.2}$　　　　**i)** $\frac{7.5 \times (9.81 - 7.18)}{6.3 - 2.1}$

18 A cinema has 33 rows of 38 seats.
A cinema ticket costs $8.50.
Estimate how much money the theatre takes when it is fully booked.

19 An equilateral triangle measures 12 cm along each side, to the nearest cm. Work out the smallest possible perimeter it could have.

20 A square measures 146 mm along each side, to the nearest mm. Work out the largest possible area it could have, giving your answer correct to 3 significant figures.

21 Last week, the number attending City's home soccer game was 25 000, correct to the nearest thousand. The week before it was 24 400, correct to the nearest hundred. Is it possible that the same number of people attended both weeks? Explain your answer.

22 In a museum there is a dinosaur bone. This notice is attached:

> **This bone came from a dinosaur that was born 75 000 000 years ago. The dinosaur lived for 20 years.**

Charlie says, 'That means the dinosaur died 74 999 980 years ago.' Is Charlie right?

1.5 Rounding calculator answers

For the IGCSE examination, a calculator should have bracket keys. Bracket keys allow you to type in the value of a bracketed expression directly, exactly as it is written. Practise the next example on your calculator to make sure you know how the bracket keys on your calculator work.

As a check, you should also work out the values of some intermediate steps in the calculation and write them down in the exam. This allows the examiner to award marks for the method you have used if your final answer is not correct.

EXAMPLE

Work out the value of $2.4 \times (3.8 - 1.1 \times 1.2)^2$, correct to 3 significant figures.

SOLUTION

$$
\begin{aligned}
2.4 \times (3.8 - 1.1 \times 1.2)^2 &= 2.4 \times 2.48^2 \\
&= 2.4 \times 6.1504 \\
&= 14.760\,96 \\
&= \underline{14.8}\ (3\ \text{s.f.})
\end{aligned}
$$

Sometimes the brackets are implied rather than written down, for example in a square root problem, or when dividing one expression by another. It is good practice to write the implied brackets into the expression before working out its value. Remember to show some of the intermediate steps to secure marks for the method you have used.

EXAMPLE

Use your calculator to find the values of:

a) $\sqrt{2.2 + 3.5 \times 4.2}$ **b)** $\dfrac{3.6 + 2.2^3}{4.8 - 1.2^2}$

Give each answer correct to 3 significant figures.

SOLUTION

a) $\sqrt{2.2 + 3.5 \times 4.2}$
$$
\begin{aligned}
\sqrt{2.2 + 3.5 \times 4.2} &= \sqrt{(2.2 + 3.5 \times 4.2)} \\
&= \sqrt{16.9} \\
&= 4.110\,960\,958 \\
&= \underline{4.11}\ (3\ \text{s.f.})
\end{aligned}
$$

b) $\dfrac{3.6 + 2.2^3}{4.8 - 1.2^2} = \dfrac{(3.6 + 2.2^3)}{(4.8 - 1.2^2)}$

$= \dfrac{14.248}{3.36}$

$= 4.240\,476\,19$

$= \underline{4.24}$ (3 s.f.)

> You can use the fraction button on your calculator to work this out more efficiently. Check you know how to do this.

EXERCISE 1.5

Use your calculator to work out the values of the following expressions. Write down all the figures on your calculator display, then round the answer to 3 significant figures, where appropriate.

1 $(16.2 - 2.8 \times 2.05) \times 2.3$

2 $(2.8 \times 3.5 - 4.9)^2$

3 $\sqrt{22.3 + 2.4 \times 1.5}$

4 $\dfrac{5.4 + 4.5}{5.4 - 4.5}$

5 $\dfrac{2.8 - 1.1^2}{3.4 \times 1.6}$

6 $\sqrt{5.4 \times 4.5}$

7 $\dfrac{6.5 - 2.3 \times 1.4}{1.4}$

8 $\dfrac{2.4 + 3 \times 2.2}{10 - 1.5^2}$

9 $2.6 \times (8.45 - 1.3^2)$

10 $(6.5 - 2.3) \times 1.4$

11 $\sqrt{2.5^2 + 3.5^2}$

12 $\dfrac{2^2 + 5^2}{3^2 + 4^2}$

13 $\dfrac{4.2 + 3.5^2}{3.5^2 - 4.6}$

14 $\sqrt{10.8^2 - 9.1^2}$

15 $\dfrac{2.4 \times 1.4 + 1.8 \times 2.3}{4.5}$

1.6 Upper and lower bounds

Suppose you are asked to find the perimeter of a rectangle that measures 12 cm by 15 cm, both measurements being correct to the nearest centimetre. A reasonable calculation for finding the perimeter is:

$$12 + 15 + 12 + 15 = 54 \text{ cm}$$

The true perimeter is unlikely to be *exactly* 54 cm, however, as the dimensions are probably not exactly 12 cm and 15 cm, since they are only correct to the nearest centimetre.

It can be helpful to establish an **upper bound** and a **lower bound** for the perimeter. These are the limits between which the true perimeter must lie.

The length of the rectangle is 15 cm to the nearest centimetre, which means it could lie anywhere between 14.5 cm and 15.5 cm. Similarly, the width given as 12 cm to the nearest centimetre could actually lie anywhere between 11.5 cm and 12.5 cm.

The upper bound for the perimeter is therefore 12.5 + 15.5 + 12.5 + 15.5 = 56 cm, and the lower bound is 11.5 + 14.5 + 11.5 + 14.5 = 52 cm. This could be written as:

$$52 \text{ cm} \leqslant \text{true perimeter} < 56 \text{ cm}$$

Note the different inequality signs at each end of the above statement. The length of the perimeter cannot actually be as high as 56 cm, since the rectangle is smaller than 12.5 by 15.5 cm – those numbers would be 13 and 16 cm correct to the nearest centimetre. You cannot, however, use a smaller limit like 12.4 or 12.49, because such values are inevitably too small. If this sounds confusing, remember that the upper bound is the same as finding the boundary for the number, even though it can never quite equal it.

EXAMPLE

A rectangle measures 18 cm by 12 cm. Find the upper and lower bound for
a) its perimeter and **b)** its area.

SOLUTION

a) Upper bounds for the dimensions are 18.5 cm and 12.5 cm, so the upper
bound for the perimeter is $18.5 + 12.5 + 18.5 + 12.5 = \underline{62\ cm}$

Lower bounds for the dimensions are 17.5 cm and 11.5 cm, so the lower
bound for the perimeter is $17.5 + 11.5 + 17.5 + 11.5 = \underline{58\ cm}$

b) Upper bounds for the dimensions are 18.5 cm and 12.5 cm, so the upper
bound for the area is $18.5 \times 12.5 = \underline{231.25\ cm^2}$

Lower bounds for the dimensions are 17.5 cm and 11.5 cm, so the lower
bound for the area is $17.5 \times 11.5 = \underline{201.25\ cm^2}$

In the examination you may be asked for the least value and the greatest value:

least value   = lower bound
greatest value = upper bound

If your calculator has a replay function, you can edit the first calculation rather
than keying the expression in again.

In the previous example you simply performed the calculation once, using all
the upper bounds for the measurements involved, and then a second time, using
all the lower bounds. Sometimes the procedure is less straightforward, as, for
example, when working with compound measures involving division.

EXAMPLE

Anita sprints along an athletics track.
The track is 100 metres long, correct to the nearest 1 metre.
Her time is measured as 12.5 seconds, to the nearest half second.
a) Treating these as exact values, work out her average speed for the sprint.
b) Calculate the upper and lower bounds for her average speed.

SOLUTION

a) Average speed $= \dfrac{\text{distance}}{\text{time}}$

$= \dfrac{100}{12.5}$

$= \underline{8\ \text{metres per second}}$

> For the *highest* answer, divide the *highest* top by the *lowest* bottom…

b) Upper bound $= \dfrac{100.5}{12.25}$

$= \underline{8.204\ \text{metres per second}}$ (4 s.f.)

> …and for the *lowest* answer, divide the *lowest* top by the *highest* bottom.

Lower bound $= \dfrac{99.5}{12.75}$

$= \underline{7.804\ \text{metres per second}}$ (4 s.f.)

EXERCISE 1.6

1 A square has sides of length 12 cm, correct to the nearest centimetre.
   a) Calculate the upper and lower bounds for the perimeter of the square.
   b) Calculate the upper and lower bounds for the area of the square.

2 A rectangle has a length of 10 cm and a width of 6 cm. Both these measurements are correct to the nearest centimetre.
   a) Calculate an upper bound for the perimeter of the rectangle.
   b) Calculate a lower bound for the area of the rectangle.

3 To the nearest centimetre, $x = 4$ cm and $y = 6$ cm.
   a) Calculate the upper bound for the value of xy.
   b) Calculate the lower bound for the value of $\dfrac{x}{y}$.

   Give your answer correct to 3 significant figures.   [Edexcel]

4 A car travels a distance of 150 miles in 2.5 hours.
   a) Taking these as exact values, work out its average speed, in miles per hour.

   In fact, the distance is correct to the nearest 10 miles and the time is correct to the nearest 0.1 hour.
   b) Work out a lower bound for the speed of the car.
   c) Work out an upper bound for the speed of the car.

5 Bill has a rectangular sheet of metal.
   The length of the rectangle is **exactly** 12.5 cm.
   The width of the rectangle is **exactly** 10 cm.

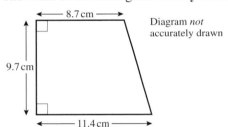

Diagram *not* accurately drawn

8.7 cm

9.7 cm

11.4 cm

   Bill cuts out a trapezium. Its dimensions, shown in the diagram, are correct to the nearest millimetre.
   He throws away the rest of the metal sheet.

   Calculate the greatest possible area of the rectangular sheet that he throws away.   [Edexcel]

REVIEW EXERCISE 1

Work out the following, without using a calculator. Show all your working.

1 $\frac{2}{5} + \frac{1}{4}$ **2** $\frac{3}{5} + \frac{2}{3}$ **3** $4\frac{5}{8} - 1\frac{1}{6}$ **4** $8\frac{3}{10} - 3\frac{3}{4}$

5 $1\frac{1}{2} \times 3\frac{1}{3}$ **6** $4\frac{2}{3} \times 5$ **7** $4\frac{1}{2} \div 1\frac{1}{5}$ **8** $3\frac{1}{2} \div \frac{1}{4}$

9 Arrange these fractions in order of size, smallest first.
   $\frac{5}{8}, \frac{3}{4}, \frac{5}{6}, \frac{7}{12}$

10 Arrange these fractions in order of size, smallest first.

$\frac{11}{15}, \frac{7}{10}, \frac{2}{3}, \frac{3}{5}$

11 Work out an estimate for the value of:

a) $\dfrac{27 \times 3.14}{0.29}$ **b)** $\dfrac{789 + 187}{9.98}$ **c)** $\dfrac{4.2 \times (3.17 + 5.21)}{11.15 - 6.2}$

12 Arrange these numbers in order of size, smallest first.

$\frac{3}{4}, 0.65, \frac{5}{8}, 0.\dot{6}$

13 a) Write the decimal 0.875 as a fraction in its lowest terms.
   b) Convert the recurring decimal $0.\dot{4}\dot{5}$ to a fraction in its simplest terms.

14 Write these numbers in order of size. Start with the smallest number.
   a) 0.56, 0.067, 0.6, 0.65, 0.605
   b) 5, -6, -10, 2, -4
   c) $\frac{1}{2}, \frac{2}{3}, \frac{2}{5}, \frac{3}{4}$ [Edexcel]

15 Simon spent $\frac{1}{3}$ of his pocket money on a computer game. He spent $\frac{1}{4}$ of his pocket money on a ticket for a football match. Work out the fraction of his pocket money that he had left.   [Edexcel]

16 Ann wins £160. She gives $\frac{1}{4}$ of £160 to Pat, $\frac{3}{8}$ of £160 to John and £28 to Peter.
   What fraction of the £160 does Ann keep?
   Give your answer as a fraction in its simplest form.   [Edexcel]

17 Write down two different fractions that lie between $\frac{1}{4}$ and $\frac{1}{2}$.   [Edexcel]

18 Nick takes 26 boxes out of his van. The weight of each box is 32.9 kg.
   Work out the *total* weight of the 26 boxes.   [Edexcel]

19 a) Work out $\frac{2}{5} + \frac{3}{8}$ **b)** Work out $5\frac{2}{3} - 2\frac{3}{4}$ [Edexcel]

20 $1.54 \times 450 = 693$

   Use this information to write down the answer to:
   a) 1.54×45 **b)** 1.54×4.5 **c)** 0.154×0.45 [Edexcel]

21 Each side of a regular pentagon has a length of 101 mm, correct to the nearest millimetre.
   a) Write down the *least* possible length of each side.
   b) Write down the *greatest* possible length of each side.   [Edexcel]

22 Using the information that: $97 \times 123 = 11\,931$

   write down the value of:
   a) 9.7×12.3
   b) $0.97 \times 123\,000$
   c) $11.931 \div 9.7$ [Edexcel]

23 Convert the recurring decimal $0.2\dot{9}$ to a fraction.   [Edexcel]

24 Use your calculator to work out the value of:

$$\frac{\sqrt{12.3^2 + 7.9}}{1.8 \times 0.17}$$

Give your answer correct to 1 decimal place.　　　　　　　　　　　　　　**[Edexcel]**

25 a) Use your calculator to find the value of:

$$\sqrt{(47.3^2 - 9.1^2)}$$

Write down all of the figures on your calculator display.

b) Write your answer to part **a)** correct to 2 significant figures.　　　　　　　**[Edexcel]**

26 a) Use your calculator to work out the value of:

$$\frac{\sqrt{(1.3^2 + 4.2)}}{5.1 - 2.02}$$ ← Check you can use your fraction button to work this out in one step.

Write down all of the figures on your calculator display.

b) Give your answer to part **a)** to an appropriate degree of accuracy.

27 a) Use your calculator to work out:

$$(2.3 + 1.8)^2 \times 1.07$$

Write down all of the figures on your calculator display.

b) Copy out the expression and then insert brackets so that its value is 45.024.

$$1.6 + 3.8 \times 2.4 \times 4.2$$　　　　　　　　　　　　　　　　　　**[Edexcel]**

28 Work out the value of:

$$\sqrt{\frac{8.35 \times 978}{1025 + 222}}$$

Give your answer correct to 3 significant figures.　　　　　　　　　　　　**[Edexcel]**

29 Change to a single fraction:
a) the recurring decimal $0.1\dot{3}$
b) the recurring decimal $0.5\dot{1}\dot{3}$　　　　　　　　　　　　　　　　**[Edexcel]**

30 a) Convert the recurring decimal $0.\dot{3}\dot{6}$ to a fraction.
b) Convert the recurring decimal $2.1\dot{3}\dot{6}$ to a mixed number.
Give your answer in its simplest form.　　　　　　　　　　　　　　　　**[Edexcel]**

31 Express the recurring decimal $2.0\dot{6}$ as a fraction.
Write your answer as a fraction in its simplest form.　　　　　　　　　　**[Edexcel]**

32 a) Express $0.\dot{2}\dot{7}$ as a fraction in its simplest form.
b) x is an integer such that $1 \leqslant x \leqslant 9$. Prove that $0.0\dot{x} = \dfrac{x}{99}$　　**[Edexcel]**

33 $x = 3$, correct to 1 significant figure. $y = 0.06$, correct to 1 significant figure.

Calculate the greatest possible value of:

$$y - \frac{x-7}{x}$$

[Edexcel]

34 Peter transports metal bars in his van. The van has a safety notice *Maximum Load 1200 kg*. Each metal bar has a label *Weight 60 kg*.

For safety reasons Peter assumes that:

1200 is rounded correct to 2 significant figures

60 is rounded correct to 1 significant figure.

Calculate the greatest number of bars that Peter can *safely* put into the van if his assumptions are correct.

[Edexcel]

35 The time period, T seconds, of a pendulum is calculated using the formula:

$$T = 6.283 \times \sqrt{\frac{L}{g}}$$

where L metres is the length of the pendulum, and g m/s^2 is the acceleration due to gravity.

$L = 1.36$ correct to 2 decimal places

$g = 9.8$ correct to 1 decimal place.

Find the difference between the lower bound of T and the upper bound of T.

[Edexcel]

Key points

1 Fractions such as $\frac{2}{3}$ and $\frac{4}{6}$ are called equivalent fractions, because they have the same values. If you have to sort a list of fractions into numerical order it can be helpful to re-write them as equivalent fractions with a common denominator.

2 Common denominators must also be used for adding or subtracting fractions.

3 To multiply two fractions, simply multiply the two top numbers (numerators) and multiply the two bottom numbers (denominators). You can make this method more efficient if you look for opportunities to cancel first.

4 To divide one fraction by another, turn the second fraction upside down and then multiply them.

   Remember, dividing by $\frac{1}{n}$ is the same as multiplying by n.

5 When multiplying or dividing mixed numbers, you must convert them to top-heavy (improper) fractions first.

6 Upper and lower bounds may be computed for quantities that have been rounded to a given level of accuracy:

   Lower bound = stated value minus half a 'step'

   Upper bound = stated value plus half a 'step'.

7 In the examination, lower bound may be referred to as least (or minimum) value, and upper bound may be referred to as greatest (or maximum) value.

   There is a subtle difference between the concepts of *maximum* and *upper bound*, but the examiner will expect you to treat them identically. Thus, if a length has been recorded as 18 cm to the nearest centimetre, and the examiner asks for the greatest possible value, you should write 18.5 cm, not 18.4 cm or 18.49 cm.

8 You will often need to convert terminating or recurring decimals into fraction form.

9 Terminating decimals may easily be expressed as fractions with denominator 10, 100, 1000, etc., and are then cancelled down where possible.

10 Recurring decimals are more awkward. Make sure you have studied the multiply/ subtract method that reduces them to fractions with denominator 9, 99, 999, etc. Once again, cancel down at the end where possible.

11 Calculations involving decimals often require rounding to a sensible number of decimal places or significant figures. Make sure you understand how to do this. Significant figures, in particular, can sometimes be a little awkward to count.

Internet Challenge 1

Egyptian fractions

It is thought that the ancient Egyptians only used unit fractions such as $\frac{1}{2}, \frac{1}{3}, \frac{1}{4}$, etc., that is, of the form $\frac{1}{n}$. They wrote other fractions as sums of (different) unit fractions. For example, the quantity we write as $\frac{5}{6}$ would have been written as $\frac{1}{2} + \frac{1}{3}$ in Egyptian fractions.

Answer the following questions about Egyptian fractions. Use the internet where appropriate.

1   Work out the value of $\frac{1}{8} + \frac{1}{24}$.

2   Work out the value of $\frac{1}{4} + \frac{1}{5} + \frac{1}{10}$.

3   Find two different Egyptian fractions that add up to $\frac{1}{3}$.

4   Find three different Egyptian fractions that add up to 1.

5   Find three different Egyptian fractions that add up to $\frac{7}{9}$.

6   Can *every* ordinary fraction be expressed as the sum of a set of different Egyptian fractions in this way?

7   Jim and Emily have 5 sacks of corn, and they want to share them out between 8 chicken coops, so that each coop gets (about) the same amount of corn. They do not have any weighing or measuring equipment available.

   Jim says 'Well just have to judge five-eights of a sack as best we can.'

   Emily says 'I know a better way, using Egyption fractions.'

   What method might Emily be planning to use?

CHAPTER 2

Ratios and percentages

In this chapter you will revise earlier work on how to:

- simplify and solve problems using simple ratios
- convert fractions to percentages and vice versa.

You will learn how to:

- use non-calculator methods to find simple percentage increases and decreases
- use multiplying factors to solve harder percentage problems
- solve reverse percentage problems efficiently
- work with compound interest and depreciation.

You will also be challenged to:

- investigate monetary inflation.

Starter: How many per cent?

Look at the hexagonal honeycomb shape below.
How many different paths can you find that spell PERCENT?
The paths must be continuous.
You could use a counting method, but a calculating method is more likely to be accurate.

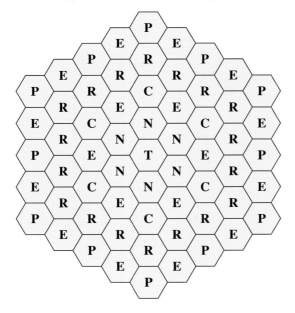

2.1 Working with ratios

Many problems on ratios are best thought of as scaling exercises. For example, suppose you have a recipe for 4 people, and you need to adapt the quantities to cater for 10 people. The method is to compute $10 \div 4 = 2.5$. This provides a multiplying **scale factor** that you can use to calculate the required amounts.

EXAMPLE

Here is a recipe for Scotch pancakes:

Plain flour	100 g
Baking power	1 teaspoon
Milk	150 ml
Eggs	3
Serves 4 people	

Roz wishes to cook Scotch pancakes for 6 people. Calculate the amount of each ingredient she requires.

SOLUTION

The scale factor is: $6 \div 4 = 1.5$

> Take two matching known amounts: 6 people and 4 people. Divide one by the other to get the scale factor, or multiplier.

Plain flour	$100 \times 1.5 = \underline{150 \text{ g}}$
Baking powder	$1 \times 1.5 = \underline{1.5 \text{ teaspoons}}$
Milk	$150 \times 1.5 = \underline{225 \text{ ml}}$
Eggs	$3 \times 1.5 = 4.5$, so $\underline{5 \text{ eggs}}$ are needed.

> You need a whole number of eggs.

EXAMPLE

Concrete mix is made by combining 1 part of cement with 2 parts sand and 4 parts gravel (by volume). A builder has 0.6 m³ of sand to use in a concrete mix.
a) Calculate the amount of the other ingredients he uses.
b) Calculate the total volume of the concrete mix.

SOLUTION

a) First, compute the multiplier: $0.6 \div 2 = 0.3$
So the parts of 1, 2, 4 can each be multiplied by 0.3, as follows:
Cement $1 \times 0.3 = \underline{0.3 \text{ m}^3}$
Sand $\quad 2 \times 0.3 = \underline{0.6 \text{ m}^3}$
Gravel $4 \times 0.3 = \underline{1.2 \text{ m}^3}$
b) The total is $0.3 + 0.6 + 1.2 = \underline{2.1 \text{ m}^3}$

In the example above, concrete was made by combining 1 part of cement with 2 parts sand and 4 parts gravel. We say that cement, sand and gravel are used **in the ratio** $1 : 2 : 4$. Like fractions, ratios can sometimes be cancelled down into simpler equivalent forms.

EXAMPLE

Express these ratios in their simplest terms.

a) $14 : 21$ **b)** $72 : 108$ **c)** $15 : 25 : 45$

SOLUTION

a) $14 : 21 = \underline{2 : 3}$

> 14 and 21 are each divisible by 7.

b) $72 : 108 = 6 : 9$
           $= \underline{2 : 3}$

> Divide 72 and 108 by 12, to give 6 and 9.
> 6 and 9 can then be cancelled again, by dividing by 3.

c) $15 : 25 : 45 = \underline{3 : 5 : 9}$

> Triple ratios simplify in exactly the same way.

EXAMPLE

Three brothers are aged 6, 9 and 15 years old. They inherit £630 between them, to be shared in the ratio of their ages. Work out the share that each brother receives.

SOLUTION

The ratio $6 : 9 : 15$ may be simplified to $2 : 3 : 5$.

$2 + 3 + 5 = 10$ shares

£630 $\div$ 10 = 63, so one share is worth £63.

The brothers receive $2 \times$ £63, $3 \times$ £63, $5 \times$ £63 respectively, that is £126, £189 and £315

> Check that these final amounts add up correctly:
> $126 + 189 + 315 = 630$ ✓

EXERCISE 2.1

Express each of these ratios in its simplest form.

1 $4 : 10$ **2** $18 : 27$ **3** $77 : 121$

4 $21 : 91$ **5** $16 : 20 : 28$ **6** $30 : 40 : 60$

7 $27 : 36 : 63$ **8** $144 : 360$ **9** $25 : 35 : 40$

10 Share £60 in the ratio $3 : 4 : 5$.   **11** Share 300 in the ratio $3 : 5 : 7$.

12 Share £450 in the ratio $1 : 3 : 5$.   **13** Share \$144 in the ratio $2 : 3 : 7$.

14 Share 132 in the ratio $2 : 4 : 5$.   **15** Share £350 in the ratio $2 : 5 : 7$.

16 Christie has £216 to spend on a birthday party. She wants to spend this on food, drinks and entertainment in the ratio $4 : 5 : 9$. Work out the actual amount she will spend on each of these three things.

17 A map has a scale of 1 : 50 000.
  a) The distance between two towns on the map is 10 cm.
  How far apart are the towns in real life? Give your answer in kilometres.
  b) Bridgetown is 16.7 km away from Littleton.
  What is the distance between these towns on the map?

18 A school's mathematics department spends money on books, photocopying and computer software in the ratio 3 : 5 : 6. The department spends £1200 on photocopying.
  a) Work out how much the department spends on books.
  b) Find the total expenditure on all three things.

19 Food for a wedding reception would cost £640 for 40 guests. Work out the cost of the food for:
  a) 60 guests   **b)** 100 guests.

20 Here are the ingredients to make dumplings to accompany a stew.

100 grams of self-raising flour
$\frac{1}{4}$ teaspoon salt
50 grams of suet
$\frac{1}{3}$ teaspoon dried mixed herbs
Serves 4 people

Ginny wants to make dumplings for 6 people. Work out the amounts of each ingredient she needs.

21 Write the scale for each of the following maps in the form 1 : n
  a) 2 cm on the map represents 5 km on the ground.
  b) 4 cm on the map represents 500 m on the ground.
  c) 5 cm on the map represents 2 m on the ground.
  d) 5 cm on the map represents 4 km on the ground.

22 Carly is planning to go travelling in her gap year.
  She converts some pounds (£) to euros (€) and dollars ($) before she travels.
  The exchange rate is €1 = £0.86 and $1 = €0.48
  a) Convert £1400 into euros.
  b) How many dollars are worth one pound?

2.2 Simple percentages

You should already be familiar with percentages. You can think of a percentage as a simple ratio out of one hundred.

EXAMPLE

Write 45% as a fraction in its lowest terms.

SOLUTION

45% means 45 out of 100, so, as a fraction:

$$45\% = \frac{\cancel{45}^{9}}{\cancel{100}^{20}} = \frac{9}{20}$$

This example reminds you how to find a percentage of an amount.

EXAMPLE

Find 24% of 350.

SOLUTION

Method 1

$$24\% \text{ of } 350 = \frac{24}{100} \times 350 = \underline{84}$$

Per cent means 'out of 100'.

To write a percentage as a decimal you divide by 100.

Method 2

$24\% \text{ of } 350 = 0.24 \times 350 = \underline{84}$

Sometimes you need to convert fractions into percentages.

EXAMPLE

A local factory employs 1200 people. 720 of these are women.
a) What fraction are women? Give your answer in its simplest terms.
b) Write your answer to part **a)** as a percentage.

SOLUTION

a) As a fraction:

$$\frac{720}{1200} = \frac{72}{120}$$

$$= \frac{6}{10}$$

$$= \frac{3}{5}$$

You can check this by entering $\frac{720}{1200}$ into your calculator, using the fraction key.

The calculator will automatically cancel the fraction down.

So $\frac{3}{5}$ of the employees are women.

b) Write the fraction as a decimal:

$$\frac{3}{5} = 3 \div 5 = 0.6$$

Then write the decimal as a percentage:

$0.6 \times 100 = \underline{60\%}$

Multiply by 100

EXERCISE 2.2

Write these percentages as fractions, in their lowest terms.

1 40%   **2** 36%   **3** 5%   **4** 30%

5 33%   **6** $33\frac{1}{3}\%$ **7** $12\frac{1}{2}\%$ **8** $\frac{1}{2}\%$

Express these fractions as percentages. You should do them by hand at first. Then use the fraction key on your calculator to check each one.

9 $\frac{3}{4}$ **10** $\frac{2}{5}$ **11** $\frac{3}{10}$ **12** $\frac{17}{20}$

13 $\frac{21}{25}$ **14** $\frac{19}{20}$ **15** $\frac{7}{8}$ **16** $\frac{3}{40}$

17 Arrange these quantities in order of size, smallest first.

$0.7, 66\%, \frac{2}{3}, 0.67, \frac{5}{7}, 69\%$

18 An examination is out of 250 marks.
  a) Sonja scores 210 marks. What percentage is this?
  b) Peter scores 62%. How many marks out of 250 did he receive?

  The marks are then converted to a UMS (uniform mark system) score. The UMS marks are out of a total of 600.
  c) Callum scores 426 UMS marks. What percentage of 600 is this?

19 Ashley has just taken two mathematics test papers. In Arithmetic he scored 56 out of 80, while in Algebra he scored 64 out of 90. His friends are talking about the results.

  Ami says 'You did *much better* in the Algebra test.'

  Marcus says 'You did *about the same* in both tests.'

  Use percentage calculations to explain who is right.

20 Naomi buys 8 bracelets at a car boot sale. She pays £2.50 for each one. She sells them on the internet for a total of £24. Work out Naomi's percentage profit.

2.3 Percentage increase and decrease

There are two standard ways of increasing an amount by a given percentage. The first method is simply to find the size of the increase, and add it on to the original amount.

EXAMPLE

Seamus buys £600 of shares in a new company. One week later, the value of the shares has increased by 15%. Find the value of the shares one week after Seamus bought them.

SOLUTION

15% of £600 is 0.15 × £600 = £90
So the value of the shares is £600 + £90 = <u>£690</u>

It is quicker to use the **multiplying factor** method. Not only is this quick but it is also applicable to a wider variety of other kinds of percentage problems.

The multiplying factor method will underpin most of the ideas encountered in the remainder of this chapter.

EXAMPLE

Martin wins an internet auction. He pays £35 for the goods, but has to add fees of 3% to this. Work out the total amount that Martin has to pay.

SOLUTION

100% + 3% = 103%, so use a multiplying factor of 1.03.

> A multiplying factor *larger* than 1 will *increase* the amount to pay.

Total amount to pay = £35 × 1.03
= <u>£36.05</u>

The multiplying factor method can be used for percentage decrease problems too.

EXAMPLE

Jeremy pays £35 000 for a new car. One year later it has lost 28% of its initial value. (This is called depreciation.) Find the value of the car one year after Jeremy bought it.

SOLUTION

100% − 28% = 72%, so use a multiplying factor of 0.72.

> Multiplying factors *smaller* than 1 will give a *decrease*, not an increase.

Value after one year = £35 000 × 0.72
= <u>£25 200</u>

EXERCISE 2.3

1 Increase 240 by 13%.

2 Decrease 420 by 9%.

> VAT stands for Value Added Tax. It is a tax that is added on to some goods and services.

3 Increase 1200 by 15%, then by a further 8% of its new value.

4 A kettle costs £29.99 plus VAT at 20%. Find the total cost of the kettle.

5 Last year the local dance club had 64 members. This year it has 73 members. Calculate the percentage increase in the membership.

6 Lance bought a new car. He paid £14 000. Each year the value of the car depreciated by 18% of its value at the beginning of that year. Work out the value of Lance's car:
   a) after one year   b) after five years.

7 Sarah travels by train to visit her grandmother. The train journey used to take $2\frac{1}{2}$ hours, but the track has now been improved, reducing the journey time by 6%. How long does Sarah's journey to her grandmother take now?

8 My salary this year, before tax, is £38 000. Next year I am getting a 4% rise, but 1% of my new salary will be contributed to a savings scheme.
   a) Work out my new salary for next year.
   b) Work out how much I will receive next year after deducting the savings contribution.
   c) The answer to part b) could have been found by multiplying £38 000 by a single number. Find the value of that single number.

9 A magazine sold 17 000 copies in June, 23 000 copies in July and 21 000 copies in August.
   a) Calculate the percentage increase in sales between June and July.
   b) Calculate the percentage decrease in sales between July and August.

10 Alan imports a camera from the USA. He has to pay Import Duty and VAT to the UK Customs and Excise authorities. Import Duty at 4.9% is charged on the cost of the camera. VAT at 20% is applied to the cost of the camera (including the Import Duty), and to the freight charges. The courier company also charges a non-taxable handling charge.

   The cost of the camera (before any taxes) is $1799 and the cost of freight (before any taxes) is $95. The company's handling charge is $15.
   a) Work out the total cost, in $, that Alan has to pay.
   b) Using an exchange rate of $1 = £0.58, work out the total cost in £.

2.4 Reverse percentage problems

Suppose a quantity increases by 10%, and you want to restore it to its original value. If you take off 10% of the new value, you would obtain the wrong answer! This is because you should be taking off 10% of the original value, not 10% of the new value.

An excellent way of solving this type of reverse percentage problem is to turn to multiplying factors. Identify the factor you would have *multiplied* by to make the increase from the original, and then just *divide* by the same factor to reverse the process.

EXAMPLE

The bill in a restaurant comes to £70.20 including a service charge of 8%.
Work out the size of the bill before the service charge is included.

SOLUTION

An 8% increase corresponds to a multiplying factor of 1.08.
Original bill × 1.08 = 70.20
So original bill = 70.20 ÷ 1.08
                 = 65
The size of the bill before service charge was £65

The same method works for percentage decrease problems, too.

EXAMPLE

In a sale, a shop reduces all its prices by 15%. Sarah buys a jacket in the sale.
She pays £68. Work out how much she saves by buying it in the sale.

SOLUTION

A 15% reduction corresponds to a multiplying factor of 0.85.

Original price $\times$ 0.85 = 68

So original price = 68 $\div$ 0.85

$\qquad\qquad\qquad$ = 80

Sarah paid £68, so she saves £80 − £68 = <u>£12</u>

EXERCISE 2.4

This exercise contains a mixture of forwards and reverse percentage problems
– so you can train yourself to spot the difference! You should use multiplying
factor methods as much as possible.

1 An electrical goods shop reduces all of its prices by 20% in a sale.
   a) Find the sale price of a television originally priced at £350.
   b) Find the original price of a radio that is £68 in the sale.

2 In a board game you can mortgage properties. To redeem them from mortgage you have to pay the
   mortgage value plus 10%.
   a) Work out the cost of redeeming a property that was mortgaged for £60.
   b) Ed redeems a set of properties, at a total cost of £715. What was the mortgage value of this set of
       properties?

3 A meal costs £91.20 inclusive of VAT at 20%. Work out the cost of the meal before VAT was applied.

4 A bookshop is having a sale. The prices of all books are reduced by 15%.
   a) Work out the sale price of a book that would normally cost £24.
   b) The shop assistant mistakenly reduces the prices by 25% instead. She charges a customer £13.50 for a
       book in the sale. Work out how much she should have charged if the sale reduction had been applied
       correctly.

5 I have upgraded my ageing computer, and it can now do tasks in only 70% of the time it used to take.
   A particular task now takes 630 milliseconds. How long did this take before the upgrade?

6 A year ago I bought a new car. It depreciates at 22% per year. My car is presently worth £7176.
   a) Work out the value of my car when I bought it.
   b) Work out how much my car will be worth three years from now.

7 Joan's diet has been successful, and she has managed to lose 15% of her body weight. She now weighs 136 pounds.

   a) Work out her weight, in pounds, before she began the diet.

   b) Rewrite your answer in stones and pounds. (1 stone = 14 pounds)

8 A galaxy was once thought to be 2.2 million light years away from Earth, but astronomers now believe its actual distance to be 30% greater than this. Calculate the distance at which it is now thought to be.

9 The average attendance at City's soccer ground has fallen by 12% from last year. This year's average is 13 552. What was the average attendance last year?

10 A graphic designer has drawn a rectangle on her computer screen. She resizes the rectangle by making it 20% smaller than it was in both length and width. After resizing, the rectangle measures 960 pixels by 620 pixels.

   a) Work out the dimensions of the rectangle before it was resized.

   b) By what percentage has the area of the rectangle been reduced when resized in this way?

2.5 Compound interest

When money is placed in a savings bank it may earn interest at a given percentage rate per year.

EXAMPLE

£1200 is invested at 4.5% interest per year. Calculate the interest earned after 1 year.

SOLUTION

4.5% of 1200 = 0.045 × 1200

$\qquad$ = £54

In reality, interest calculations are usually more complicated than this. The rate of interest might change over time, for example. Perhaps even more importantly, interest is generated on the previous interest as well as the original capital. So, if £100 were invested at 10% per annum, you would earn £10 in the first year. This is added to the principal amount, so in the second year you earn interest on £110, which would be £11. This type of interest is called **compound interest**.

Compound interest problems are best solved using multiplying factors.

EXAMPLE

£1200 is invested at 4.5% per annum. Calculate the compound interest earned over 5 years.

SOLUTION

The multiplying factor for a 4.5% increase is 1.045.

At the end of 5 years, the money invested will have grown to:

1200 × 1.045 × 1.045 × 1.045 × 1.045 × 1.045 = 1200 × 1.045^5

$\qquad\qquad\qquad\qquad\qquad\qquad\qquad$ = 1495.42

So the compound interest earned is £1495.42 − £1200 = £295.42

33

EXAMPLE

£800 is invested at $r\%$ per annum. After 6 years it has grown to £955.24.
Use a trial and improvement method to find the value of r.

SOLUTION

After 6 years the investment has a value of $800 \times \left(1 + \dfrac{r}{100}\right)^6$.

Try $r = 10$: $800 \times 1.10^6 = 1417.25$ too big

Try $r = 5$: $800 \times 1.05^6 = 1072.08$ too big

Try $r = 2$: $800 \times 1.02^6 = 900.23$ too small

Try $r = 3$: $800 \times 1.03^6 = 955.24$ correct

> Show full details of all the trials, not just the final one.

Hence $r = 3$

EXERCISE 2.5

1   Work out the compound interest on £450 invested at 3% per annum for 4 years.

2   Work out the compound interest on £3000 invested at 5.5% per annum for 8 years.

3   Bill invests some money at a compound interest rate of 3.5% per annum. After 6 years his investment is worth £921.94. How much did Bill invest?

4   On his 16th birthday Anu puts £160 in a savings account. The account pays compound interest at 3% per annum. How old will Anu be when his savings have increased to over £200?

5   A ten-year savings bond pays 4% interest for the first year, then 5% per annum compound interest after that. Work out the total final value after ten years of an initial investment of £500.

6   Michelle invests some money in a savings account. It pays compound interest at a rate of 4% per annum. Michelle says, 'My money will double in 25 years.' Michelle is wrong.
    a)   Explain how you think Michelle arrived at a figure of 25 years.
    b)   Work out the correct figure.

REVIEW EXERCISE 2

1   Write the ratio 18 : 27 in its lowest terms.

2   Work out 15% of 360°.

3   Simplify the ratio 16 : 20 : 30.

4   Alan scores 38 out of 40 in a mathematics test. What percentage is this?

5   A computer costs £320 plus VAT at 20%. Work out the cost of the computer including VAT.

6 Jack buys a box of 20 pens for £3.00. He sells the pens for 21p each. He sells all the pens.
Work out his percentage profit. [Edexcel]

7 A customer who cancels a holiday with Funtours has to pay a cancellation charge. The cancellation charge depends on the number of days before the departure date the customer cancels the holiday.

Number of days before the departure date the customer cancels the holiday	Cancellation charge as a percentage of the cost of the holiday
29–55	40%
22–28	60%
15–21	80%
4–14	90%
3 or fewer	100%

The cancellation charge is a percentage of the cost of the holiday. The table shows the percentages.

The cost of Amy's holiday was £840. She cancelled her holiday 25 days before the departure date.

a) Work out the cancellation charge she had to pay.

The cost of Carol's holiday was £600. She cancelled her holiday and had to pay a cancellation charge of £480.

b) Work out £480 as a percentage of £600.

Ravi cancelled his holiday 30 days before the departure date. He had to pay a cancellation charge of £272.

c) Work out the cost of his holiday. [Edexcel]

8 Brass is made up from copper and zinc. Every 100 grams of brass contains 20 grams of zinc.

a) Work out the weight of zinc in 60 grams of brass.

Brass contains 4 parts by weight of copper to 1 part by weight of zinc.

b) Work out the weight of copper in 350 grams of brass. [Edexcel]

9 In a sale all the prices are reduced by 30%. The sale price of a jacket is £28. Work out the price of the jacket before the sale. [Edexcel]

10 There are 800 students at Prestfield School. 45% of these students are girls.

a) Work out 45% of 800.

There are 176 students in Year 10.

b) Write 176 out of 800 as a percentage. [Edexcel]

11 Ben bought a car for £12 000. Each year the car depreciated by 10%. Work out the value of the car 2 years after he bought it. [Edexcel]

12 A photocopier makes copies that are 40% larger than the original.

a) An original is 12 cm long. Find the length of the copy.

b) A copy is 35 cm long. Find the length of the original.

13 Calculate the compound interest earned when $360 is invested at 6% per annum for 3 years.

14 The price of a new smart television is £960 including VAT at 20%.
Work out the price of the television before VAT was added.

15 A travel shop increases the prices of all its weekend city breaks by 8%.

Before the increase, the price of a weekend city break to Paris was £222.

a) Work out the price of a city break to Paris after the increase.

The price of a city break to London increased by £26.

b) Work out the price of a city break to London after the increase.

16 Toby invites a group of his friends to a party, but only 80% of the people invited turn up.

68 people turn up to the party.

How many people did Toby invite?

17 Each year the value of a car falls by 15% of its value at the beginning of that year.

Sally bought a new car for £9600 on the 1st January one year.

Calculate the value of the car after 3 years.

18 This is a list of the ingredients for making a pear and almond crumble for 4 people.

> Ingredients for 4 people.
> 80 g plain flour
> 60 g ground almonds
> 90 g soft brown sugar
> 60 g butter
> 4 ripe pears

Work out the amount of each ingredient needed to make a pear and almond crumble for **10** people. [Edexcel]

19 £5000 is invested for 3 years at 4% per annum **compound** interest. Work out the **total interest** earned over the three years.

20 In a sale, all the normal prices are reduced by 15%. The normal price of a jacket is £42. Syreeta buys the jacket in the sale.

a) Work out the sale price of the jacket.

In the same sale, Winston pays £15.64 for a shirt.

b) Calculate the normal price of the shirt. [Edexcel]

21 The price of a telephone is £36.40 plus VAT. VAT is charged at a rate of 17.5%.

a) Work out the amount of VAT charged.

In a sale, normal prices are reduced by 12%. The normal price of a camera is £79.

b) Work out the sale price of the camera. [Edexcel]

22 Wasim opened an account with £650 at the London Bank. After one year the bank paid him interest. He then had £676 in his account.

a) Work out, as a percentage, London Bank's interest rate.

Holly opened an account at the Anglia Bank. Anglia Bank's interest rate was 5%. After one year, the bank paid her interest. The total amount in her account was then £1029.

b) Work out the amount with which she opened her account. [Edexcel]

23 A company bought a van that had a value of £12 000. Each year the value of the van depreciates by 25%.

Work out the value of the van at the end of three years. [Edexcel]

24 Harvey invests £4500 at a compound interest rate of 5% per annum. At the end of n complete years the investment has grown to £5469.78. Find the value of n. [Edexcel]

Key points

1 Simple ratios may be expressed using whole numbers, such as 4 : 6, or 1 : 2 : 4.

2 Ratios may be cancelled down in a similar way to fractions, so 4 : 6 is equivalent to 2 : 3.

3 Percentages may be thought of as ratios out of 100. To convert a percentage into a fraction, simply write the given percentage over a denominator of 100, then cancel down if appropriate.

4 Percentage increase or decrease may be calculated very efficiently using a multiplying factor. This may either be greater than 1 (increase) or less than 1 (decrease). For a 3% increase you would use 1.03; for a 3% decrease, 0.97.

5 Reverse percentage problems, such as 'find the price before VAT was applied', are best solved using multiplying factors. To reverse the calculation, simply divide by the multiplying factor instead (1.2 in the case of VAT at 20%).

6 Compound interest requires a fresh calculation to be done for each year. These problems are best solved using multiplying factors.

Internet Challenge 2

Investigating inflation

Each year the price of things goes up, and so does the amount people earn. This is called inflation, and is usually measured using percentages, to give an annual rate of inflation.

Here are some questions about inflation. You will need internet access to help you research the answers.

1 What is the meaning of the Retail Price Index (RPI)?

2 How often is it calculated?

3 What is the value of the present UK yearly rate of inflation, based on the RPI?

4 What is the meaning of the Bank of England base rate?

5 How often is it calculated?

6 What is the present value of the Bank of England base rate?

There was a four-year period in Germany's history when the purchasing power of printed banknotes fell to a purchasing power only one trillionth of what it had been before the inflation set in – in other words, prices soared a trillion times.

7 In which years did this happen?

8 Write the number one trillion in figures. There is an older, imperial definition of a trillion, and a newer, metric one, which are not the same. Give both answers.

> 'In October of 1993 the government created a new currency unit. One new dinar was worth one million of the old dinars. In effect, the government simply removed six zeroes from the paper money. This of course did not stop the inflation and between 1 October 1993 and 24 January 1995 prices increased by 5 quadrillion per cent. This number is a 5 with 15 zeros after it.'

9 To which country does this text refer?

10 What name do economists give to excessively high inflation like this?

CHAPTER 3

Powers and roots

In this chapter you will learn how to:

- work with simple powers and roots using mental methods
- use a calculator to compute harder powers and roots
- work with fractional and negative powers, and reciprocals
- use the laws of indices to simplify numerical expressions
- write large and small numbers using standard form
- calculate using standard form
- decompose integers into prime factors
- calculate highest common factors (HCFs) and lowest common multiples (LCMs).

You will also be challenged to:

- investigate astronomical numbers.

Starter: Roman numerals

The Romans used letters for numbers.

$$I = 1$$
$$V = 5$$
$$X = 10$$
$$L = 50$$
$$C = 100$$
$$D = 500$$
$$M = 1000$$

To add two letters, write the smaller one after the larger one. For example, VI = 5 + 1 = 6. To subtract two letters, write the smaller one before the larger one. For example, XL = 50 − 10 = 40.

Task 1
Turn these Roman numbers into ordinary numbers.
a) XVII   b) XIV   c) XLV
d) LXX   e) XCII   f) DCIX

Task 2
Write these ordinary numbers as Roman numerals.
a) 21   b) 24   c) 39
d) 212   e) 319   f) 47

Task 3
Film makers often use Roman numerals in their credits. *Star Wars* was originally released in MDCCCCLXXVII. *The Lion King* was released in MDCCCCXCIV. Write these two years as ordinary numbers.

3.1 Basic powers and roots

When a number is multiplied by itself, the result is called the **square** of the original number. For example, 5 squared is $5 \times 5 = 25$.

Reversing this process gives the **square root**. The square root of 25 is 5. This can be written using the $\sqrt{\ }$ symbol as $\sqrt{25} = 5$.

In fact, 25 has two different square roots, because -5 times -5 also makes 25, so you could say that -5 is another square root of 25. To distinguish between these two cases, we say that 5 is the **positive square root** of 25 and -5 is the **negative square root** of 25.

When a number is multiplied by itself and then by itself again, the result is called the **cube** of the original number. For example, 5 cubed is $5 \times 5 \times 5 = 125$.

Reversing this process gives the **cube root**. The cube root of 125 is 5. This can be written using the $\sqrt[3]{\ }$ symbol as $\sqrt[3]{125} = 5$. (There is no negative cube root for 125.)

You are expected to know the squares and cubes of some basic whole numbers, and to recognise the corresponding square roots and cube roots. Here are the squares you should learn.

$1^2 = 1$	$6^2 = 36$	$11^2 = 121$
$2^2 = 4$	$7^2 = 49$	$12^2 = 144$
$3^2 = 9$	$8^2 = 64$	$13^2 = 169$
$4^2 = 16$	$9^2 = 81$	$14^2 = 196$
$5^2 = 25$	$10^2 = 100$	$15^2 = 225$

You should also learn these cubes:

$1^3 = 1$	$3^3 = 27$	$5^3 = 125$
$2^3 = 8$	$4^3 = 64$	$10^3 = 1000$

The squares, cubes, square roots and cube roots of other numbers will normally be found using a calculator. Make sure that you know how to use the keys for this on your calculator; most calculators use similar keys but their locations on the keypad vary between different models.

EXAMPLE

Without using a calculator, obtain the values of:

a) 13^2 **b)** 5^3 **c)** $\sqrt{121}$ **d)** $\sqrt[3]{27}$

SOLUTION

a) $13^2 = \underline{169}$

> Make sure that you have learnt the basic squares and cubes so you can spot these answers by eye.

b) $5^3 = \underline{125}$

c) $\sqrt{121} = \underline{11}$

> The symbol $\sqrt{\ }$ means the positive square root only.

d) $\sqrt[3]{27} = \underline{3}$

EXAMPLE

Fred is answering an algebra problem. He has worked out that $x^2 = 81$. Give
two possible values of x that would complete Fred's answer.

SOLUTION

$x^2 = 81$

So $x = \sqrt{81}$ or $-\sqrt{81}$

$\quad = 9$ or -9

> You will meet this idea again later in the book,
> when you are working with quadratic equations.

EXAMPLE

Use your calculator to work out the values of:
a) 9.4^2 **b)** 2.5^3 **c)** $\sqrt{109}$ **d)** $\sqrt[3]{44.8}$

Round your answer to a sensible level of accuracy where appropriate.

SOLUTION

a) $9.4^2 = 88.36$

b) $2.5^3 = 15.625$

c) $\sqrt{109} = 10.440\ 306\ 51$

$\quad = 10.44$ (4 s.f.)

d) $\sqrt[3]{44.8} = 3.551\ 616\ 007$

$\quad = 3.552$ (4 s.f.)

> **a)** and **b)** are calculated using the x^2 and x^3 keys
> on a calculator. No rounding is needed.

> **c)** and **d)** are calculated using the $\sqrt{\ }$ and $\sqrt[3]{\ }$ keys on a calculator.
> The calculator generates a full screen of decimals.

> A good exam habit is to show your full calculator result…
> …as well as the rounded answer.

Take care when you work with negative numbers; you must use brackets if you
use your calculator.

The square of -3 is $(-3)^2 = (-3) \times (-3) = +9$ and not -9.

EXERCISE 3.1

Find the values of the following, without using a calculator.

1 5^2 **2** 2^3 **3** 7^2 **4** 3^3

5 $(-9)^2$ **6** 4^3 **7** $(-12)^2$ **8** 10^3

9 $\sqrt{144}$ **10** $\sqrt{225}$ **11** $\sqrt[3]{64}$ **12** $\sqrt{36}$

13 $\sqrt{196}$ **14** $\sqrt[3]{125}$ **15** $\sqrt{81}$ **16** $\sqrt[3]{1000}$

Use your calculator to work out the values of the following expressions. Round
your answers to 3 significant figures where appropriate.

17 19^2 **18** $(-1.8)^2$ **19** 14.6^2 **20** $(-9)^3$

21 16.3^3 **22** $(-1.2)^3$ **23** $\sqrt{13}$ **24** $\sqrt{300}$

25 $\sqrt[3]{24}$ **26** $\sqrt[3]{50}$ **27** $\sqrt{2.5}$ **28** $\sqrt[3]{6.8}$

29 Find x if $x^2 = 72$. Give your answers to 3 s.f.   **30** Find y if $y^3 = 38$. Give your answer to 4 s.f.

3.2 Higher powers and roots

Although squares and cubes occur frequently in mathematics, other (higher) whole number powers and roots may also be used. The notation x^n represents n factors of x multiplied together so, for example, 6^4 means $6 \times 6 \times 6 \times 6 = 1296$.

Similarly, higher roots may be obtained too, using this idea in reverse.
The 5th root of 32 is 2, because $2 \times 2 \times 2 \times 2 \times 2 = 32$.
Roots are denoted using fractional powers, so you would write this as $32^{\frac{1}{5}} = 2$.
The notation $x^{\frac{1}{n}}$ represents the nth root of x.

As with basic powers, you will solve simple problems by sight, but may use a calculator for harder ones. Make sure that you know how to use your calculator's power and root keys.

EXAMPLE

Without using a calculator, obtain the values of:

a) 7^3 **b)** 2^8 **c)** $81^{\frac{1}{4}}$ **d)** $125^{\frac{1}{3}}$

SOLUTION

a) $7^3 = 7 \times 7 \times 7$
$= 49 \times 7$
$= \underline{343}$

b) $2^8 = 2 \times 2 \times 2 \times 2 \times 2 \times 2 \times 2 \times 2$
$= 4 \times 4 \times 4 \times 4$
$= 16 \times 16$
$= \underline{256}$

c) Since $3 \times 3 \times 3 \times 3 = 81$,
$$81^{\frac{1}{4}} = \underline{3}$$

d) Since $5 \times 5 \times 5 = 125$
$$125^{\frac{1}{3}} = \underline{5}$$

EXAMPLE

Use your calculator to obtain the values of:

a) 14^4 **b)** 1.5^6 **c)** $1045^{\frac{1}{4}}$ **d)** $125^{\frac{1}{6}}$

Round your answers to 3 significant figures where appropriate.

SOLUTION

a) $14^4 = \underline{38\ 416}$

b) $1.5^6 = 11.390\ 625$
$= \underline{11.4}$ (3 s.f.)

c) $1045^{\frac{1}{4}} = 5.685\ 636\ 266$ ← | Remember to show the full calculator values as well as your final rounded answer.
$= \underline{5.69}$ (3 s.f.)

d) $125^{\frac{1}{6}} = 2.236\ 067\ 977$
$= \underline{2.24}$ (3 s.f.)

EXERCISE 3.2

Without using a calculator, find the exact values of:

1 3^4 **2** 100^3 **3** $216^{\frac{1}{3}}$ **4** 10^4

5 9^3 **6** 2^5 **7** $32^{\frac{1}{5}}$ **8** 2^{10}

9 $1000^{\frac{1}{3}}$ **10** $400^{\frac{1}{2}}$ **11** 12^3 **12** $16^{\frac{1}{4}}$

Use your calculator to find the value of each expression. Round your answers as indicated.

13 12^5 (3 s.f.)   **14** 9.8^4 (4 s.f.)   **15** 1.3^7 (3 d.p.)   **16** 0.95^4 (4 d.p.)

17 $6^{\frac{1}{3}}$ (4 s.f.)   **18** $12^{\frac{1}{4}}$ (3 s.f.)   **19** $6.3^{\frac{1}{5}}$ (3 d.p.)   **20** $41.6^{\frac{1}{3}}$ (3 s.f.)

3.3 Fractional (rational) indices

Some expressions contain fractional indices, for example $8^{\frac{2}{3}}$. These require two processes to be applied together – you need to raise to a power, and also apply a root.

The top of the fraction tells you what power to apply – squaring in this case.

The bottom of the fraction tells you what root to apply – cube root in this case.

EXAMPLE

Find the value of $8^{\frac{2}{3}}$, without using a calculator.

SOLUTION

Method 1
$8^{\frac{2}{3}} = 8$ squared and then cube rooted
$= 64$ cube rooted
$= \underline{4}$

Method 2

$8^{\frac{2}{3}}$ = 8 cube rooted and then squared

$\quad$ = 2 squared

$\quad$ = $\underline{4}$

Notice that the order of the two processes did not affect the final result. Method 2 is probably more efficient, because the intermediate numbers you are working with are smaller.

EXERCISE 3.3

Without using a calculator, write the following expressions as simply as possible.

1 $4^{\frac{3}{2}}$

2 $27^{\frac{2}{3}}$

3 $25^{\frac{3}{2}}$

4 $9^{\frac{5}{2}}$

5 $36^{\frac{3}{2}}$

6 $64^{\frac{3}{2}}$

7 $64^{\frac{2}{3}}$

8 $100^{\frac{5}{2}}$

9 $81^{\frac{3}{4}}$

10 $125^{\frac{4}{3}}$

11 $16^{\frac{3}{2}}$

12 $16^{\frac{3}{4}}$

3.4 Negative powers

So far you have used *positive* powers: for example 10^3 tells you to *multiply* by 10, then by 10 again, and then by 10 again, so $10^3 = 1000$.

Negative powers are also used for numerical expression. For example, 10^{-3} tells you to *divide* by 10, then by 10 again, and then by 10 again, so $10^{-3} = 1/1000$.

Here is a general rule for negative powers:

$$x^{-n} = \frac{1}{x^n}$$

(This rule may be used as long as x is not zero; if x is zero then $\dfrac{1}{x^n}$ is not defined.)

EXAMPLE

Work out the values of:

a) 2^{-3} $\qquad$ **b)** 10^{-4} $\qquad$ **c)** 6^{-2}

SOLUTION

a) $2^{-3} = \dfrac{1}{2^3}$

$\qquad = \dfrac{1}{2 \times 2 \times 2}$

$\qquad = \underline{\dfrac{1}{8}}$

b) $10^{-4} = \dfrac{1}{10^4}$

$= \dfrac{1}{10 \times 10 \times 10 \times 10}$

$= \dfrac{1}{10000}$

c) $6^{-2} = \dfrac{1}{6^2}$

$= \dfrac{1}{6 \times 6}$

$= \dfrac{1}{36}$

The **reciprocal** of a whole number is 1 divided by that number. For example, the reciprocal of 2 is $\frac{1}{2}$, and the reciprocal of 4 is $\frac{1}{4}$.

Using power notation, reciprocals are indicated by a power of -1. So $2^{-1} = \frac{1}{2}$, and $4^{-1} = \frac{1}{4}$.

EXAMPLE

Work out the values of:

a) 8^{-1} **b)** 25^{-1} **c)** 3^{-1}

SOLUTION

a) $8^{-1} = \frac{1}{8}$ (or 0.125)

> Fractions are usually preferable to decimals in this type of question, since they are exact; decimals might not be.

b) $25^{-1} = \frac{1}{25}$ (or 0.04)

> Note that $\frac{1}{3}$ does not have an exact terminating decimal form, so this answer is best given as a fraction.

c) $3^{-1} = \frac{1}{3}$

Fractions, too, have reciprocals. To find the reciprocal of a fraction, simply interchange the top and bottom of the fraction.

To raise a fraction to a negative power, use the equivalent positive power and then interchange the top and bottom.

EXAMPLE

Work out the values of:

a) $\left(\dfrac{2}{3}\right)^{-1}$ **b)** $\left(\dfrac{5}{7}\right)^{-2}$ **c)** $\left(\dfrac{16}{25}\right)^{-\frac{1}{2}}$

SOLUTION

a) $\left(\dfrac{2}{3}\right)^{-1} = \dfrac{3}{2}$

b) $\left(\dfrac{5}{7}\right)^{-2} = \left(\dfrac{7}{5}\right)^{2}$

$$= \dfrac{7^2}{5^2}$$

$$= \underline{\dfrac{49}{25}}$$

c) $\left(\dfrac{16}{25}\right)^{-\frac{1}{2}} = \left(\dfrac{25}{16}\right)^{\frac{1}{2}}$

$$= \dfrac{25^{\frac{1}{2}}}{16^{\frac{1}{2}}}$$

$$= \underline{\dfrac{5}{4}}$$

EXERCISE 3.4

Work out the values of these, leaving your answers as exact fractions.

1 3^{-2} **2** 10^{-3} **3** 5^{-2} **4** 4^{-1}

5 9^{-2} **6** 4^{-2} **7** 2^{-5} **8** 10^{-1}

9 5^{-1} **10** 20^{-2}

Evaluate these expressions, giving your answers as exact fractions.

11 $\left(\dfrac{3}{5}\right)^{-1}$ **12** $\left(\dfrac{4}{3}\right)^{-1}$ **13** $\left(\dfrac{25}{4}\right)^{-\frac{1}{2}}$ **14** $\left(\dfrac{4}{5}\right)^{-2}$

15 $\left(\dfrac{2}{3}\right)^{-3}$ **16** $\left(\dfrac{1}{2}\right)^{-1}$ **17** $\left(\dfrac{9}{64}\right)^{-\frac{1}{2}}$ **18** $\left(\dfrac{5}{3}\right)^{-2}$

19 $\left(\dfrac{100}{49}\right)^{-\frac{3}{2}}$ **20** $\left(\dfrac{27}{64}\right)^{-\frac{2}{3}}$

3.5 The laws of indices

Another name for a power is an **index**, so powers are often called **indices**.
There are several laws of indices that can help you to simplify index problems.

EXAMPLE

Write $10^3 \times 10^5$ as a single power of 10.

SOLUTION

$10^3 = 10 \times 10 \times 10$

and $\quad 10^5 = 10 \times 10 \times 10 \times 10 \times 10$

so $\quad 10^3 \times 10^5 = (10 \times 10 \times 10) \times (10 \times 10 \times 10 \times 10 \times 10)$

$\qquad\qquad = 10 \times 10 \times 10 \times 10 \times 10 \times 10 \times 10 \times 10$

$\qquad\qquad = \underline{10^8}$

You could have solved this example much more efficiently just by adding the indices to give the final result: $10^3 \times 10^5 = 10^{3+5} = 10^8$.

EXAMPLE

Write $2^9 \div 2^6$ as a single power of 2.

SOLUTION

$2^9 = 2 \times 2 \times 2 \times 2 \times 2 \times 2 \times 2 \times 2 \times 2$

and $\quad 2^6 = 2 \times 2 \times 2 \times 2 \times 2 \times 2$

so $\quad 2^9 \div 2^6 = \dfrac{2 \times 2 \times 2 \times 2 \times 2 \times 2 \times 2 \times 2 \times 2}{2 \times 2 \times 2 \times 2 \times 2 \times 2}$

$\qquad\qquad = \dfrac{\cancel{2}^1 \times \cancel{2}^1 \times \cancel{2}^1 \times \cancel{2}^1 \times \cancel{2}^1 \times \cancel{2}^1 \times 2 \times 2 \times 2}{\cancel{2}^1 \times \cancel{2}^1 \times \cancel{2}^1 \times \cancel{2}^1 \times \cancel{2}^1 \times \cancel{2}^1}$

$\qquad\qquad = \dfrac{2 \times 2 \times 2}{1}$

$\qquad\qquad = \underline{2^3}$

Once again, there is a more efficient method. You could have just subtracted the indices to give the final result: $2^9 \div 2^6 = 2^{9-6} = 2^3$.

EXAMPLE

Write $(5^4)^2$ as a single power of 5.

SOLUTION

$5^4 = 5 \times 5 \times 5 \times 5$

so $\quad (5^4)^2 = (5 \times 5 \times 5 \times 5) \times (5 \times 5 \times 5 \times 5)$

$\qquad\qquad = 5 \times 5 \times 5 \times 5 \times 5 \times 5 \times 5 \times 5$

$\qquad\qquad = \underline{5^8}$

Again, it would be quicker to multiply the indices to give the final result:
$(5^4)^2 = 5^{4 \times 2} = 5^8$.

The three examples above illustrate three general laws of indices, which may be expressed symbolically like this:

$$x^a \times x^b = x^{a+b} \qquad \text{when } \textit{multiplying, add} \text{ the indices}$$
$$x^a \div x^b = x^{a-b} \qquad \text{when } \textit{dividing, subtract} \text{ the indices}$$
$$(x^a)^b = x^{ab} \qquad \text{when } \textit{raising to a power, multiply} \text{ the indices}$$

You should look for opportunities to use these rules whenever you are simplifying numerical expressions involving indices.

The laws of indices allow you to assign a meaning to a power 0, such as 7^0. For example, using the laws of indices, $7^5 \div 7^5 = 7^{5-5} = 7^0$, but since $7^5 \div 7^5 = 1$ the value of 7^0 must be 1. More generally, $x^0 = 1$, for any value of x (provided x is not 0). Do not confuse this with x^1, which is just x.

This is loosely stated in words as 'anything to the power zero equals 1'. This rule covers all cases except 0^0, which is not defined to have a value. To summarise:

$$x^1 = x$$
$$x^0 = 1 \qquad \text{provided } x \text{ is not 0}$$
$$0^0 \qquad \text{is not defined}$$

EXAMPLE

Use the laws of indices to write these expressions as simply as possible.

a) $8^3 \times 8^4$ **b)** $5^{10} \div 5^8$ **c)** $(4^3)^2$ **d)** 8^0

SOLUTION

a) $8^3 \times 8^4 = 8^{3+4}$
$$= \underline{8^7}$$

b) $5^{10} \div 5^8 = 5^{10-8}$
$$= \underline{5^2}$$

Although you can do each of these in your head, it is good discipline to write down the steps of the simplification as shown here.

This helps you to master the laws of indices, and also lets your teacher follow your reasoning clearly.

c) $(4^3)^2 = 4^{3 \times 2}$
$$= \underline{4^6}$$

d) $8^0 = \underline{1}$

EXERCISE 3.5

Simplify each of these expressions, giving your answer as a number to a single power.

1 $2^3 \times 2^4$ **2** $5^4 \times 5^3$ **3** $8^2 \times 8^7$ **4** $6^5 \div 6^2$

5 $9^4 \div 9^3$ **6** $3^{10} \div 3^9$ **7** $(2^3)^4$ **8** $(3^2)^3$

9 $3^2 \times 3^0$ **10** $2^{\frac{3}{2}} \times 2^{\frac{5}{2}}$ **11** $3^{\frac{1}{2}} \times 3^{\frac{3}{2}}$ **12** 6×2^0

Work out each of these, giving your answer as an ordinary number.

13 $2^3 \times 2^2$ **14** $3^6 \div 3^5$ **15** $10^3 \times 10^3$ **16** $4^8 \div 4^6$

17 $(2^2)^3$ **18** $3^2 \times 3^3$ **19** $2^3 \times 2$ **20** $7^6 \div 7^4$

21 $3^{12} \div 3^9$ **22** $(3^0)^4$ **23** $(10^2)^3$ **24** $(5^2)^0$

3.6 Standard index form

Standard index form, or **standard form** as it is often called, is a very convenient way of writing very large or very small quantities. You start with a number between 1 and 10, and multiply (or divide) by a suitable number of powers of 10. For example, the number 3 000 000 could be written as 3×10^6, meaning that the 3 has to be multiplied by six powers of 10.

EXAMPLE

Write these numbers in standard form.
a) 4 000 000 000   **b)** 36 000   **c)** 14 300 000

SOLUTION

a) 4 000 000 000 = $\underline{4 \times 10^9}$

b) 36 000 = $\underline{3.6 \times 10^4}$ ←   Note that 36×10^3 would not be right here, because 36 does not lie between 1 and 10.

c) 14 300 000 = $\underline{1.43 \times 10^7}$

For numbers smaller than 1 you divide by powers of 10, instead of multiplying. This gives rise to a negative index of 10, instead of a positive one.

EXAMPLE

Write these numbers in standard form.
a) 0.0006   **b)** 0.000 000 25   **c)** 0.000 000 000 001 8

SOLUTION

Count the number of hops needed to restore the 2.05 to the original number:
0.0000002.05
There are 7 altogether.

a) 0.0006 = $\underline{6 \times 10^{-4}}$

b) 0.000 000 205 = $\underline{2.05 \times 10^{-7}}$

c) 0.000 000 000 001 8 = $\underline{1.8 \times 10^{-12}}$

Remember that numbers in standard form *always* have the decimal point after the first non-zero digit.

EXERCISE 3.6

Write these numbers in standard index form.

1 350 000   **2** 40 000   **3** 352 000 000

4 19 300 000   **5** 765   **6** 0.0045

7 0.8   **8** 0.002 03   **9** 0.000 000 000 827

10 0.000 33

Write these as ordinary numbers

11 7.4×10^6 **12** 2.15×10^7 **13** 1.05×10^5

14 2×10^9 **15** 8.4×10^3 **16** 5×10^{-3}

17 2.5×10^{-6} **18** 1.004×10^{-7} **19** 8.3×10^{-11}

20 5.05×10^{-4}

3.7 Calculating with numbers in standard form

In order to add or subtract two numbers in standard form, you have to make sure that the digits line up in their correct place values. You can do this either by converting them into ordinary numbers, or adjusting them so they both have the same power of 10. This latter method means that your working contains index numbers that are not 'standard', but this does not matter provided the final answer is expressed correctly.

EXAMPLE

Add together 4.2×10^4 and 7.3×10^5. Write your answer in standard form.

SOLUTION

Method 1
$4.2 \times 10^4 = 42\ 000$ and $7.3 \times 10^5 = 730\ 000$

so $4.2 \times 10^4 + 7.3 \times 10^5 = 42\ 000 + 730\ 000$
$$= 772\ 000$$
$$= \underline{7.72 \times 10^5}$$

Method 2
Write both numbers using the lower power (10^4):

$7.3 \times 10^5 = 73 \times 10^4$

Thus $4.2 \times 10^4 + 7.3 \times 10^5 = 4.2 \times 10^4 + 73 \times 10^4$
$$= 77.2 \times 10^4$$
$$= \underline{7.72 \times 10^5}$$

Standard form numbers can be used in multiplication or division quite easily.

EXAMPLE

Multiply 3×10^5 and 2.5×10^7.

SOLUTION

$$3 \times 10^5 \times 2.5 \times 10^7 = 3 \times 2.5 \times 10^5 \times 10^7$$
$$= 7.5 \times 10^{5+7}$$
$$= 7.5 \times 10^{12}$$

> Multiply the two number parts, and multiply the two power terms.

Sometimes the final answer is not in standard index form, however, and needs a little adjustment.

EXAMPLE

Work out $(4 \times 10^8) \div (5 \times 10^2)$.

SOLUTION

$$(4 \times 10^8) \div (5 \times 10^2) = (4 \div 5) \times (10^8 \div 10^2)$$
$$= 0.8 \times 10^6$$
$$= 8.0 \times 10^{-1} \times 10^6$$
$$= 8 \times 10^5$$

> Divide the two number parts, and divide the two power terms.

> Note the adjustment into standard form here.

A good calculator will allow you to type in the numbers using standard form. Some older models require the use of an `EXP` key to do this, but on the latest models you will find a `×10ˣ` key that allows a more natural entry notation. Make sure you know how your calculator works!

EXAMPLE

If $a = 3.55 \times 10^8$ and $b = 2.065 \times 10^9$ use your calculator to work out the values of each of these expressions. Give your answer in standard form, correct to 3 significant figures.

a) ab **b)** $\dfrac{a}{b}$ **c)** $\sqrt{a + 2b}$

SOLUTION

a) $ab = 3.55 \times 10^8 \times 2.065 \times 10^9$
$$= 7.330\,75 \times 10^{17}$$
$$= 7.33 \times 10^{17} \text{ (3 s.f.)}$$

b) $\dfrac{a}{b} = (3.55 \times 10^8) \div (2.065 \times 10^9)$

$= 0.171\,912\,8329$

$= \underline{1.72 \times 10^{-1}}$ (3 s.f.)

c) $a + 2b = 3.55 \times 10^8 + 2 \times (2.065 \times 10^9)$

$= \underline{4\,485\,000\,000}$

Thus $\sqrt{a + 2b} = \sqrt{4\,485\,000\,000}$

$= 66\,970.1426$

$= 67\,000$ (3 s.f.)

$= \underline{6.70 \times 10^4}$ (3 s.f.)

> First work out just $a + 2b$...
> ...then square root it.

EXERCISE 3.7

Work out the answers to these calculations without using a calculator.

Give your answers in standard form.

1 $(4 \times 10^6) + (7 \times 10^8)$

2 $(2.4 \times 10^5) + (1.8 \times 10^6)$

3 $(7 \times 10^7) - (5 \times 10^5)$

4 $(3 \times 10^5) - (8 \times 10^4)$

5 $(1.2 \times 10^{10}) \times (5 \times 10^7)$

6 $(7 \times 10^6) \times (6 \times 10^7)$

7 $(4 \times 10^{10}) \div (8 \times 10^7)$

8 $(3 \times 10^6) \div (4 \times 10^{-3})$

9 In the year 2004 a total of 2.17×10^8 passengers passed through UK airports. 6.7×10^7 of these passengers passed through Heathrow Airport. How many of the UK passengers did not pass through Heathrow? Give your answer as an ordinary number.

Use your calculator to evaluate these expressions.

Give your answers in standard form, correct to 3 significant figures.

10 $(2.45 \times 10^7) \div (8.22 \times 10^{11})$

11 $\dfrac{(3.5 \times 10^7) + (4.8 \times 10^9)}{(8.4 \times 10^6)}$

12 $184\,000 \div 0.0023$

13 $(1.5 \times 10^7)^2$

14 $\dfrac{(1.2 \times 10^6) - (4.8 \times 10^4)}{7 \times 10^{-5}}$

15 $(2.8 \times 10^{10}) \div 0.15$

16 $x = 4 \times 10^{2n}$ where n is a positive integer.

Find an expression for

a) $\sqrt{x}$ **b)** x^3

Simplify each expression and write your answers in standard form.

17 Light takes 3.33×10^{-9} seconds to travel 1 metre.

Pluto is 4.4 billion kilometres from the Sun.

How long does light from the Sun take to reach Pluto?

Give your answer in hours and minutes, to the nearest minute.

3.8 Factors, multiples and primes

A **multiple** of a number is the result of multiplying it by a whole number.

The multiples of 4 are 4, 8, 12, 16,…

A **factor** of a number is a whole number that divides exactly into it, with no remainder.

The factors of 12 are 1, 2, 3, 4, 6, 12.

A **prime** number is a whole number with exactly two factors, namely 1 and itself. The number 1 is not normally considered to be prime, so the prime numbers are 2, 3, 5, 7, 11,…

If a large number is not prime, it can be written as the product of a set of prime factors in a unique way. For example, 12 can be written as $2 \times 2 \times 3$.

A **factor tree** is a good way of breaking a large number into its prime factors. The next example shows how this is done.

EXAMPLE

Write the number 180 as a product of its prime factors.

SOLUTION

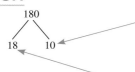

Begin by splitting the 180 into a product of two parts. You could use 2 times 90, or 4 times 45, or 9 times 20, for example. The result at the end will be the same in any case. Here we begin by using 18 times 10.

Since neither 18 nor 10 is a prime number, repeat the factorising process.

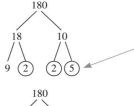

18 has been broken down into 9 times 2, and 10 into 2 times 5. The 2s and the 5 are prime, so they are circled and the tree stops there.

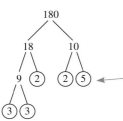

The 9 is not prime, so the process can continue.

The factor tree stops growing when all the branches end in circled prime numbers.

So $180 = 2 \times 2 \times 3 \times 3 \times 5$
$= 2^2 \times 3^2 \times 5$

2^2 means the factor 2 is used twice (two squared). If it had been used three times, you would write 2^3 (two cubed).

EXERCISE 3.8

1 List all the prime numbers from 1 to 40 inclusive.
You should find that there are 12 such prime numbers altogether.

2 Use your result from question **1** to help answer these questions:
  a) How many primes are there between 20 and 40 inclusive?
  b) What is the next prime number above 31?
  c) Find two prime numbers that multiply together to make 403.
  d) Write 91 as a product of two prime factors.

3 Use the factor tree method to obtain the prime factorisation of:
  a) 80   **b)** 90   **c)** 450.

4 Use the factor tree method to obtain the prime factorisation of:
  a) 36   **b)** 81   **c)** 144.
  What do you notice about all three of your answers?

5 When 56 is written as a product of primes, the result is $2^a \times b$ where a and b are positive integers. Find the values of a and b.

3.9 Highest common factor, HCF

Consider the numbers 12 and 20. The number 2 is a factor of 12, and 2 is also a factor of 20. Thus 2 is said to be a **common factor** of 12 and 20.

Likewise, the number 4 is also a factor of both 12 and 20, so 4 is also a common factor of 12 and 20.

It turns out that 12 and 20 have no common factor larger than this, so 4 is said to be the **highest common factor (HCF)** of 12 and 20. You can check that 4 really is the highest common factor by writing 12 as 4×3 and 20 as 4×5; the 3 and 5 share no further factors.

EXAMPLE

Find the highest common factor (HCF) of 30 and 80.

SOLUTION

By inspection, it looks as if the highest common factor may well be 10.

Check: $30 = 10 \times 3$, and $80 = 10 \times 8$

> By inspection means that you can just spot the answer by eye, without any formal working.

and clearly 3 and 8 have no further factors in common.
So <u>HCF of 30 and 80 is 10</u>

There is an alternative, more formal, method for finding highest common factors. It requires the use of **prime factorisation**.

EXAMPLE

Use prime factorisation to find the highest common factor of 30 and 80.

SOLUTION

By the factor tree method: $30 = 2 \times 3 \times 5$

Similarly, $80 = 2^4 \times 5$

So HCF of 30 and 80 $= 2 \times 5$
$= \underline{10}$

> Look at the 2's: 30 has one of them, 80 has four.
> Pick the lower number: one 2
>
> Look at the 3's: 30 has one of them, but 80 has none.
> Pick the lower number: no 3s
>
> Look at the 5's: 30 has one of them, and 80 has one.
> Pick the lower number: one 5

The prime factorisation method involves a lot of steps, but it is particularly effective when working with larger numbers, as in this next example.

EXAMPLE

Use prime factorisation to find the highest common factor of 96 and 156.

SOLUTION

By the factor tree method: $96 = 2^5 \times 3$
and $156 = 2^2 \times 3 \times 13$

HCF of 96 and 156 $= 2^2 \times 3$
$= 4 \times 3$
$= \underline{12}$

EXERCISE 3.9

1 Use the method of inspection to write down the highest common factor of each pair of numbers. Check your result in each case.

    a) 12 and 18   **b)** 45 and 60   **c)** 22 and 33

    d) 27 and 45   **e)** 8 and 27   **f)** 26 and 130

2 Write each of the following numbers as the product of prime factors. Hence find the highest common factor of each pair of numbers.

    a) 20 and 32   **b)** 36 and 60   **c)** 80 and 180

    d) 72 and 108   **e)** 120 and 195   **f)** 144 and 360

3 Find the highest common factor of each pair of numbers.

    a) 12 and 30   **b)** 24 and 36   **c)** 96 and 120

    d) 90 and 140   **e)** 78 and 102   **f)** 48 and 70

3.10 Lowest common multiple, LCM

Consider the numbers 15 and 20.

The multiples of 15 are 15, 30, 45, 60, 75,…

The multiples of 20 are 20, 40, 60, 80,…

Any multiple that occurs in both lists is called a **common multiple**.

The smallest of these is the **lowest common multiple (LCM)**. In this example, the LCM is 60.

There are several methods for finding lowest common multiples. As with highest common factors, one of these methods is based on prime factorisation.

EXAMPLE

Find the lowest common multiple of 48 and 180.

SOLUTION

First, find the prime factors of each number using a factor tree if necessary.

$$48 = 2^4 \times 3$$
$$180 = 2^2 \times 3^2 \times 5$$

Look at the powers of 2:

$$48 = 2^4 \times 3$$
$$180 = 2^2 \times 3^2 \times 5$$

There are 4 factors of 2 in 48, but only 2 in 180. Pick the higher of these: 4

Next, the powers of 3:

$$48 = 2^4 \times 3$$
$$180 = 2^2 \times 3^2 \times 5$$

There is 1 factor of 3 in 48, but 2 in 180. Pick the higher of these: 2

Finally, the powers of 5:

$$48 = 2^4 \times 3$$
$$180 = 2^2 \times 3^2 \times 5$$

There is no factor of 5 in 48, but 1 in 180. Pick the higher of these: 1

Putting all of this together:

$$
\begin{aligned}
\text{LCM of 48 and 180} &= 2^4 \times 3^2 \times 5 \\
&= 16 \times 9 \times 5 \\
&= 144 \times 5 \\
&= \underline{720}
\end{aligned}
$$

An alternative method is based on the fact that the product of the LCM and the HCF is the same as the product of the two original numbers. This gives the following result:

$$\text{LCM of } a \text{ and } b = \frac{a \times b}{\text{HCF of } a \text{ and } b}$$

This can be quite a quick method if the HCF is easy to spot.

EXAMPLE

Find the lowest common multiple of 70 and 110.

SOLUTION

By inspection, HCF is 10

So:

$$\text{LCM} = \frac{70 \times 110}{10}$$
$$= 7 \times 110$$
$$= \underline{770}$$

It is also possible to find the HCF and LCM of three (or more) numbers. The prime factorisation method remains valid here, but other shortcut methods can fail. This example shows you how to adapt the factorisation method when there are three numbers.

EXAMPLE

Find the HCF and LCM of 16, 24 and 28.

SOLUTION

Write these as products of prime factors:

$16 = 2^4$

$24 = 2^3 \times 3$

$28 = 2^2 \times 7$

HCF of 16, 24 and 28 is $2^2 = \underline{4}$

LCM of 16, 24 and 28 is $2^4 \times 3 \times 7 = 16 \times 21 = \underline{336}$

> The lowest number of 2s from 2^4 or 2^3 or 2^2 is 2^2

> The highest number of 2s from 2^4 or 2^3 or 2^2 is 2^4

EXERCISE 3.10

Find the lowest common multiple (LCM) of each of these pairs of numbers.

1 12 and 20	**2** 16 and 26	**3** 18 and 45
4 25 and 40	**5** 36 and 48	**6** 6 and 20
7 14 and 22	**8** 30 and 50	**9** 36 and 60
10 44 and 55	**11** 16 and 36	**12** 28 and 42
13 18 and 20	**14** 14 and 30	**15** 27 and 36
16 33 and 55		

17 a) Write 60 and 84 as products of their prime factors.
  b) Hence find the LCM of 60 and 84.

18 a) Write 66 and 99 as products of their prime factors.
   b) Hence find the LCM of 66 and 99.
   c) Find also the HCF of 66 and 99.

19 a) Write 10, 36 and 56 as products of their prime factors.
   b) Work out the Highest Common Factor, HCF, of 10, 36 and 56.
   c) Work out the Lowest Common Multiple, LCM, of 10, 36 and 56.

20 a) Write 40, 48 and 600 as products of their prime factors.
   b) Work out the Highest Common Factor, HCF, of 40, 48 and 600.
   c) Work out the Lowest Common Multiple, LCM, of 40, 48 and 600.

21 Virginia has two friends who regularly go round to her house. Joan goes round once every 4 days and India goes round once every 5 days. How often are both friends at Virginia's house together?

22 Eddie owns three motorcycles. He cleans the Harley once every 8 days, the Honda once every 10 days and the Kawasaki once every 15 days. Today he cleaned all three motorcycles. When will he next clean all three motorcycles on the same day?

REVIEW EXERCISE 3

1 Write down the values of:
   a) 5^3 **b)** $\sqrt{144}$ **c)** 15^2

2 Work out the values of:
   a) 10^{-2} **b)** 4^{-3} **c)** $8^{\frac{2}{3}}$

3 Write each of these using a single power.
   a) $6^3 \times 6^2$ **b)** $3^{10} \div 3^7$ **c)** $(4^3)^2$

4 Use a factor tree to find the prime factorisation of:
   a) 70 **b)** 124 **c)** 96 **d)** 240

5 a) Find the Highest Common Factor (HCF) of 24 and 56.
   b) Find the Lowest Common Multiple (LCM) of 24 and 56.

6 a) Write down the Highest Common Factor (HCF) of 20 and 22.
   b) Hence find the Lowest Common Multiple (LCM) of 20 and 22.

7 a) Write 360 in the form $2^a \times 3^b \times 5^c$.
   b) Write $2^4 \times 3^2 \times 5$ as an ordinary number.

8 Who is right? Explain carefully.

Chuck says 'If the HCF of two numbers is 1, then they must both be primes.'

Lilian says 'Not necessarily true.'

9 Arrange these numbers in order of size, smallest first:
3.2×10^8, 7.6×10^{-2}, 1.4×10^9, 15 300

10 Find the value of $\sqrt{(2 \times 2 \times 3 \times 3 \times 5 \times 5)}$. [Edexcel]

11 Evaluate:
   a) 3^{-2} b) $36^{\frac{1}{2}}$ c) $27^{\frac{2}{3}}$ d) $\left(\dfrac{16}{81}\right)^{-\frac{3}{4}}$ [Edexcel]

12 Work out:
   a) 4^0 b) 4^{-2} c) $16^{\frac{3}{2}}$ [Edexcel]

13 Work out the values of
   a) $(2^2)^3$ b) $(\sqrt{3})^2$ c) $\sqrt{2^4 \times 9}$ [Edexcel]

14 a) Write 84 000 000 in standard form.
   b) Work out:

   $$\frac{84\,000\,000}{4 \times 10^{12}}$$

   Give your answer in standard form. [Edexcel]

You may use your calculator for the remaining questions.

15 a) Work out the value of 5^3.
   b) Work out the value of
   i) $\sqrt{(4.5^2 - 0.5^2)}$
        Write down all the figures on your calculator display.
   ii) Write your answer correct to 2 decimal places. [Edexcel]

16 Calculate the value of

   $$\frac{5.98 \times 10^8 + 4.32 \times 10^9}{6.14 \times 10^{-2}}$$

   Give your answer in standard form correct to 3 significant figures. [Edexcel]

17 420 000 carrot seeds weigh 1 gram

   a) Write the number 420 000 in standard form.

   b) Calculate the weight, in grams, of one carrot seed.

   Give your answer in standard form, correct to 2 significant figures. [Edexcel]

18 $y^2 = \dfrac{ab}{a+b}$

   $a = 3 \times 10^8$
   $b = 2 \times 10^7$
   Find y. Give your answer in standard form correct to 2 significant figures. [Edexcel]

19 Pens cost 25p each. Mr Smith spends £120 on pens.
   Work out the number of pens he gets for £120. [Edexcel]

20 The number 1104 can be written as $3 \times 2^c \times d$, where c is a whole number and d is a prime number. Work out the value of c and the value of d.

[Edexcel]

21 a) Express 72 and 96 as products of their prime factors.

b) Use your answer to part **a)** to work out the Highest Common Factor of 72 and 96.

[Edexcel]

Key points

1 Powers tell you how many times a number is to be multiplied by itself.
For example, $2^4 = 2 \times 2 \times 2 \times 2$.
Powers are also called indices.

2 The reverse of raising to a power is taking a root.
So, for example, the fourth root of 16 is 2.
This may be written as $\sqrt[4]{16}$ or $16^{\frac{1}{4}}$.

3 Fractional powers indicate a combination of raising to a power and taking a root.
The power $\frac{3}{2}$, for example, tells you to cube and also square root the number.

4 n^{-1} indicates the reciprocal of n and takes the value $\frac{1}{n}$.
The reciprocal of $\frac{a}{b}$ is $\frac{b}{a}$.

5 Negative powers indicate reciprocals, thus $3^{-2} = \frac{1}{3^2}$.

6 There are three key laws of indices:

$x^a \times x^b = x^{a+b}$ when multiplying, add the indices
$x^a \div x^b = x^{a-b}$ when dividing, subtract the indices
$(x^a)^b = x^{ab}$ when raising to a power, multiply the indices

7 Remember also that:

$x^1 = x$
$x^0 = 1$ (provided x is not 0)
0^0 is not defined.

8 Very large or very small numbers may be written in the form $a \times 10^n$,
where n is a whole number and $1 \leqslant a < 10$.
This is called standard (index) form.

9 Non-prime whole numbers may be written as a product of primes, using the factor tree method. This leads to a powerful method of working out the Highest Common Factor or Lowest Common Multiple of two numbers.

10 Sometimes you may be able to spot HCFs or LCMs by inspection. This result might help you to check them:

$$\text{LCM of } a \text{ and } b = \frac{a \times b}{\text{HCF of } a \text{ and } b}$$

Internet Challenge 3

Astronomical numbers

Astronomers work with very large numbers, so they often use standard index form, sometimes alternatively called scientific notation.

Here are some astronomical statements with missing values. The values are given, in jumbled-up order, to the right. Use the internet to help you decide which answer belongs with which statement.

1 Astronomers have calculated that the mass of the Sun is about ▢▢▢▢▢ kg.

2.998×10^8

2 The Sun is thought to have formed about ▢▢▢▢▢▢ years ago.

10^{-9}

3 Each second the Sun's mass decreases by about ▢▢▢▢▢▢ tonnes.

6×10^3

4 The surface temperature of the Sun has been measured to be about ▢▢▢▢▢▢ degrees C.

10^{11}

5 It takes our solar system about ▢▢▢▢▢▢ years to make one revolution around the Milky Way galaxy.

2×10^{30}

6 Light travels through space at a speed of ▢▢▢▢▢▢ metres per second.

2.25×10^8

7 Visible light has a wavelength of about ▢▢▢▢▢▢ metres.

4×10^6

8 X-rays can have wavelengths as short as ▢▢▢▢▢▢ metres.

2.8×10^6

9 The Andromeda galaxy is so remote that light from it takes ▢▢▢▢▢▢ years to reach us.

5×10^9

10 It is thought that the Universe contains about ▢▢▢▢▢▢ individual galaxies.

5.5×10^{-7}

CHAPTER 4

Working with algebra

In this chapter you will revise and extend earlier work on how to:

- substitute numbers into formulae and expressions
- work with indices
- expand brackets and collect like terms.

You will learn how to:

- factorise algebraic expressions
- generate formulae
- change the subject of a formula.

You will also be challenged to:

- investigate the language of algebra.

Starter: **Right or wrong?**

Each question is followed by two possible answers. Identify the right one – and explain what slip might have caused the wrong one in each case.

1 3×0 3   or   0

2 $6 - 5 + 1$ 0   or   2

3 5^2 10   or   25

4 $2 + 3 \times 4$ 20   or   14

5 $4 + 10 \div 2$ 9   or   7

6 $(-3)^2$ 9   or -9

4.1 Substituting numbers into formulae and expressions

In this chapter you will be using letters in expressions to stand for mathematical quantities. Sometimes you are told the numerical value of each letter. It is then possible to work out the value of the corresponding expression.

For example, $x + 3$ is an algebraic expression. If x has the value 5 then, clearly, $x + 3$ has the value 8.

Care must be taken when substituting given numbers into more intricate expressions. In particular, be careful with minus signs and squares, cubes, etc. Remember the order of priorities described by BIDMAS – you should work out Brackets first, then Indices (squares, cubes, etc.). Next come Division and Multiplication, together, and finally Addition and Subtraction, also together.

EXAMPLE

If $p = 5$, $q = 2$ and $r = -4$, work out the values of

a) $3p + 4q$ **b)** $2p^2$ **c)** $pq - 5r$

SOLUTION

a) $3p + 4q = 3 \times (5) + 4 \times (2)$ ← Multiply before adding: BID**MA**S
$= 15 + 8$
$= \underline{23}$

b) $2p^2 = 2 \times (5)^2$ ← Indices before multiplying: B**ID**M**A**S
$= 2 \times 25$
$= \underline{50}$

c) $pq - 5r = (5) \times (2) - 5 \times (-4)$ ← Multiply before subtracting: BID**MAS**
$= 10 - - 20$
$= 10 + 20$
$= \underline{30}$

EXAMPLE

If $x = 4$, $y = -1$ and $z = 3$, work out the values of

a) $3(x + 2z)^2$ **b)** $\dfrac{2x + 3y}{z + y}$

SOLUTION

a) $3(x + 2z)^2 = 3 \times (4 + 2 \times 3)^2$ ← Brackets first, then indices, then multiplying: B**ID**M**A**S
$= 3 \times (4 + 6)^2$
$= 3 \times (10)^2$
$= 3 \times 100$
$= \underline{300}$

b) $\dfrac{2x + 3y}{z + y} = \dfrac{2 \times 4 + 3 \times (-1)}{3 + (-1)}$ ← In an algebraic fraction, you should evaluate the top and bottom separately first.

Then do the division.

$= \dfrac{8 + -3}{3 + - 1}$

$= \dfrac{5}{2}$

$= \underline{2\frac{1}{2}}$

Sometimes you need to substitute values into a **formula** rather than an **expression**. The only difference is that a formula contains an *equals sign*, so the final numerical answer can tell you the value of another algebraic letter, as in the next example.

EXAMPLE

The distance s, travelled by a particle is given by the formula $s = ut + \frac{1}{2}at^2$.
a) Use your calculator to work out the value of s when $a = 9.8$, $t = 3.5$ and $u = 2.4$. Write down all the figures from your calculator display.
b) Round your answer to part a) correct to 2 significant figures.

SOLUTION

a) $s = ut + \frac{1}{2}at^2$
$= 2.4 \times 3.5 + (1 \div 2) \times 9.8 \times 3.\boxed{x^2}$
$= \underline{68.425}$

b) Correct to 2 significant figures, $s = \underline{68}$

EXERCISE 4.1

If $p = 6$, $q = 5$ and $r = 2$, find the value of:

1 $3p - 2q$ **2** $5pq + 10q$ **3** $r^2 - 3p$ **4** $(2p - 3r)^2$

If $x = 4$, $y = 6$ and $z = -2$, find the value of:

5 $x^2 + z^2$ **6** $3x - z$ **7** $2z - 3y$ **8** $(y - 2z)^2$

If $f = 3$, $g = -1$ and $h = 2$, find the value of:

9 $4f^2$ **10** $5(f + g)$ **11** $f^2 + gh$ **12** $3g^2$

13 The number of bacteria N in a colony is modelled by the formula $N = 2500(1 + kt)$ where t is the time and k is a growth factor.
a) Find the value of N when $k = 0.3$ and $t = 1.5$.
b) Write down the number of bacteria at time $t = 0$.

14 The velocity v of a particle is given by the formula $v^2 = u^2 + 2as$.
a) Find the value of v^2 when $u = 5$, $a = -3$ and $s = 2$.
b) Find the corresponding value of v, correct to 3 s.f.

15 The voltage V in an electronic circuit is given by the formula $V = IR$.
a) Find the exact value of V when $I = 13$ and $R = 20$.
b) Find the value of R when $V = 240$ and $I = 13$, giving your answer to 3. s.f.

4.2 Working with indices

Indices are used in algebra as a short notation when a quantity is multiplied by itself repeatedly. In the previous section you used x^2 (said as 'x squared') to mean $x \times x$. In a similar way, x^3 (said as 'x cubed') is a short way of writing $x \times x \times x$. The number 2 or 3 is a power, or index. Sometimes you will meet higher indices too.

EXAMPLE

Simplify:
a) $f \times f \times f$ **b)** $t \times t \times t \times t$ **c)** $2 \times m \times m$

SOLUTION

a) $f \times f \times f = \underline{f^3}$ **b)** $t \times t \times t \times t = \underline{t^4}$ **c)** $2 \times m \times m = \underline{2m^2}$

You have probably already met the idea of simplifying algebraic expressions using indices, which are powers such as squares or cubes. For example, you may have been asked to simplify $x^2 \times x^3$, or, perhaps, $(y^4)^3$.

The rules for simplifying such expressions are quite straightforward, but they can appear confusing at first, so you should practise in order to be able to apply them confidently and correctly.

Here are the rules for multiplication and division.

Multiplication	Division
$x^a \times x^b = x^{a+b}$	$x^a \div x^b = x^{a-b}$

EXAMPLE

Simplify these expressions:
a) $x^2 \times x^3$ **b)** $x^7 \div x^3$

SOLUTION

a) $x^2 \times x^3 = x^{2+3}$
$$= \underline{x^5}$$
b) $x^7 \div x^3 = x^{7-3}$
$$= \underline{x^4}$$

Sometimes the algebraic terms will have whole number multiples in front of them – these are known as **coefficients**. If these are present, you simply multiply or divide the coefficients in the usual way, and then multiply or divide the algebraic terms as well.

EXAMPLE

Simplify these expressions:
a) $5x^2 \times 4x^3$ **b)** $12x^7 \div 4x^3$ $\leftarrow$ $\dfrac{12x^7}{4x^3}$

SOLUTION

a) $5x^2 \times 4x^3 = 5 \times 4 \times x^{2+3}$

> Multiply the two number coefficients…
> …and multiply the two algebraic terms.

$= 20 \times x^5 \leftarrow$

$= \underline{20x^5}$

b) $12x^7 \div 4x^3 = (12 \div 4) \times x^{7-3}$

> Divide the two number coefficients…
> …and divide the two algebraic terms.

$= 3 \times x^4 \leftarrow$

$= \underline{3x^4}$

Take care with an expression such as $(x^3)^2$ – it does not simplify to x^5. This is because $(x^3)^2$ means x^3 times x^3 which gives x^{3+3}, in other words, x^6. This gives us another rule for indices:

$$\boxed{(x^a)^n = x^{a \times n}}$$

EXAMPLE

Simplify these expressions:
a) $(x^2)^5$ **b)** $(2x^3)^4$

SOLUTION

a) $(x^2)^5 = x^{2 \times 5}$

$= \underline{x^{10}}$

b) $(2x^3)^4 = 2^4 \times (x^3)^4$

> Remember to raise 2 to the power 4…
> …as well as working out x^3 to the power 4.

$= 16 \times x^{12}$

$= \underline{16x^{12}}$

EXAMPLE

Simplify these expressions:
a) $2a^0$ **b)** $4a^3b^{-2} \times 2a^{-2}b^5$ **c)** $\dfrac{10a^2b}{2a^3\sqrt{b}}$

SOLUTION

a) $2a^0 = 2 \times 1$

> Anything to the power of 0 equals 1.

$= 2$

b) $4 \times 2 = 8 \leftarrow$

> Multiply the two number coefficients…

$a^3 \times a^{-2} = a^{3+(-2)} = a^1 = a$

> …then add the powers of a…

$b^{-2} \times b^5 = b^{-2+5} = b^3 \leftarrow$

> …and then b.

So $4a^3b^{-2} \times 2a^{-2}b^5 = \underline{8ab^3}$

c) $\dfrac{10a^2b}{2a^3\sqrt{b}} = \dfrac{10a^2b^1}{2a^3b^{\frac{1}{2}}}$ ← b is the same as b^1...

← ...and $\sqrt{b}$ is $b^{\frac{1}{2}}$.

$10 \div 2 = 5$

$a^2 \div a^3 = a^{2-3} = a^{-1}$ ← When you divide you subtract the powers.

$b^1 \div b^{\frac{1}{2}} = b^{1-\frac{1}{2}} = b^{\frac{1}{2}}$

So $\dfrac{10a^2b}{2a^3b^{\frac{1}{2}}} = 5a^{-1}b^{\frac{1}{2}}$

This is the same as $\dfrac{5\sqrt{b}}{a}$.

EXERCISE 4.2

Write these expressions using indices.

1 $k \times k \times k$

2 $u \times u$

3 $x \times x \times x \times x \times x \times x$

4 $n \times n \times n \times n$

5 $2 \times g \times g$

6 $5 \times t \times t \times t$

Simplify these expressions using the index laws $x^a \times x^b = x^{a+b}$ and $x^a \div x^b = x^{a-b}$.

7 $x^3 \times x^5$

8 $y^{10} \div y^3$

9 $z^4 \times z^3 \times z^2$

10 $2x^3 \times 5x^4$

11 $4x \times 6x^5$

12 $12y^6 \div 6y^3$

13 $4y^2 \times 2y^4$

14 $18z^6 \div 3z$

15 $2x^2 \times 3x^3 \times x^4$

16 $10z^6 \div 20z^4$

17 $x^3 \times x^{-3}$

18 $x^3 \div x^{-3}$

Simplify these expressions using the index law $(x^a)^n = x^{a \times n}$.

19 $(x^4)^2$

20 $(y^2)^3$

21 $(3z^3)^2$

22 $(4x^5)^2$

23 $(y^{10})^3$

24 $(2z^2)^4$

25 $(2x^4)^2$

26 $(5x^2)^3$

27 $(4xy)^3$

28 $(6x^2y)^2$

29 $\left(2\sqrt{x}\right)^4$

30 $\left(y^{-1}\right)^{-3}$

Simplify these expressions.

31 $3x^2 \times 5x^3$

32 $4y^2 \div 2y$

33 $(3z^2)^3$

34 $12y^{10} \div 12y^9$

35 $4x^3 \times 10x^2$

36 $48x^4 \div 16x$

37 $10x^7 \times 10x^3$

38 $(x^3y)^2$

39 $(x^3y^2)^3$

40 $4x^4 \times 3x$

41 $\dfrac{12x^2y^3}{3xy^4}$

42 $\dfrac{18x\sqrt{y}}{\sqrt{(9x^2y^3)}}$

4.3 Expanding brackets

In this section you will learn how to expand and simplify brackets. The instruction 'simplify' tells you to collect together like terms where possible. The next example reminds you how to collect like terms.

EXAMPLE

Simplify

a) $8x + 3y + 5x + 2y$ **b)** $3x + 2y + 9x - 6y$ **c)** $5c - 3e - 8c + 4e$

SOLUTION

a) $8x + 3y + 5x + 2y = 8x \qquad\qquad + 5x$
$$\qquad\qquad\qquad + 3y \qquad\quad + 2y$$
$$= 13x + 5y$$

b) $3x + 2y + 9x - 6y = 3x \qquad\qquad + 9x$
$$\qquad\qquad\qquad + 2y \qquad\quad - 6y$$
$$= 12x - 4y$$

c) $5c - 3e - 8c + 4e = 5c \qquad\qquad - 8c$
$$\qquad\qquad\qquad - 3e \qquad\quad + 4e$$
$$= -3c + e$$

Some algebraic expressions are written with brackets. It may be possible to 'expand' the brackets, which means that you multiply them out and rewrite the result without using brackets.

EXAMPLE

Expand $3(2x + 5)$.

> *'Expand'* means *'clear away the brackets'*.

SOLUTION

$3(2x + 5) = 3 \times (2x) + 3 \times 5$
$$= 6x + 15$$

> The bracket contains two terms, namely $2x$ and $+5$. Each term gets multiplied by the 3.

EXAMPLE

Expand:

a) $4y(5 - 3y)$

b) $2z(3z^2 - 4z + 1)$

SOLUTION

a) $4y(5 - 3y) = 4y \times 5 - 4y \times 3y$ ← | Multiply each term inside the bracket by $4y$
$= \underline{20y - 12y^2}$

b) $2z(3z^2 - 4z + 1) = 2z \times 3z^2 - 2z \times 4z + 2z \times 1$ ← | Multiply each term inside the bracket by $2z$
$= \underline{6z^3 - 8z^2 + 2z}$

Examination questions often require you to do this twice, and then collect like terms to write the result in a neater form. In such a case you will be told to 'expand and simplify', as in the next example.

EXAMPLE

Expand and simplify $4(3x + 7) + 5(x + 2)$.

First you multiply out the brackets…

SOLUTION

$4(3x + 7) + 5(x + 2) = 12x + 28 + 5x + 10$

…then collect up $12x + 5x$ to make $17x$… and $+ 28$ and $+ 10$ to make $+38$.

$= \underline{17x + 38}$ ←

Sometimes there are minus signs inside one or more of the brackets.

EXAMPLE

Expand and simplify $4(3x - 7) + 5(2 - x)$.

Again, multiply out the brackets …

SOLUTION

$4(3x - 7) + 5(2 - x) = 12x - 28 + 10 - 5x$

… then collect up $12x - 5x$ to make $7x$ … and $- 28$ and $+ 10$ to make -18.

$= \underline{7x - 18}$ ←

Watch carefully when there is a minus sign *in front of* one of the brackets, because the multiplication is much more tricky.

EXAMPLE

Expand and simplify $5(4x + 3) - 2(x + 3)$.

Multiply the terms in the second bracket by -2.

SOLUTION

$5(4x + 3) - 2(x + 3) = 20x + 15 - 2x - 6$

Then collect up $20x - 2x$ to make $18x$ … and $+ 15$ and $- 6$ to make $+ 9$.

$= \underline{18x + 9}$

Notice that the minus outside the bracket ends up changing all the signs inside the bracket. Finally, watch for a double minus multiplying to give a positive term, as in this example.

EXAMPLE

Expand and simplify $2(4x - 1) - 3(x - 2)$.

SOLUTION

$$2(4x - 1) - 3(x - 2) = 8x - 2 - 3x + 6$$
$$= 5x + 4$$

The -3 multiplies with -2 to give $+6$.

EXERCISE 4.3

In questions **1** to **16** you are to multiply out the brackets and simplify the results. These are straightforward questions, without any awkward sign problems.

1 $x(3x+2)$

2 $5x(4x+1)$

3 $3x(2-4x)$

4 $-3x(x-7)$

5 $2x(x^2+2x-1)$

6 $3x(2x^2-5x-3)$

7 $2(x + 5) + 5(x + 2)$

8 $3(2x + 1) + 2(x + 5)$

9 $4(2x + 5) + 3(x + 3)$

10 $2(x + 5) + 3(2x + 5)$

11 $10(x + 1) + 6(2x + 1)$

12 $3(x + 5) + 4(x - 1)$

13 $2(x - 1) + 7(2x + 3)$

14 $3(x + 2) + 2(2x - 1)$

15 $5(x + 1) + 4(3x - 1)$

16 $6(3x - 2) + 5(4x + 3)$

In questions **11** to **20**, expand and simplify the result. Take special care when there is a negative number in front of the second bracket.

17 $2x(x-1)+3x(4-x)$

18 $3x(5x-2)+2x(3-2x)$

19 $6(2x - 1) + 3(3x - 1)$

20 $4(x + 3) - 2(x + 1)$

21 $6(2x + 1) - 3(3x + 1)$

22 $8(2x - 5) + 5(3x - 2)$

23 $16(10x - 5) + 5(3x - 2)$

24 $12(x + 2) - 3(2x + 4)$

25 $5(2x + 5) - 4(x - 2)$

26 $6(x + 1) - 2(x - 3)$

27 $7(x - 1) - 2(2x + 1)$

28 $4x + 3(2x - 1) - 5x$

29 $3x(2x-3)-2x(5-2x)$

30 $5x(1-2x)-7x(2x-1)$

4.4 Multiplying brackets together

It is possible to expand the product of a pair of brackets multiplied together. There are several possible methods, all leading to the same end result. These include 'smiles and eyebrows', 'FOIL' and a grid method. They are demonstrated in the following examples.

EXAMPLE

Expand and simplify $(x + 3)(2x + 5)$.

SOLUTION

$(x + 3)(2x + 5) \quad = 2x^2 + 6x + 5x + 15$

Each term in the first bracket is multiplied by each term in the second one. The 'smiles and eyebrows' show which pairs of terms are multiplying at each stage.

$= 2x^2 + 11x + 15$

EXAMPLE

Expand and simplify $(2x + 3)(3x - 1)$.

SOLUTION

$(2x + 3)(3x - 1) = 6x^2 - 2x + 9x - 3$
$= 6x^2 + 7x - 3$

Here we are using 'FOIL'.
First $2x$ times $3x$ gives $6x^2$
Outside: $2x$ times -1 gives $-2x$
Inside: $+ 3$ times $3x$ gives $+ 9x$
Last: $+ 3$ times -1 gives -3

EXAMPLE

Expand and simplify $(4x - 1)(2x - 5)$.

SOLUTION

	$4x$	-1
$2x$	$8x^2$	$-2x$
-5	$-20x$	$+5$

The two terms from the first bracket are written along one edge of the grid, and the terms from the other bracket down the other edge. The grid is then filled in by multiplying corresponding pairs of terms, for example $4x$ times -5 gives $-20x$.

$(4x - 1)(2x - 5) = 8x^2 - 2x - 20x + 5$
$= 8x^2 - 22x + 5$

You may use whichever of these methods you prefer – or even a combination of them. They are different ways of obtaining the same list of terms prior to collecting like terms.

You also need to be able to expand the product of three brackets.

EXAMPLE

Expand and simplify $(x - 3)(x + 4)(x - 1)$

SOLUTION

$(x - 3)(x + 4)(x - 1)$

Choose two of the brackets and expand them:

$(x+4)(x-1) = x^2 - x + 4x - 4$

Multiply everything in the second bracket by x and then by $+4$

$= x^2 + 3x - 4$ — Simplify

Then multiply your simplified expression by the 3rd bracket.

$(x-3)(x+4)(x-1) = (x-3)(x^2 + 3x - 4)$

	x^2	$+3x$	-4
x	x^3	$+3x^2$	$-4x$
-3	$-3x^2$	$-9x$	$+12$

Multiply everything in the second bracket by x and then by -3

$(x-3)(x+4)(x-1) = x^3 + 3x^2 - 3x^2 - 4x - 9x + 12$ — Simplify

$= x^3 - 13x + 12$

In the above example the x^2 terms cancelled out. When you expand three brackets you should expect up to 4 terms: one in x^3, one in x^2, one in x and a constant term.

EXAMPLE

Expand and simplify $(2x - 1)(x + 3)^2$.

SOLUTION

Expand $(x+3)^2$

$(x+3)^2 = (x+3)(x+3)$

Multiply everything in the second bracket by x and then by $+3$

$= x^2 + 3x + 3x + 9$

$= x^2 + 6x + 9$

So $(2x-1)(x+3)^2 = (2x-1)(x^2 + 6x + 9)$

Multiply everything in the second bracket by $2x$ and then by -1

	x^2	$+6x$	$+9$
$2x$	$2x^3$	$+12x^2$	$+18x$
-1	$-x^2$	$-6x$	-9

$(2x-1)(x^2 + 6x + 9) = 2x^3 + 12x^2 - x^2 + 18x - 6x - 9$ — Simplify

$= 2x^3 + 11x^2 + 12x - 9$

EXERCISE 4.4

Expand and simplify these products of brackets. You may use any valid method of your choice, but you should show all the steps in your working.

1 $(x + 3)(3x + 4)$ **2** $(x + 2)(4x + 5)$ **3** $(x + 4)(2x + 1)$

4 $(x + 5)(2x - 1)$ **5** $(x - 3)(2x + 2)$ **6** $(2x + 11)(2x + 1)$

7 $(3x + 4)(x + 2)$ **8** $(x - 6)(6x + 1)$ **9** $(2x + 5)(2x - 3)$

10 $(x + 13)(4x - 1)$ **11** $(3x + 2)(2x + 3)$ **12** $(4x - 1)(2x + 5)$

13 $(7x + 3)(2x - 3)$ **14** $(x - 3)(x - 4)$ **15** $(2x + 3)(3x - 2)$

16 $(x - 3)(2x - 5)$ **17** $(x + 7)(x - 7)$ **18** $(2x - 3)(2x + 3)$

19 $(x + 3)^2$ **20** $(1 - x)^2$ **21** $(3x - 4)^2$

> Note that, in question 19, $(x + 3)^2$ means $(x + 3)(x + 3)$.

22 $(x+1)(x+2)(x+3)$ **23** $(x+2)(x+3)(x+4)$ **24** $(x+5)(x-2)(x+1)$

25 $(x-2)(x-4)(x-5)$ **26** $(x-2)(x+1)^2$ **27** $(x-2)^3$

28 $(x+2)(x-3)(2x+1)$ **29** $(x+2)(2x+1)(3x-1)$ **30** $(3x+2)^3$

4.5 Factorising – common factors

Sometimes it is desirable to apply the reverse of expanding brackets – this is known as factorising. The idea is to take an expression that does not contain brackets and rewrite it as some kind of product of factors, so that brackets are required in the final answer. There are several different types of factorisation on the IGCSE specification, and you should learn to recognise when it is appropriate to use each type.

The simplest type of factorisation is to extract a common factor. You examine the terms of the expression one at a time, and look for the highest numerical and/or algebraic factors of each term, as in these two examples.

EXAMPLE

Factorise $16x + 20y$.

SOLUTION

> $16x$ and $20y$ are both multiples of 4, so you divide out 4 as a common factor.

$16x + 20y = \underline{4(4x + 5y)}$

EXAMPLE

Factorise $18x + 24x^2$.

SOLUTION

$18x + 24x^2 = \underline{6x(3 + 4x)}$

> Although $18x$ and $24x^2$ are both multiples of 2, you can do better – they are both multiples of 6. Also, it is possible to factor x out of both $18x$ and $24x^2$, so the highest common factor is $6x$.

Even if there are more terms, and more letters, the same overall principle applies. Find the highest common factor of the numerical coefficients first, then the highest common factor of the x parts, then the y parts, and so on.

EXAMPLE

Factorise $22x^2y^3 + 33x^3y^2 - 44x^4y$.

SOLUTION

The highest common factor of all three terms is $11x^2y$.

Thus $22x^2y^3 + 33x^3y^2 - 44x^4y = 11x^2y(\ldots + \ldots - \ldots)$
$$= \underline{11x^2y(2y^2 + 3xy - 4x^2)}$$

> Look at 22, 33, 44 to select 11
> Next, look at x^2, x^3 and x^4 to select x^2
> Finally, look at y^3, y^2 and y to select y
> Thus the HCF is $11x^2y$

EXERCISE 4.5

Factorise these expressions. They may all be done using the common factor method.

1 $x^2 + 6x$

2 $2x^2 + 6x$

3 $2x^2 + 6xy$

4 $y^2 - 10y$

5 $2y^2 - 10y$

6 $6x + 9x^2$

7 $12y^2 + 8$

8 $12y^2 + 8y$

9 $fg + 3g^2$

10 $9y^2 + 12y$

11 $5x^5 - 4x^4$

12 $12x^2 - 6x^3$

13 $14a^2 + 21ab$

14 $5xy + 10y$

15 $14 + 10y$

16 $15xy - 9x^2y$

17 $8y^2 - 20y^3$

18 $12y^2 - 8y$

19 $6 + 18x^2$

20 $12pq^3 - 12pq^2 + 15pq$

4.6 Factorising – quadratic expressions

Earlier in this chapter you practised multiplying out the products of two brackets. For example, $(x + 1)(x + 3)$ could be multiplied out to make $x^2 + 4x + 3$. It is possible to reverse this process, in order to factorise some kinds of algebraic expressions, known as **quadratics**.

EXAMPLE

Factorise $x^2 + 7x + 6$.

SOLUTION

$$x^2 + 7x + 6 = (x + \ldots)(x + \ldots)$$
$$= (x + 6)(x + 1)$$

Check:
$$(x + 6)(x + 1) = x^2 + 6x + x + 6$$
$$= x^2 + 7x + 6 \text{ as required.}$$

Each bracket must contain an x, to give a product of x^2 …

… and there must be two numbers in here which multiply together to make $+6$.

They cannot be $+2$ and $+3$, since these would contribute $2x$ and $3x$, which do not combine to make $7x$.

They could be $+6$ and $+1$, since these would contribute $+6x$ and $+x$, which do combine to make $+7x$.

This method of factorising can involve some experimentation before you find the right solution, especially when there are minus signs involved too. It is a good idea to check your final answer by multiplying the brackets back out again.

Here is another way in which the signs sometimes occur.

EXAMPLE

Factorise $x^2 - 7x + 12$.

SOLUTION

$$x^2 - 7x + 12 = (x - \ldots)(x - \ldots)$$
$$= (x - 3)(x - 4)$$

Check:
$$(x - 3)(x - 4) = x^2 - 3x - 4x + 12$$
$$= x^2 - 7x + 12 \text{ as required.}$$

To give a product of x^2, each bracket must contain an x.

Both the signs must be negative, in order to generate $-7x$ but multiply to $+12$.

You could try 12 and 1, or 6 and 2, but 4 and 3 look more promising.

Sometimes the final number term is negative, indicating that one of the factors is positive and the other negative. Take care to match them the right way round.

EXAMPLE

Factorise $x^2 + 3x - 28$.

SOLUTION

$$x^2 + 3x - 28 = (x - \ldots)(x + \ldots)$$
$$= (x - 4)(x + 7)$$

Check:
$$(x - 4)(x + 7) = x^2 - 4x + 7x - 28$$
$$= x^2 + 3x - 28 \text{ as required.}$$

One sign is positive, and one negative, in order to generate -28 at the end.

Factors of 4 and 7 look good, since they multiply to make 28, and they differ by 3.

Try $(x - 4)(x + 7)$ and $(x - 7)(x + 4)$. They both give -28, but one gives $-3x$ and the other $+3x$.

EXERCISE 4.6

Factorise these quadratic expressions.

1 $x^2 + 8x + 7$ **2** $x^2 + 9x + 14$ **3** $x^2 + 5x + 6$

4 $x^2 + 11x + 30$ **5** $x^2 + 10x + 16$ **6** $x^2 - 4x + 3$

7 $x^2 - 7x + 10$ **8** $x^2 - 11x + 30$ **9** $x^2 - 3x + 2$

10 $x^2 - 7x + 12$ **11** $x^2 + 3x - 4$ **12** $x^2 + x - 6$

13 $x^2 - x - 6$ **14** $x^2 - 4x - 5$ **15** $x^2 - x - 12$

16 $x^2 - 8x + 12$ **17** $x^2 + 12x + 32$ **18** $x^2 - x - 72$

19 $x^2 + 7x + 12$ **20** $x^2 - 7x - 44$

4.7 Factorising – harder quadratic expressions

Suppose you need to factorise a quadratic such as $2x^2 + x - 3$. This is a little harder than the examples you have tried so far. Exactly the same methods are used, but there are more possibilities to consider, because the $2x^2$ can factorise as $2x$ in one bracket and x in the other.

EXAMPLE

Factorise $2x^2 + x - 3$.   Means $1x$.

To give a product of x^2, one bracket must contain an x, and the other, $2x$.

SOLUTION

$2x^2 + x - 3 = (2x \ldots)(x \ldots)$

$\qquad\qquad\;\; = (2x + 3)(x - 1)$

One of the signs must be positive, and the other negative, to get a product of -3.

Check:
$(2x + 3)(x - 1) = 2x^2 + 3x - 2x - 3$
$\qquad\qquad\qquad = 2x^2 + x - 3$ as required.

The numbers must be 3 and 1, and after some experimentation, this combination of $+ 3$ and $- 1$ is seen to work.

EXERCISE 4.7

Factorise these quadratic expressions.

1 $2x^2 + 3x + 1$ **2** $2x^2 + 5x + 3$ **3** $2x^2 + 5x + 2$

4 $3x^2 - 5x + 2$ **5** $3x^2 - 2x - 1$ **6** $5x^2 + 4x - 1$

7 $2x^2 - x - 1$ **8** $5x^2 - 9x - 2$ **9** $3x^2 - 8x + 4$

10 $2x^2 + 11x - 6$ **11** $2x^2 - 9x + 9$ **12** $6x^2 + x - 1$

13 $6x^2 - 5x - 25$

14 $12x^2 + 8x + 1$

15 $15x^2 + 19x + 6$

16 $4x^2 - 4x + 1$

17 $6x^2 - 13x + 2$

18 $2x^2 + 9x + 7$

19 $4x^2 + 12x + 9$

20 $2x^2 - 3x - 9$

4.8 Factorising – difference of two squares

Finally, you may meet a quadratic expression with no middle term, such as $x^2 - 25$. This is equal to having $x^2 + 0x - 25$. In this case the two factors are symmetric, one with a positive sign and one negative, to give the result $x^2 - 25 = (x + 5)(x - 5)$.

More generally,

$$x^2 - a^2 = (x + a)(x - a)$$

This is a result known as the **difference of two squares.**

EXAMPLE

Factorise $x^2 - 144$.

SOLUTION

$x^2 - 144 = \underline{(x + 12)(x - 12)}$

EXAMPLE

Factorise $10x^2 - 360$.

SOLUTION

$10x^2 - 360 = 10(x^2 - 36)$
$\qquad\qquad = \underline{10(x + 6)(x - 6)}$

> First, take out a common factor …
>
> … then apply the difference of two squares.

EXERCISE 4.8A

Factorise these expressions, using the difference of two squares method.

1 $x^2 - 1$

2 $y^2 - 121$

3 $x^2 - 81$

4 $y^2 - 400$

5 $3x^2 - 75$

6 $2x^2 - 18$

7 $7y^2 - 63$

8 $10x^2 - 40$

9 $3x^2 - 27$

10 $4y^2 - 100$

The next exercise contains a mixture of all the different factorising methods you have learnt so far.

EXERCISE 4.8B

Factorise these expressions.

1 $x^2 + 6x + 5$

2 $x^2 + 8x$

3 $y^2 + 15y + 44$

4 $x^2 + 11x + 30$

5 $x^2 + 7x$

6 $y^2 + 3y - 10$

7 $4x^2 - 9x + 2$

8 $y^2 - y - 30$

9 $x^2 - 3x + 2$

10 $x^2 - 8x + 15$

11 $y^2 - 16$

12 $5xy - 10y^2$

13 $4x^2 - 8x + 3$

14 $7y^2 - 700$

15 $x^2 + 2x - 24$

16 $2y^2 + y - 10$

17 $4z^2 - 4z$

18 $2x^2 + 3x + 1$

19 $3x^2 - 12$

20 $2x^2 + 5x - 3$

4.9 Generating formulae

Formulae can be generated from information given in words. You may also generate them from information given by a diagram, or even another formula.

EXAMPLE

A factory produces handbags by cutting and shaping rectangles of material. To make one bag, a rectangle of material p cm by q cm is needed. The factory finds that an area A of material is just sufficient to make n bags.

Obtain a formula for A in terms of p, q and n.

SOLUTION

The area required for one bag is found by multiplying p and q together: pq.

For n such bags, the area of material must be n times bigger: npq.

Thus the required formula is

$A = npq$

EXAMPLE

A square of side x cm is removed from each corner of a rectangle measuring a cm by b cm.
a) Draw a sketch to show this information.
b) Find a formula for the area, A cm^2, remaining after the removal of the four squares.
c) The sides are now folded up to make a rectangular tray of depth x cm. Find a formula for the volume, V cm^3, of the tray.

SOLUTION

a)

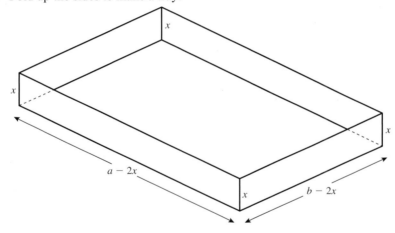

b) The original rectangle has an area of ab.

Each square has an area of x^2, and there are four, giving a total of $4x^2$.

Thus the area remaining is:

$$A = ab - 4x^2$$

c) Fold up the sides to make a tray:

The base is now a rectangle of dimensions $(a - 2x)$ and $(b - 2x)$,
so the area of the base rectangular base of the tray is $(a - 2x)(b - 2x)$.
To obtain the volume, this is multiplied by the depth, x, to give:

$$V = x(a - 2x)(b - 2x)$$

EXERCISE 4.9

1 An equilateral triangle has sides of length x cm. Obtain a formula for the perimeter P of the triangle.

2 A helicopter consumes n litres of fuel per minute. Obtain a formula for the total number of litres T of fuel consumed during a flight lasting half an hour.

3 Each month I have to pay £x for my house mortgage, and each year I have to pay £y for buildings insurance. Find a formula for the total £T I have to pay in mortgage and buildings insurance over a period of 5 years.

4 Ginny is given £500 on her 18th birthday, which she saves in a building society account. She then adds £10 per month to her savings. Obtain a formula for the amount £P she will have in the account after m months of saving.

5 A rectangle of area A has length l cm. Find a formula for its width, w cm, in terms of A and l.

6 My brother is 11 years older than me.
   a) If my present age is n years, write down my brother's present age.
   b) Obtain a formula for T, the total obtained by adding our two ages together.

7 Pencils cost 15 pence each, and pens 25 pence each.
   a) Write an expression for the cost of x pencils.
   b) Write a formula for the total cost, T pence, of x pencils and y pens.

8 The cost of hiring a bicycle is £5 plus a daily charge of £2 per day.
   a) Find the cost of hiring the bicycle for 5 days.
   b) Obtain a formula for the cost, £C, of hiring the bicycle for n days.

9 Digital photos are stored as files on a memory card. Each photo takes up 0.3 MB of space on the card. The card can hold 128 MB of data.
   a) Find the amount of space occupied by 60 photos.
   b) Obtain a formula for the amount of space S remaining on the card when n photos have been stored on it.
   c) What is the maximum value of n?

10 A rectangle of dimensions a cm by b cm has a 1 cm square cut out from each of its four corners. The sides thus formed are then folded up to make a rectangular tray.
   a) Find an expression for the area of the base of the tray.
   b) Find a formula for the volume, V cm^3, of the rectangular tray.

4.10 Changing the subject of a formula

A formula usually has a single letter term on the left-hand side of the equals sign. This is called the subject of the formula. For example, the formula $C = 2\pi r$ has C as its subject.

Sometimes you will want to rearrange the formula so that one of the other letters becomes the subject instead.

EXAMPLE

Make r the subject of $C = 2\pi r$.

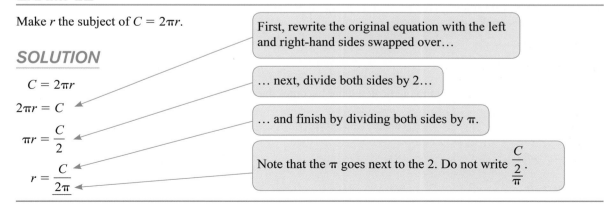

SOLUTION

$$C = 2\pi r$$

$$2\pi r = C$$

$$\pi r = \frac{C}{2}$$

$$r = \frac{C}{2\pi}$$

First, rewrite the original equation with the left and right-hand sides swapped over…

… next, divide both sides by 2…

… and finish by dividing both sides by π.

Note that the π goes next to the 2. Do not write $\dfrac{\frac{C}{2}}{\pi}$.

Some problems require a mixture of addition/subtraction and multiplication/division. You need to think carefully about the appropriate order in which to do these.

EXAMPLE

Make x the subject of $y = mx + c$.

SOLUTION

$$y = mx + c$$

$$mx + c = y$$

$$mx = y - c$$

$$x = \frac{y - c}{m}$$

> Again, begin by swapping over the left- and right-hand sides…

> …. next, subtract c from both sides….

> … and finish by dividing both sides by m.

The quantity to be made the new subject might appear as a squared term in the initial formula, for example, r^2. In this case, just make r^2 the subject of the new formula, then square root at the end.

EXAMPLE

Make r the subject of $V = \frac{1}{3}\pi r^2 h$.

SOLUTION

$$V = \tfrac{1}{3}\pi r^2 h$$

$$\tfrac{1}{3}\pi r^2 h = V$$

$$\pi r^2 h = 3V$$

$$r^2 = \frac{3V}{\pi h}$$

> Start by swapping the left and right-hand sides over …

> … then multiply both sides by 3 …

> … and divide both sides by πh.

Square rooting both sides, we obtain:

$$r = \sqrt{\frac{3V}{\pi h}}$$

Harder examination questions might be set where the new subject appears twice in the original equation, or the new subject appears as part of an algebraic fraction. You will find some examples of these harder types of problems in Chapter 23 of this book.

EXERCISE 4.10

Rearrange these formulae so that the indicated letter becomes the subject.

1 $A = \pi r l$ (make r the subject)

2 $v = u + at$ (u)

3 $v = u + at$ (a)

4 $V = \frac{1}{3}\pi r^2 h$ (h)

5 $E = mc^2$ (m)

6 $y = 4x + 3$ (x)

7 $y = \frac{x}{5} + 3$ (x)

8 $y = \frac{x + 3}{5}$ (x)

9 $A = \frac{1}{2}bh$ (h)

10 $E = mc^2$ (c)

11 $A = 4xy + x^2$ (y)

12 $P = I^2 R$ (R)

13 $y = m(x - a)$ (x)

14 $v^2 = u^2 + 2as$ (a)

15 $A = 4\pi r^2$ (r)

16 $y = x^2 - 9$ (x)

17 $x^2 = y^2 + z^2$ (y)

18 $V = abc$ (b)

19 $V = \frac{4}{3}\pi r^3$ (r)

20 $v^2 = u^2 + 2as$ (u)

REVIEW EXERCISE 4

1 If $p = 4$, $q = 2$ and $r = -5$, find the values of:
   a) $3pq$ **b)** $2p^2$ **c)** $4p - 3r$ **d)** pqr

2 If $x = 3$, $y = -2$ and $z = 10$, find the values of:
   a) $2x^2$ **b)** y^3 **c)** $3z - xy$ **d)** $z(x + y)$

3 If $s = 1$, $t = 4$ and $u = -1$, find the values of:
   a) su **b)** $t^2 + 3u$ **c)** $2s + 3t + 4u$

Simplify each of these algebraic expressions:

4 $x^2 \times x^5$ **5** $3x^4 \times x^2$ **6** $4x^3 \times 3x^2$

7 $10y^8 \div 2y^5$ **8** $8z^5 \div 2z^4$ **9** $12xy^8 \div 4y^5$

10 $(x^2)^3$

11 $(5xy^2)^2$

12 $(3xy)^2 \div y^2$

13 $(3x^2)^3 + 3x^6$

14 $\dfrac{5x^4 \times 6x^2}{3x^3}$

15 $\dfrac{(3xy) \times (4x^2y^3)}{6x^3y^2}$

Expand and simplify the following expressions:

16 $5(x + 2) + 2(x + 3)$

17 $2(y + 5) + 3(y + 1)$

18 $3(z + 1) + 5(z - 2)$

19 $7(x - 1) + 6(x + 2)$

20 $4(2x - 2) + 2(x - 3)$

21 $2(2x + 5) - 2(x + 1)$

22 $7(3x + 1) + 9(x - 1)$

23 $4(x - 2) - 2(x + 4)$

24 $3(2x + 4) + 4(x - 3)$

25 $5(2x + 2) - 2(5x - 1)$

26 $2x(x - 1) + 3x(2 - x)$

27 $4x(x - 2) - 2x(5 - 3x)$

Expand and simplify these expressions:

28 $(x + 5)(x + 1)$

29 $(y + 5)(y + 7)$

30 $(z + 4)(2z + 1)$

31 $(x - 5)(x + 4)$

32 $(2x - 3)(x + 5)$

33 $(2x - 1)(x - 1)$

34 $(3x + 2)(x - 3)$

35 $(2x - 3)(2x + 3)$

36 $(x + 4)(x - 4)$

37 $5(x + 1)(x - 1)$

38 $(x - 1)(x - 2)(x - 3)$

39 $(x + 2)(x + 5)(x - 3)$

40 $(x + 1)(x + 3)(2x - 5)$

41 $(2x - 1)(x + 2)^2$

Factorise the following expressions. You might need to use common factors, quadratic factorisation, or the difference of two squares.

42 $24x^2 + 10x$

43 $x^2 + 10x + 21$

44 $y^2 + 2y + 1$

45 $z^2 - 64$

46 $2y^2 + 9y - 5$

47 $2x^2 - 9x + 4$

48 $12x^2 - 10x$

49 $2y^2 + 7y + 6$

50 $4x^2 - 36$

51 a) Expand   **i)** $(x + 1)^2$ **ii)** $(x + 2)^2$

   b) Hence expand and simplify $(x + 1)^2(x + 2)^2$.

52 Simplify:

   a) $2a^0$ **b)** $(b^{-\frac{3}{2}})^2$ **c)** $\dfrac{2c^{-3} \times 6c^2}{3c}$

53 A Post Office sells x 26 pence stamps and y 19 pence stamps during one day. The total income from the stamps is T pence. Write a formula expressing T in terms of x and y.

54 A theatre charges £5 for adult tickets and £3 for children. Altogether a group of x adults and y children pays a total of £T.
   a) Find a formula for T in terms of x and y.
   b) What can you say about the values of x and y if the average ticket price for the group turned out to be £4?

55 In a dice game you score either 5 points or 2 points each time you play. Fred plays 10 times, and wins 5 points on n of the 10 games.
  a) Write an expression for the total number of points Fred scores in all 10 games.
  b) Simplify your expression as much as possible.

56 Rearrange the formula $C = 2\pi r$ to make r the subject.

57 Make a the subject of the formula $s = ut + \frac{1}{2}at^2$.

58 Rearrange the formula $A = \pi r^2$ to make r the subject.

59 Make l the subject of the formula $T = 2\pi \sqrt{\dfrac{l}{g}}$.

60 Lisa packs pencils in boxes. She packs 12 pencils in each box. Lisa packs x boxes of pencils.
  a) Write an expression, in terms of x, for the number of pencils Lisa packs.
  Lisa also packs pens in boxes. She packs 10 pens into each box. Lisa packs y boxes of pens.
  b) Write down an expression, in terms of x and y, for the total number of pens and pencils Lisa packs.
          [Edexcel]

61 Sharon earns p pounds per hour. She works for h hours. She also earns a bonus of b pounds.
  Write down a formula for the total amount she earns, w pounds.   [Edexcel]

62 Daniel buys n books at £4 each. He pays for them with a £20 note.
  He receives C pounds change. Write down a formula for C in terms of n.   [Edexcel]

63 **a)** Simplify $5p - 4q + 3p + q$.
  b) Simplify $\dfrac{x^7}{x^2}$
  c) Factorise $4x + 6$.
  d) Multiply out and simplify $(x + 3)(x - 2)$.
  e) Simplify $2x^3 \times x^5$.   [Edexcel]

64 **a)** Simplify $y^3 \times y^4$.
  b) Expand and simplify $5(2x + 3) - 2(x - 1)$.
  c) **(i)** Factorise $4a + 6$.
     (ii) Factorise completely $6p^2 - 9pq$.   [Edexcel]

65 Tayub said, 'When $x = 3$, then the value of $4x^2$ is 144.'
  Bryani said, 'When $x = 3$, then the value of $4x^2$ is 36.'
  a) Who was right? Explain why.
  b) Work out the value of $4(x + 1)^2$ when $x = 3$.   [Edexcel]

66 Simplify:
  a) $3a^2b \times 4a^3b^2$ **b)** $\left(\dfrac{5p^3}{q}\right)^3$ **c)** $\dfrac{12t^5}{u^4} \times \dfrac{u^3}{3t^2}$ [Edexcel]

67 **a)** Expand and simplify $(x + 5)(x - 3)$.
  b) Factorise completely $6a^2 - 9ab$.   [Edexcel]

68 a) Expand and simplify $(x + y)^2$.

    b) Hence or otherwise find the value of $3.47^2 + 2 \times 3.47 \times 1.53 + 1.53^2$.   **[Edexcel]**

69 Simplify fully:

    a) $(p^3)^3$ **b)** $\dfrac{3q^4 \times 2q^5}{q}$ **[Edexcel]**

70 Make x the subject of the formula $y = \dfrac{x^2 + 4}{5}$.   **[Edexcel]**

71 a) Simplify $k^5 \div k^2$.

    b) Expand and simplify:

        (i) $4(x + 5) + 3(x - 7)$

        (ii) $(x + 3y)(x + 2y)$.

    c) Factorise $(p + q)^2 + 5(p + q)$.

    d) Simplify $(m^{-4})^{-2}$.

    e) Simplify $2t^2 \times 3r^3t^4$.   **[Edexcel]**

Key points

1. When substituting numbers into expressions, remember the BIDMAS sequence – Brackets, then Indices, followed by Division/Multiplication and, finally, Addition/ Subtraction.

2. In particular, remember that, for example, the value of $2x^2$ when $x = 3$, is 2 times $9 = 18$, and not $6^2 = 36$; the squaring must be done before the multiplication by 2.

3. There are three algebraic laws of indices:

$$x^a \times x^b = x^{a+b}$$
$$x^a \div x^b = x^{a-a}$$
$$(x^a)^b = x^{ab}$$

4. When expanding brackets, watch for a minus sign in front of a bracket – this will change the sign of all the terms inside the bracket, for example:

$$-3(2x + 5) = -6x - 15$$

5. In algebra, factorising is the reverse process of expanding. It can be confusing because there are several different methods, so make sure you have studied them all. You need to know the common factor method, the quadratic method and the difference of two squares, and when it is appropriate to apply each approach.

6. Rearranging the subject of a formula is, perhaps, the most awkward topic in this chapter, and you may need to practise some more questions in order to master it. If you still find it tricky, ask your teacher to explain the *reverse flow diagram* method; this can be easy to use, but it only works for certain types of question.

Internet Challenge 4

The language of algebra

The wordsearch below contains 20 algebraic words for you to find. Once you have located them, use the internet to check the precise meaning of each word.

T	E	Q	U	A	T	I	O	N	N	D	P	O	R	I
V	X	S	P	G	F	E	N	Y	M	Z	S	E	R	Y
B	P	O	W	E	R	F	T	L	A	C	I	D	A	R
Q	R	X	O	Y	U	O	M	K	P	P	M	S	T	L
U	E	S	N	E	O	A	N	T	P	N	P	C	I	H
O	S	L	D	R	U	S	A	O	I	S	L	D	O	S
T	S	A	O	D	E	R	L	U	N	F	I	V	N	J
I	I	N	O	E	X	Y	E	A	G	I	F	A	A	Y
E	O	J	E	L	N	C	D	U	H	C	Y	R	L	T
N	N	D	N	O	I	T	C	N	U	F	E	I	Q	I
T	E	R	M	G	E	O	W	L	A	L	H	A	F	T
A	R	I	L	T	C	U	D	O	R	P	N	B	A	N
V	A	D	J	H	Y	T	E	D	W	T	X	L	B	E
L	Y	T	S	F	A	C	T	O	R	I	S	E	L	D
C	I	T	A	R	D	A	U	Q	F	V	H	I	Y	I

Here are the words to find. They may run left, right, up, down or diagonally.

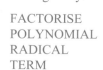

EQUATION	EXPAND	EXPRESSION	FACTORISE	FUNCTION
IDENTITY	INDEX	MAPPING	POLYNOMIAL	POWER
PRODUCT	QUADRATIC	QUOTIENT	RADICAL	RATIONAL
ROOT	SIMPLIFY	SURD	TERM	VARIABLE

CHAPTER 5

Algebraic equations

In this chapter you will **revise and extend earlier work on**:

- the language of algebra.

You will **learn how to**:

- solve simple equations
- solve harder linear equations
- solve linear equations involving brackets and fractions.

You will also be **challenged to**:

- investigate the mathematics of Carl Friedrich Gauss.

Starter: Triangular arithmagons

In a triangular arithmagon, the number along each side of the triangle is obtained by adding up the numbers in the adjacent corners. The first arithmagon has been filled in, to show you how this works. See if you can complete the other three arithmagons.

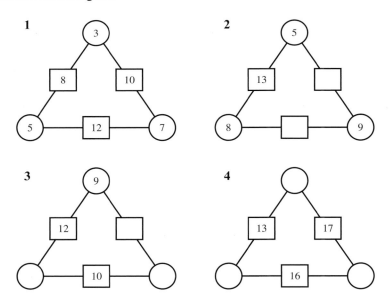

Now try devising some arithmagon puzzles of your own.

5.1 Expressions, equations and identities

In your work on algebra, you have met the words *expression*, *equation* and *formula*. They all have different meanings.

An **expression** is a statement usually containing one or more algebraic symbols.

$\dfrac{x^2 + 7}{4x}$ is an expression.

An **equation** also contains algebraic symbols and numbers, but contains an equals sign.

$\dfrac{x^2 + 7}{4x} = 2$ is an equation.

If an equation is designed to solve a problem then it is often called a **formula**. $V = IR$ is a formula used to calculate electrical voltage, V.

You can think of equations as statements that are true for some values of the quantities involved, and false for others. For example, the equation $\dfrac{x^2 + 7}{4x} = 2$ is true when $x = 1$, since $\dfrac{1^2 + 7}{4 \times 1} = \dfrac{8}{4} = 2$ but false when $x = 3$, since $\dfrac{3^2 + 7}{4 \times 3} = \dfrac{16}{12} = 1\frac{1}{3}$.

Some equations, however, are true for *all* values of x, and are called identical equations, or **identities**.

$3(x + 5) = 3x + 15$ is an identity.

Strictly speaking, identities should be written with a special symbol $3(x + 5) \equiv 3x + 15$, but in practice this is often overlooked.

EXERCISE 5.1

Look at the various algebraic statements labelled A to J:

A	$(1 + x)^3$	B	$A = \pi r^2$
C	$x^2 + 10x - 3$	D	$x(x + 4) = x^2 + 4x$
E	$x(x + 4) = 12$	F	$x + 5 = 17 - x$
G	$(x + 3)(x + 4) = 20$	H	$(x + 3)(x + 4) = x^2 + 7x + 12$
I	$4x^2 = 9$	J	$s = ut + \frac{1}{2}at^2$

1 Which ones are expressions?

2 Which ones are equations?

3 Which ones would you call formulae?

4 Pick out any identities, and rewrite them using the identity sign $\equiv$.

5.2 Simple equations

Some simple algebraic equations can be solved in just one step, or even by inspection. For example, if $x + 2 = 9$ you can spot, by inspection, that x must be 7. A more formal approach would be to subtract 2 from both sides, so that $x = 9 - 2 = 7$.

In this section you will be solving simple equations of this kind. Although you may well be able to spot some of the solutions by inspection, it is better to solve them formally, since this equips you with the skills needed for harder equations where the solutions cannot be spotted by inspection.

EXAMPLE

Solve these equations:
a) $3x = 12$ **b)** $x + 5 = 7$ **c)** $x - 5 = 1$ **d)** $\frac{1}{2}x = 12$ **e)** $\frac{x}{2} = \frac{4}{9}$

SOLUTION

a) $3x = 12$

$$x = \frac{12}{3}$$ ← Divide both sides by 3 ….

$$x = 4$$

b) $x + 5 = 7$

$$x = 7 - 5$$ ← Subtract 5 from both sides …

$$x = 2$$

c) $x - 5 = 1$

$$x = 1 + 5$$ ← Add 5 to both sides …

$$x = 6$$

d) $\frac{1}{2}x = 12$

$$x = 12 \times 2$$ ← $\frac{1}{2}x$ means 'x divided by 2'. To solve the equation, multiply both sides by 2.

$$x = 24$$

e) $\frac{x}{2} = \frac{4}{9}$

Cross-multiplying:

$$9 \times x = 4 \times 2$$ ← Cross-multiplying is a handy technique when two fractions are equal to each other.
The top of one fraction gets multiplied by the bottom of the other one, and vice versa.

$$9x = 8$$

$$x = \frac{8}{9}$$

If you need to use the square root process, remember to allow for both positive and negative options. For example, if $x^2 = 9$ then x could be 3 or -3.

EXAMPLE

Solve the equation $4y^2 = 9$.

SOLUTION

$4y^2 = 9$

Square rooting both sides,

$2y = 3$ or -3

$y = \frac{3}{2}$ or $-\frac{3}{2}$

EXERCISE 5.2

Solve these algebraic equations. Use a formal method, and show the steps of your working.

1 $5t = 20$ **2** $y + 3 = 10$ **3** $x - 2 = 7$

4 $\frac{1}{10}x = 3$ **5** $\frac{x}{2} = 6$ **6** $2y = 4$

7 $t - 3 = 1$ **8** $x + 13 = 4$ **9** $\frac{x}{5} = \frac{1}{2}$

10 $\frac{x}{2} = \frac{3}{5}$

Find the values of the letters in each of these equations.

11 $3t = 48$ **12** $u + 7 = 4$ **13** $7p = 4$

14 $3q = 8$ **15** $14 = 20 - r$ **16** $13 = 30 - u$

17 $98g = 7$ **18** $3 - z = 3$ **19** $16x^2 = 25$

20 $y^2 = 144$

5.3 Harder linear equations

Some equations contain many stages, not just one or two. You may have to reorganise the equation to collect all the x terms on one side and all the numbers on the other. The like terms are then collected together and simplified, before performing the final step of the solution.

When reorganising the terms, you can move a term from one side of the equation to the other – but it must then change its sign. These examples show you how this works.

EXAMPLE

Solve the equation $5x + 3 = 3x + 17$.

SOLUTION

$$5x + 3 = 3x + 17$$

$$5x - 3x + 3 = 17$$

First, subtract $3x$ from both sides. This causes the $3x$ to disappear from the right-hand side, and appear as $-3x$ on the left-hand side.

$$2x + 3 = 17$$

$$2x = 17 - 3$$

Similarly, subtract 3 from both sides.

$$2x = 14$$

$$x = \frac{14}{2}$$

Finally, divide both sides by 2.

$$\underline{x = 7}$$

Here is another example, this time with some minus signs. The principle is exactly the same.

EXAMPLE

Solve the equation $3x - 11 = 3 - x$.

SOLUTION

$$3x - 11 = 3 - x$$

$$3x + x - 11 = 3$$

First, add x to both sides.

$$4x - 11 = 3$$

$$4x = 3 + 11$$

Next, add 11 to both sides.

$$4x = 14$$

$$x = \frac{14}{4}$$

Finally, divide both sides by 4.

$$\underline{x = 3\tfrac{1}{2}}$$

If the overall coefficient of x looks like being negative, it may be more convenient to collect the x terms on the right-hand side instead, as in this final example.

EXAMPLE

Solve the equation $17 + x = 12 + 5x$.

SOLUTION

$$17 + x = 12 + 5x$$

$$17 = 12 + 5x - x$$

First, take x from both sides.

$$17 = 12 + 4x$$

$$17 - 12 = 4x$$

Next, take 12 from both sides.

$$5 = 4x$$

$$x = \frac{5}{4}$$

Finally, divide both sides by 4.

EXERCISE 5.3

Solve these algebraic equations, showing the steps of your working clearly. All the answers should be integers, but some may be negative.

1 $4x + 5 = x + 14$

2 $6x + 1 = 8 - x$

3 $10t + 3 = 8t + 11$

4 $x + 3 = 7x + 15$

5 $12m + 5 + m = 44$

6 $15 - 2u = 30 - 5u$

7 $14 + 2x = 9x$

8 $10 + 5k = 3k + 6$

9 $x + 3 = 3 - x$

10 $4x = 55 - x$

Solve these algebraic equations, showing the steps of your working clearly. Answers should be given as top-heavy fractions or mixed numbers, rather than decimals.

11 $6t + 2 = 2t + 5$

12 $4u + 3 = 10 - u$

13 $8p + 13 = 10p + 4$

14 $5q + 7 = 12 - 3q$

15 $10r - 4 = 4r - 7$

16 $10u - 1 = 11u + 1$

17 $3x + 31 = x + 30$

18 $4y - 1 = y + 4$

19 $5x - 4 = 1 + 6x$

20 $10y - 3 = 4 + 7y$

5.4 Equations and brackets

You may encounter brackets in an equation. It is usually a good idea to expand the brackets, so that you can gather together like terms in order to solve the equation.

EXAMPLE

Solve the equation $3(x + 2) + 2(2x - 5) = 5(x - 1) + 9$.

SOLUTION

$$3(x + 2) + 2(2x - 5) = 5(x - 1) + 9$$
$$3x + 6 + 4x - 10 = 5x - 5 + 9$$
$$7x - 4 = 5x + 4$$
$$7x = 5x + 4 + 4$$
$$7x = 5x + 8$$
$$7x - 5x = 8$$
$$2x = 8$$
$$x = 4$$

Some word problems can be formulated using brackets, to obtain an equation that can then be solved.

EXAMPLE

I am thinking of a number. If I add 20 on to my number I get twice as much as if I only add 6. What number am I thinking of?

SOLUTION

Suppose the number I am thinking of is x.

Then $x + 20$ is twice as much as $x + 6$.

$$x + 20 = 2(x + 6)$$
$$x + 20 = 2x + 12$$
$$20 = 2x + 12 - x$$
$$20 = x + 12$$
$$20 - 12 = x$$
$$8 = x$$

So the number I am thinking of is 8.

EXERCISE 5.4

Multiply out the brackets, and hence solve these equations. Show each step of your working.

1 $5(x + 5) + 3 = 13$ **2** $4(x - 1) + 3x = 45$ **3** $3(y + 5) + y = 23$

4 $2(n - 4) + 3 = 7$ **5** $3(2p + 7) - 38 = 4p + 3$ **6** $x + 13 = 6(x - 2) + 5$

7 $5(2x - 1) - 2(3x + 4) = 3$ **8** $4(r - 2) - 2(r - 3) = 6$ **9** $2(3s + 14) + 11 = 3(s + 9)$

10 $3(2x + 3) = 2(x + 1) + 23$ **11** $2d - 9 = 5d - 3(3d + 2)$ **12** $5(x + 4) - 2(x - 1) = 43$

Write an equation, involving brackets, to formulate each of the problems below. Then expand your brackets and solve the equation, to obtain the answer to the problem.

13 I think of a number, add 12, and then multiply the new total by 2. I get the same answer as if I had just multiplied the original number by 4. What number did I think of?

14 At the moment Ravi is n years old. He is five years older than his brother. In four years' time, Ravi will be exactly twice as old as his brother is now. Find out how old Ravi is now.

15 Nat and Marina each think of the same number. Nat multiplies the number by 7, and then adds 5. Marina adds 7 to the number, and then multiplies by 5. They both end up with the same answer.
   a) Write this information as an equation.
   b) Solve your equation, to find the number they both thought of.

5.5 Equations with fractional coefficients

Some equations contain fractional coefficients. A good approach is to multiply the whole equation by a positive whole number, to clear the fractions away. You would normally use the lowest common denominator of the fractions.

EXAMPLE

Solve the equation $\dfrac{3x}{5} = \dfrac{x+1}{10}$.

SOLUTION

The lowest common multiple of 5 and 10 is 10, so multiply both fractions by 10:

$$\frac{10 \times 3x}{5} = \frac{10 \times (x+1)}{10}$$

Cancelling, and then simplifying:

$$\frac{\cancel{10}^{2} \times 3x}{\cancel{5}^{1}} = \frac{\cancel{10}^{1} \times (x+1)}{\cancel{10}^{1}}$$

$$6x = x + 1$$
$$6x - x = 1$$
$$5x = 1$$
$$\underline{x = \tfrac{1}{5}}$$

Harder examples may involve more terms.

EXAMPLE

Solve $\dfrac{5y+8}{6} - \dfrac{3y+2}{4} = 1$.

SOLUTION

The LCM of 6 and 4 is 12, so multiply all three terms by 12:

$$\frac{12 \times (5y+8)}{6} - \frac{12 \times (3y+2)}{4} = 12 \times 1$$

Cancelling, and then simplifying:

$$\frac{\cancel{12}^{2} \times (5y+8)}{\cancel{6}^{1}} - \frac{\cancel{12}^{3} \times (3y+2)}{\cancel{4}^{1}} = 12 \times 1$$

$$2(5y + 8) - 3(3y + 2) = 12$$
$$10y + 16 - 9y - 6 = 12$$
$$y + 10 = 12$$
$$y = 12 - 10$$
$$\underline{y = 2}$$

It is a good idea to check your answer, by substitution:

$$\frac{5y+8}{6} - \frac{3y+2}{4} = \frac{5 \times 2 + 8}{6} - \frac{3 \times 2 + 2}{4}$$
$$= \frac{18}{6} - \frac{8}{4}$$
$$= 3 - 2$$
$$= 1 \quad \text{as required}$$

EXERCISE 5.5

Solve these equations.

1 $\dfrac{x-1}{2} = \dfrac{x+1}{3}$

2 $\dfrac{7x-5}{10} = \dfrac{x}{2}$

3 $\dfrac{3x+2}{7} = \dfrac{x+2}{3}$

4 $\dfrac{x+8}{2} + \dfrac{x+6}{4} = 7$

5 $\dfrac{7x-5}{10} = \dfrac{3x}{5}$

6 $\dfrac{x+1}{2} + \dfrac{2x+1}{3} = 9$

7 $\dfrac{x+1}{2} = \dfrac{x}{3}$

8 $\dfrac{x+1}{2} + \dfrac{3x-1}{4} = 4$

9 $\dfrac{x+13}{2} - \dfrac{12-3x}{3} = 1$

10 $\dfrac{3x-1}{7} = \dfrac{9-x}{2}$

REVIEW EXERCISE 5

1 Insert the best word (expression, equation, formula or identity) into the missing space in each of the following sentences.

   a) The volume of a cuboid may be found by using the ⬜⬜⬜⬜⬜⬜⬜ $V = abc$.

   b) My age n years after my 21st birthday is given by the ⬜⬜⬜⬜⬜⬜⬜⬜⬜⬜ $21 + n$.

   c) The ⬜⬜⬜⬜⬜⬜⬜⬜ $3x + 5 = 17$ has a solution at $x = 4$.

   d) The result $x(x + 12) = x^2 + 12x$ is an example of an ⬜⬜⬜⬜⬜⬜⬜⬜.

Solve these equations. You may solve them by inspection, which means by sight, and write the answer down.

2 $x + 5 = 11$

3 $y + 2 = 0$

4 $x - 5 = 2$

5 $3y = 1$

6 $4x = 7$

7 $25y^2 = 64$

8 $16 - 2x = 8$

9 $x^2 = 81$

10 $\dfrac{z}{4} = 5$

Solve these linear equations. Show all the steps in your working.

11 $4x + 3 = x + 15$

12 $7x - 5 = 2x + 15$

13 $x - 1 = 5 - 2x$

14 $13 + x = 7x + 25$

15 $\dfrac{x}{5} = \dfrac{7}{20}$

16 $2x + 6 = 9x + 13$

17 $10 - x = 15 - 2x$

18 $16 + 3x = 5x + 20$

19 $8 - x = 8 + x$

20 $\dfrac{2x}{3} = \dfrac{10}{9}$

Expand the brackets, and hence solve these equations.

21 $5(y - 1) = 2y + 7$

22 $9z = 4(z - 2) + 3$

23 $2x = 5(12 - x) + 3$

24 $9(x + 7) = 7(x + 9)$

25 $7(2x - 3) = 59 + 4(x - 5)$

26 $15(x - 1) + 3 = 2(x + 1) + 3(x + 2)$

Solve these equations.

27 $\dfrac{3x + 4}{5} = x$

28 $\dfrac{8 - x}{6} = \dfrac{x - 3}{4}$

29 $\dfrac{x}{2} + \dfrac{x}{3} = 5$

30 $\dfrac{x + 1}{3} + \dfrac{x}{4} = 5$

31 Glenn has done a homework exercise about solving equations. He got 19 of the 20 questions right. Here is the one he got wrong.

$$5(2x - 1) - 2(x + 4) = 19$$
$$10x - 5 - 2x + 8 = 19$$
$$8x + 3 = 19$$
$$8x = 16$$
$$x = 2$$

Look carefully at Glenn's work, and see if you can spot what he has done wrong. Then write out the corrected answer, showing all the lines of working.

32 Marco and Seyi are solving the equation $3(2x - 1) + 4(x + 8) = 19$.

Marco says 'I reckon $x = 1$.'

Seyi says 'I reckon that $x = -1$.'

Work out which one of them was right.

33 a) Solve $7p + 2 = 5p + 8$.
  b) Solve $7r + 2 = 5(r - 4)$. [Edexcel]

34 Solve the equation $7(x - 1) = 2x - 1$. [Edexcel]

35 a) Solve $7x + 18 = 74$.
  b) Solve $4(2y - 5) = 32$.
  c) Solve $5p + 7 = 3(4 - p)$. [Edexcel]

36 a) Solve the equation $5p - 4 = 11$.
  b) Solve the equation $7(q + 5) = 21$.
  c) Solve the equation $\dfrac{21 + x}{6} = x$. [Edexcel]

37 Nassim thinks of a number. When he multiplies his number by 5 and subtracts 16 from the result he gets the same answer as when he adds 10 to his number and multiplies that result by 3. Find the number Nassim is thinking of. [Edexcel]

38 a) Solve $20y - 16 = 18y - 9$.
  b) Solve $\dfrac{40 - x}{3} = 4 + x$. [Edexcel]

39 The diagram represents a garden in the shape of a rectangle.

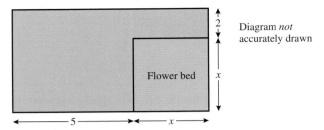

Diagram *not* accurately drawn

All measurements are given in metres.

The garden has a flower bed in one corner.

The flower bed is a square of side x.

a) Write an expression, in terms of x, for the shortest side of the garden.

b) Find an expression, in terms of x, for the perimeter of the garden.

Give your answer in its simplest form.

The perimeter of the garden is 20 metres.

c) Find the value of x.

[Edexcel]

Key points

1. An expression is a mathematical phrase, such as $x^2 + 5x$.

2. An equation is a mathematical sentence, containing an $=$ sign, such as $y = x^2 + 5x$.

3. An equation designed for a purpose, for example to calculate an area, is called a formula.

4. An equation that is always true is called an identity.

5. Simple equations may be solved by inspection. If square rooting, remember to allow for both positive and negative options.

6. Harder equations may include several stages and, possibly, the expansion of brackets. With this kind of problem you should always show each step of your working carefully.

7. Some algebraic equations involve fractions. You can clear the fractions away by multiplying through by the lowest common denominator of the fractions.

Internet Challenge 5

Carl Friedrich Gauss

Much pioneering work on the theory of equations was done by Gauss. Use the internet to help answer these questions about him:

1 What nationality was Carl Friedrich Gauss?

2 When and where was he born?

3 How long did he live?

4 Gauss solved a difficult geometric construction problem while he was still a teenager. What was this?

5 '*Work out 1 + 2 + 3 + ... + 100 in your head.*' How did Gauss do this when he was 9 years old?

6 Which university did Gauss enter in 1795?

7 What astronomical discovery is jointly credited to Gauss and the Italian astronomer Guiseppe Piazzi?

8 '*Every (positive) whole number is the sum of at most three triangular numbers.*'
   Is this statement true or false?

9 What is the Fundamental Theorem of Algebra?

10 It used to be necessary to *degauss* a computer's CRT monitor. What does this mean?

11 A central idea in modern statistics is the *Gaussian distribution*, but this name is misleading, as it was not originated by Gauss. Which mathematician was responsible for first introducing this distribution, and by what other name is it often known?

12 What are complex numbers?

13 Gauss allegedly said words to the effect '*Tell her to wait a minute until I've finished*' on what occasion?

14 By what regal nickname is Gauss sometimes known?

15 When and where was Gauss buried?

16 What mathematical shape did Gauss want inscribed on his gravestone? Was this done?

17 What is the significance of the number of questions in this exercise?

CHAPTER 6

Graphs of straight lines

In this chapter you will revise earlier work on:

- using coordinates in all four quadrants.

You will learn how to:

- plot graphs of linear functions defined implicitly or explicitly
- use gradient and intercept to sketch linear graphs
- recognise the equation of a linear graph by looking at its gradient and intercept
- calculate the midpoint of a line segment
- use properties of parallel and perpendicular lines.

You will also be challenged to:

- investigate parallels.

Starter: Matchstick puzzles

1  Starting with these twelve matches, remove two matches so that only two squares remain.

2 These four matches make a cocktail glass containing a cherry. Move two matches so that the cherry is outside the glass.

3 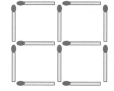 Move three matches so that the fish swims in the opposite direction.

4

Move one match to make a square.

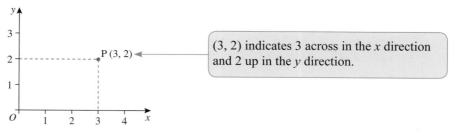

6.1 Coordinates in all four quadrants

You will already be familiar with the idea of using x and y coordinates like this:

(3, 2) indicates 3 across in the x direction and 2 up in the y direction.

These are sometimes called *Cartesian* coordinates, after the French mathematician and philosopher René Descartes, although he was not the first mathematician to use them.

You can extend the basic Cartesian coordinate system into four regions, or quadrants, by using negative coordinates, like this:

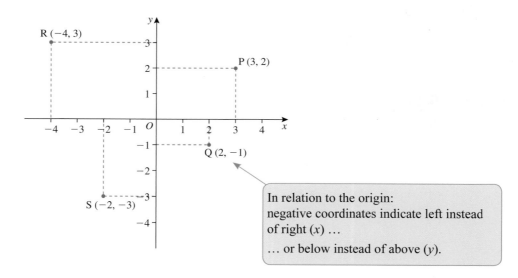

In relation to the origin:
negative coordinates indicate left instead of right (x) …

… or below instead of above (y).

EXAMPLE

A is the point (2, 5) and B is the point (7, 9).
Find the coordinates of the midpoint of the line AB.

SOLUTION

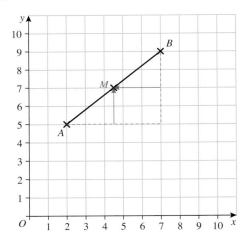

The midpoint M is at the point $\left(4\frac{1}{2}, 7\right)$.

You can find the midpoint without a diagram by finding:

- the value halfway between the x co-ordinates: $\frac{2+7}{2} = 4\frac{1}{2}$

- the value halfway between the y co-ordinates: $\frac{5+9}{2} = 7$

So the midpoint is at $\left(4\frac{1}{2}, 7\right)$.

EXERCISE 6.1

1 Using the diagram below, write down the coordinates of A, B, C, D and E.

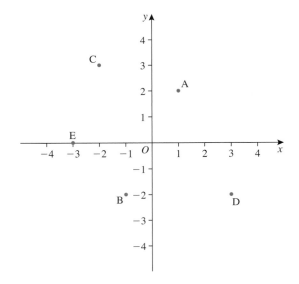

2 The questions refer to the diagram below:
   a) Which point is at $(-3, 1)$?
   b) What are the coordinates of E?
   c) Which point has the same x and y coordinates?
   d) Which point is midway between $(-2, 5)$ and $(4, 3)$?
   e) What are the coordinates of H?
   f) Which point has the largest y coordinate?
   g) Which point has the smallest x coordinate?

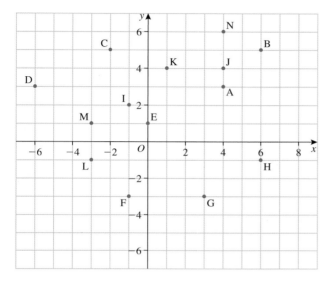

3 A is the point $(4, 9)$ and B is the point $(-2, -3)$.

Find the coordinates of the midpoint of the line.

4 A circle has a diameter AB where A has coordinates $(6, -2)$ and B has coordinates $(-3, -4)$. Work out the coordinates of the centre of the circle.

6.2 Graphs of linear functions

Expressions such as $3x + 5$ and $4 - 2x$ are called linear expressions. They must not contain any terms such as x^2, x^3 or $\frac{1}{x}$. Linear expressions are always of the form $ax + b$, where a and b are numbers. Although a and b often take positive whole number values, this is not always the case – they may be fractional, negative or even zero.

A relation of the form $y = ax + b$ is called a linear function. Linear functions are so-called because, when you plot their graphs, the result is a straight line.

EXAMPLE

Plot the graph of $y = 2x + 3$ for values of x from -5 to 5.

SOLUTION

When $x = -5$, $y = 2 \times (-5) + 3 = -10 + 3 = -7$.

When $x = 0$, $y = 2 \times (0) + 3 = 0 + 3 = 3$.

When $x = 5$, $y = 2 \times (5) + 3 = 10 + 3 = 13$.

> Use the formula to work out matching x and y values for a low value of x (-5), a middle value (0) and a high value (5).

x	-5	0	5
y	-7	3	13

> It is convenient to store these values in a table, like this.

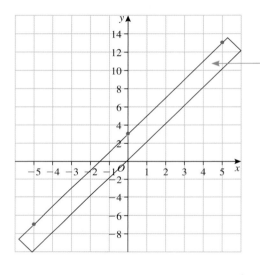

> After plotting the points, use a transparent ruler to check that the points form a straight line.
>
> In fact you need only two points to define a straight line …
>
> … but the third point acts as a check.

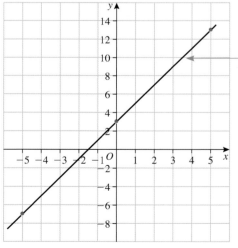

> Now join them with a single neat line.
>
> Note that the line continues slightly beyond the end-points at $(5, 13)$ and $(-5, -7)$.

Strictly speaking, the line $y = 2x + 3$ is infinitely long, since it extends indefinitely in both directions. The portion of this line cut off between $x = -5$ and $x = 5$ is more correctly known as a *line segment*.

EXERCISE 6.2

For questions 1 to 8 you are given a linear function and an incomplete table of values. Work out the missing values to complete the table, and then plot the graph of the corresponding line segment. You may use either graph paper or squared paper.

1 $y = 2x + 1$

x	−4	0	4
y	−7	1	

2 $y = x + 4$

x	−5	0	5
y	−1		

3 $y = 3x - 1$

x	−4	0	5
y			

4 $y = 2x - 3$

x	−2	0	4
y			

5 $y = \frac{1}{2}x + 4$

x	−6	0	4
y			

6 $y = x + 1$

x	−5	0	5
y			

7 $x + y = 10$

x	0	5	10
y			

8 $2x + y = 5$

x	−2	0	4
y			

9 Draw up a set of coordinate axes so that x can run from −10 to 10 and y from −25 to 25.
   a) Calculate the coordinates of three points that lie on the line $y = 2x$. Hence plot the line $y = 2x$ on your coordinate axes.
   b) Now calculate the coordinates of three points that lie on the line $y = 2x - 1$. Plot the line $y = 2x - 1$ on the same set of coordinate axes.
   c) Look at your two graphs. What do you notice?

10 Draw up a set of coordinate axes so that x can run from 0 to 10 and y from −5 to 10.
   a) Calculate the coordinates of three points that lie on the line $x + y = 8$. Hence plot the line $x + y = 8$ on your coordinate axes.
   b) Now calculate the coordinates of three points that lie on the line $x + y = 5$. Plot the line $x + y = 5$ on the same set of coordinate axes.
   c) Look at your two graphs. What do you notice?

6.3 Gradient and intercept of linear functions

In question **9** of the previous exercise you were asked to plot the graph of $y = 2x$. Your graph should have looked like this:

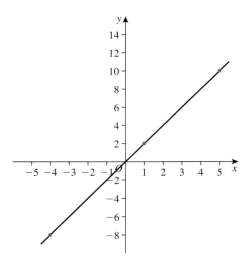

You can measure the *gradient* of the line by constructing a triangle underneath it – the exact size of the triangle is unimportant – and measuring the horizontal and vertical changes. These are sometimes referred to as 'rise' and 'run'. You can choose any two points on the line.

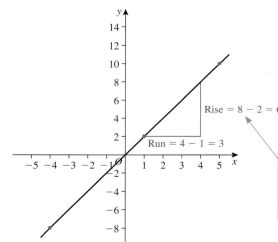

It is dangerous just to count squares.

You must read the values off the graph carefully: the x and y axes may have different scales, as here.

Then the **gradient** m is defined as:

$$\text{gradient } m = \frac{\text{rise}}{\text{run}} = \frac{6}{3} = 2$$

The next diagram shows a family of three graphs, all with gradient 2. They are distinguished by the fact that each one crosses the y axis at a different position – this point is known as the **intercept** (or y intercept, to give it its full name).

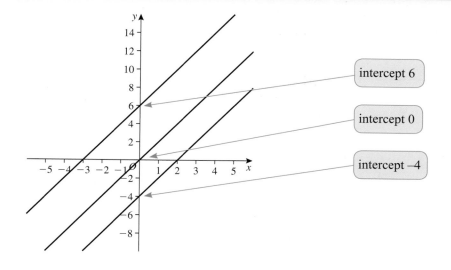

EXAMPLE

A straight line passes through the points (0, 5) and (3, 14). Find its gradient *m* and intercept *c*.

SOLUTION

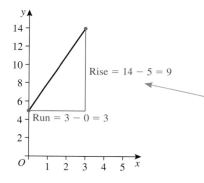

You do not need the graph to be plotted accurately: a sketch to show which numbers are being used is enough.

$$\text{Gradient} = \frac{\text{rise}}{\text{run}} = \frac{9}{3} = 3$$

Intercept = 5

Thus $\underline{m = 3}$ and $\underline{c = 5}$

Some linear graphs have negative gradients. This simply means that the graph slopes *down*, as you move to the right, not up.

EXAMPLE

A straight line passes through the points $(1, 6)$ and $(3, 2)$. Find its gradient m and intercept c.

SOLUTION

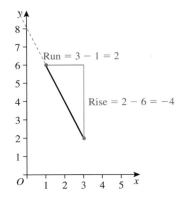

$$\text{Gradient} = \frac{\text{rise}}{\text{run}} = \frac{-4}{2} = -2.$$

By extending the line segment to the left the intercept may be read from the y axis:

Intercept $= 8$.

Thus $m = -2$ and $c = 8$

EXERCISE 6.3

Find the gradient m and the intercept c for each of the lines marked in questions **1** to **8** below.

1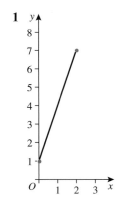

2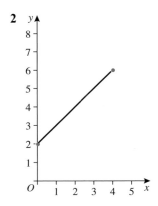

3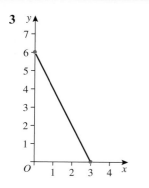

4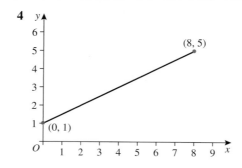

5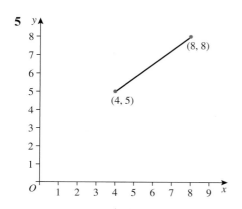

6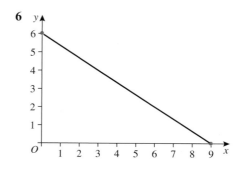

7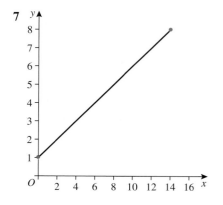

8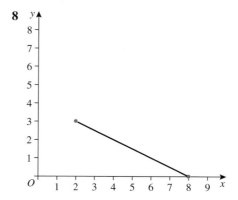

6.4 Equations and graphs

Look again at this graph of $y = 2x + 3$ that was used on page 102.

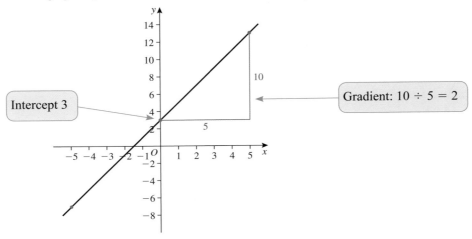

Intercept 3

Gradient: $10 \div 5 = 2$

Notice that the line has gradient 2, and intercept 3, and these also happen to be the values of the two coefficients that appear in the equation of the line.

This illustrates an important general result:

The graph of the function $y = mx + c$ has gradient m and intercept c.

You can use this principle to help sketch graphs of linear functions.

EXAMPLE

Sketch the graph corresponding to the function $y = 3x + 1$.

SOLUTION

The intercept is $c = 1$, so the graph must cross the y axis at $(0, 1)$.
The gradient is $m = 3$, so the graph rises by 3 units for each 1 unit to the right.
Thus the graph will look like this:

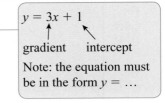

$y = 3x + 1$

gradient   intercept

Note: the equation must be in the form $y = \dots$

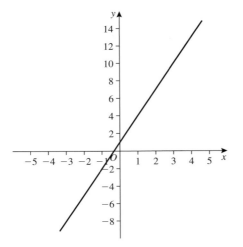

You can use this idea the other way round, to find the equation of a given straight-line graph.

EXAMPLE

Find the equation of this straight line:

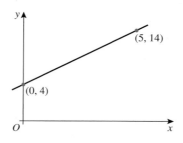

SOLUTION

The intercept is $c = 4$.

The gradient is $\dfrac{14-4}{5-0} = \dfrac{10}{5} = 2$.

Thus the equation of the line is $\underline{y = 2x + 4}$

EXERCISE 6.4

1 to 8 Write down the equations of the straight lines whose gradients and intercepts you found in Exercise 6.3, questions **1** to **8**.

9 The diagram shows the graph corresponding to a linear function of x.
   a) Write down the coordinates of the points P and Q on the line.
   b) Find the gradient and intercept of the line.
   c) Hence write down the equation of the straight line.

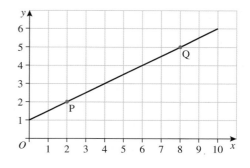

10 The diagram shows the graph of a linear function of x.
   a) Find the gradient and intercept of the line.
   b) Hence write down the equation of the straight line.

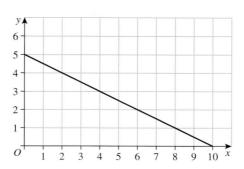

109

6.5 Parallel and perpendicular lines

Here are the graphs of $y = 2x - 4$, $y = 2x$ and $y = 2x + 6$ from page 105.

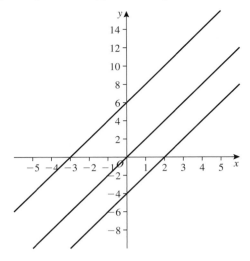

All three graphs have the same gradient, namely 2.

Geometrically, this means that all three lines are parallel.

In general, two lines will be parallel if, and only if, their gradients are equal.

EXAMPLE

Find the equation of the line passing through (5, 13) that is parallel to the line $y = 2x - 4$.

SOLUTION

Suppose the required line has equation $y = mx + c$.

Since it is parallel to $y = 2x - 4$ then the gradient must be 2, that is, $m = 2$.

Thus the required line has equation $y = 2x + c$.

Since it passes through the point (5, 13), we may substitute $x = 5$ and $y = 13$ to obtain:

$$13 = (2 \times 5) + c$$
$$13 = 10 + c$$
$$c = 3$$

Thus the required line has equation $\underline{y = 2x + 3}$

EXAMPLE

Investigate whether any of these lines is parallel to any of the others:

A $y = 3x + 5$
B $y = 2x + 5$
C $y = x + 5$
D $x - y = 9$
E $2y - 4x = 7$

SOLUTION

Clearly neither A nor B nor C is parallel to another since they have gradients of 3, 2, 1 respectively.

Equations D and E need to be rearranged to make y the subject before any further comparison is possible.

D may be written as $y = x - 9$, which has gradient 1, so <u>C and D are parallel.</u>

E may be written as $y = 2x + 3.5$, which has gradient 2, so <u>B and E are parallel.</u>

When two lines are perpendicular the product of their gradients is −1.

This is sometimes written as

$$m_1 \times m_2 = -1 \text{ or } m_1 = \frac{-1}{m_2}$$

where m_1 and m_2 are the gradients of two perpendicular lines.

Below is a short proof of this important result.

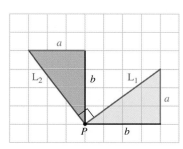

The blue triangle can be rotated through 90° anticlockwise to give the red triangle.
The angle between the lines L_1 and L_2 is 90° so the lines are perpendicular.

The gradient of the blue line, L_1, is $\dfrac{a}{b}$.

The gradient of the purple line, L_2, is $-\dfrac{b}{a}$. ← The purple line slopes downwards, so its gradient is negative.

Multiplying these gradients gives:

$$\frac{\cancel{a}}{\cancel{b}} \times \left(-\frac{\cancel{b}}{\cancel{a}} \right) = -1$$

So the product of the gradients is -1.

EXAMPLE

Find the equation of the line perpendicular to $y = \frac{1}{2}x + 1$ that passes through the point $(1, 4)$.

SOLUTION

The gradient of $y = \frac{1}{2}x + 1$ is $\frac{1}{2}$.

So the gradient of a line perpendicular to this line is $\dfrac{-1}{\frac{1}{2}} = -2$.

Using $m_1 = \dfrac{-1}{m_2}$

Check: $-2 \times \frac{1}{2} = -1$ ✓

The line you need passes through $(1, 4)$ and has gradient -2.
Now substitute $x = 1$, $y = 4$ and $m = -2$ into $y = mx + c$ to find the value of c.
$4 = -2 \times 1 + c$
So $c = 4 + 2 = 6$
So the equation of the line is
$$y = -2x + 6$$

Check that $(1, 4)$ lies on this line:
$4 = -2 \times 1 + 6$ ✓

EXERCISE 6.5

1 Rearrange each of these equations into the form $y = ax + b$. Then pick out the two that represent a pair of parallel lines.

    a) $x - y = 6$ **b)** $2x + y + 5 = 0$ **c)** $y - 1 = \frac{1}{2}x$ **d)** $x - 2y + 5 = 0$

2 The line $y = ax + b$ is parallel to $y = 5x + 1$, and passes through the point $(1, 0)$.
    a) Write down the value of a.
    b) Work out the value of b, and hence obtain the equation of the line.

3 The line $y = mx + c$ is parallel to the line $y = 4x - 1$, and passes through the point $(0, 3)$.
    a) Find the values of m and c, and write down the equation of the line.
    b) The line also passes through the point $(3, p)$. Find p.

4 A line has equation $\dfrac{y - 2}{3} = x$.
    a) Rearrange the equation into the form $y = mx + c$.
    b) Find the equation of the parallel line that passes through the point $(2, 1)$.

5 Find the gradient of a line that is perpendicular to $2y - x + 6 = 0$.

6 Find the equation of the line perpendicular to $y = x - 2$ that passes through the point $(0, 4)$.

7 Find the equation of the line perpendicular to $y = 3x + 1$ that passes through the point $(9, -5)$.

    Give your answer in the form $ax + by + c = 0$.

REVIEW EXERCISE 6

For questions **1** to **4** you are given a linear function and an incomplete table of values. Copy and complete the table, and then plot the graph of the corresponding line segment.

1 $y = x + 4$

x	-5	0	2
y	-1		6

2 $y = \frac{1}{2}x + 1$

x	-2	0	6
y	0		

3 $y = 2x - 5$

x	-6	0	6
y			

4 $x + y = 20$

x	0	8	20
y			

5 Work out the gradient and intercept of each of the lines shown below. Hence obtain their equations.

a)

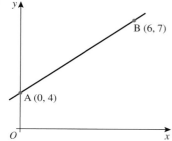

b)

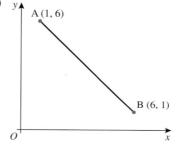

6 Find the equation of a line parallel to $y = 4x + 3$ but with a y-intercept of 7.

7 Find the equation of the line parallel to $y = 3x - 5$ that passes through $(2, 8)$.

8 Show that the lines $3y = 2x + 1$ and $2y + 3x + 2 = 0$ are perpendicular.

9 The line $\mathbf{L_1}$ passes through the points $A(2, 5)$ and $B(5, 11)$.
   a) Find the gradient of $\mathbf{L_1}$.

   The line $\mathbf{L_2}$ is perpendicular to $\mathbf{L_1}$ and passes through the point A.
   b) Find the equation of $\mathbf{L_2}$.

   Give your answer in the form $ax + by + c = 0$.

10 The diagram shows three points A $(-1, 5)$, B $(2, -1)$ and C $(0, 5)$.

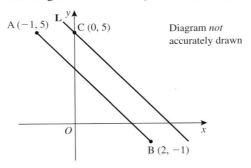

Diagram *not* accurately drawn

The line **L** is parallel to AB and passes through C.
Find the equation of the line **L**.

[Edexcel]

11 A straight line has equation $y = \frac{1}{2}x + 1$. The point P lies on the straight line. P has a y-coordinate of 5.
   a) Find the x-coordinate of P.
   b) Write down the equation of a different straight line that is parallel to $y = \frac{1}{2}x + 1$.
   c) Rearrange $y = \frac{1}{2}x + 1$ to make x the subject.

[Edexcel]

12 The line with equation $x + 2y = 6$ has been drawn on the grid below.

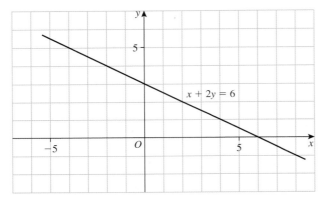

   a) Rearrange the equation $x + 2y = 6$ to make y the subject.
   b) Write down the gradient of the line with equation $x + 2y = 6$.
   c) Write down the equation of the line that is parallel to the line with equation
       $x + 2y = 6$ and passes through the point with coordinates $(0, 7)$.

[Edexcel]

13 ABCD is a rectangle. A is the point $(0, 1)$. C is the point $(0, 6)$.
   The equation of the straight line through A and B is $y = 2x + 1$

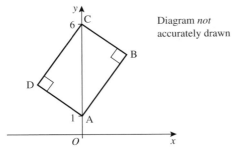

Diagram *not* accurately drawn

Find the equation of the straight line through D and C.

[Edexcel]

Key points

1 Points in a 2-D (two-dimensional) plane may be described using a coordinate system in which x runs from left to right and y from bottom to top. The coordinate axes cross over at the origin. x values to the left of the origin are negative; so, too, are y values below the origin.

2 Linear functions such as $y = 3x + 1$ may be plotted accurately by drawing a table of values. Although two points are sufficient to define a line, it is customary to plot three points as this helps detect errors caused by a slip in the working.

3 The gradient of a linear function is defined as the ratio of the height gained to the horizontal distance covered, or 'rise over run' for short. Graphs that go down as you move to the right will have negative gradients.

4 The intercept (or y intercept) of a linear function tells you where it crosses the y axis.

5 A linear graph with gradient m and y intercept c will have equation $y = mx + c$. This principle allows you to sketch linear functions, and to recognise the equation of a given straight line graph. In order to compare the gradients of two linear functions, it is best to rearrange them (if necessary) into the form $y = mx + c$.

6 Two lines will be parallel if, and only if, their gradients have the same value.

7 When two lines are perpendicular the product of their gradients is -1.

8 Finally, the methods in this chapter apply to linear functions containing both x and y terms. You will occasionally encounter linear graphs that are purely vertical (equation $x = $ a constant) or purely horizontal ($y = $ a constant), like these:

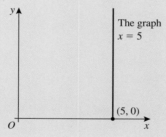

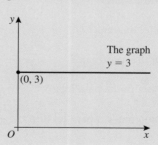

Internet Challenge 6

Parallels

Use the internet to help you answer these questions about parallels.

1 What name is given to a quadrilateral with two sets of parallel sides?

2 What name is given to a quadrilateral with only one set of parallel sides?

3 What is a parallelepiped? How do you draw one?

4 Which iconic rock group recorded the album 'Parallel Lines' in 1978?

5 What is the 49th parallel?

6 What are parallel universes?

7 What is the parallel postulate?

8 Where might you find a parallel port?

9 Where might you make a parallel turn?

10 'Parallel lines never meet.' True or false?

11 Is it possible for two curves to be parallel?

12 Who might choose to place things in parallel rather than in series?

13 Look at the picture at the top of this page.
   a) How many of the lines running from left to right are parallel? Now check your answer with a ruler or straight edge.
   b) The picture is called 'Café Wall'. Find out the location of the café that inspired this picture, and the name of the mathematician who first described it.

CHAPTER 7

Simultaneous equations

In this chapter you will learn how to:

- solve simple simultaneous equations by inspection
- solve harder simultaneous equations by algebraic elimination
- solve simultaneous equations by graphical methods
- solve problems using simultaneous equations.

You will also be challenged to:

- investigate magic squares.

Starter: **Fruity numbers**

Each fruit symbol stands for a missing number – and has the same value each time it occurs. Work out the value of each fruit.

7.1 Simultaneous equations

Sometimes you need to solve a pair of equations such as:

$$5x + 2y = 19$$
$$5x + 3y = 21$$

These are called **simultaneous equations**. The idea is to find a value for x and a matching value for y so that both equations are true together.

The method of **inspection** requires you to look at the two equations and spot any obvious slight differences between them. It should be used only for simple problems.

EXAMPLE

Solve the simultaneous equations:

$$5x + 2y = 19$$
$$5x + 3y = 21$$

SOLUTION

First, label the two equations as (1) and (2). Then compare them.

$$5x + 2y = 19 \qquad (1)$$
$$5x + 3y = 21 \qquad (2)$$

You can see that equation (2) has an extra y on the left, and a total of 2 more on the right $(21 - 19)$

By inspection, $y = 2$

Now substitute this value into equation (1):

$$5x + 2 \times 2 = 19$$
$$5x + 4 = 19$$
$$5x = 19 - 4$$
$$5x = 15$$
$$x = 3$$

Your answer should give values for both x and y.

So the final solution pair is $x = 3$ and $y = 2$

You can check your answer by substituting these values into the other equation, i.e. number (2):

Checking is a very good habit.

$$5x + 3y = 5 \times 3 + 3 \times 2 = 15 + 6 = 21 \qquad \text{as required.}$$

In the exam you will need to show all of your working and explain your reasoning dearly.

EXAMPLE

Solve the simultaneous equations: $3x + y = 7$
$5x + y = 5$

SOLUTION

$3x + y = 7$ (1)

$5x + y = 5$ (2)   Equation (2) has an extra $2x$ on the left but is 2 less ($7 - 5$) on the right, so $2x = -2$.

By inspection, $2x = -2$, so $x = -1$.

Now substitute this value into equation (1) to obtain:

$3 \times (-1) + y = 7$
$-3 + y = 7$
$y = 7 + 3$
$y = 10$

So the final solution pair is $x = -1$ and $y = 10$.

Check by substituting these values into equation (2):
$5x + y = 5 \times (-1) + 10 = -5 + 10 = 5$ as required.

EXERCISE 7.1

Solve these problems using the method of inspection. Write out all the stages clearly, as in the examples above.

1 $3x + 4y = 16$
    $3x + 5y = 17$

2 $x + 4y = 15$
    $x + 5y = 18$

3 $3x + y = 3$
    $4x + y = 2$

4 $4x + 2y = 6$
    $5x + 2y = 5$

5 $x + 8y = 4$
    $x + 10y = 6$

6 $6x - y = 9$
    $5x - y = 7$

7 $5x - 3y = 47$
    $7x - 3y = 67$

8 $2x - y = 7$
    $4x - y = 13$

9 $x + 2y = 3$
    $x + 3y = 1$

10 $5x + 3y = 20$
     $5x + 4y = 20$

7.2 Solving simultaneous equations by algebraic elimination

This method is used for most problems, if the answer is not obvious by inspection. The idea is to multiply one, or both, of the equations by a suitable multiplier, until they have a matching number of x's (or y's).

There are two variants of the elimination method, depending on the signs involved.

If the matching terms are the same, but one is positive and the other is negative, then you use the **addition** method. If, however, they are both positive or both negative, then you use the **subtraction** method instead. A useful rule is DASS: Different, Add; Same, Subtract!

EXAMPLE

Solve the simultaneous equations:

$$2x - y = 8$$
$$x + 3y = 11$$

SOLUTION

$2x - y = 8$	(1)
$x + 3y = 11$	(2)

> Look at the y terms. If you multiply equation (1) by 3 then they will both contain $3y$.

(1) ×3:	$6x - 3y = 24$	(3)
(2) ×1:	$x + 3y = 11$	(4)
Adding:	$7x = 35$	
	$x = 35 \div 7$	
	$x = 5$	

> The matching parts are $-3y$ and $+3y$.
> One of these is positive and the other negative, so you use the addition method.
> When you add $-3y$ and $+3y$ together there are no y's left at all.

Now substitute this value into equation (1) to obtain:

$$2 \times (5) - y = 8$$
$$10 - y = 8$$
$$2 - y = 0$$
$$y = 2$$

So the solution is $\underline{x = 5 \text{ and } y = 2}$

Check by substituting these values into equation (2):

$$x + 3y = (5) + 3 \times (2) = 5 + 6 = 11 \qquad \text{as required.}$$

EXAMPLE

Solve the simultaneous equations:

$$7x + 2y = 24$$
$$5x + 3y = 25$$

SOLUTION

$7x + 2y = 24$	(1)
$5x + 3y = 25$	(2)

> Look at the y terms. If you multiply equation (1) by 3 and equation (2) by 2 then they will both contain $6y$.

(1) ×3:	$21x + 6y = 72$	(3)
(2) ×2:	$10x + 6y = 50$	(4)
Subtracting:	$11x = 22$	
	$x = 22 \div 11$	
	$x = 2$	

> The matching parts are $+6y$ and $+6y$.
> These are both positive, so you use the subtraction method.
> When you subtract $+6y$ from $+6y$ there are no y's left at all.

Now substitute this value into equation (1) to obtain:

$$7 \times (2) + 2y = 24$$
$$14 + 2y = 24$$
$$2y = 24 - 14$$
$$2y = 10$$
$$y = 5$$

So the solution is $x = 2$ and $y = 5$

Check by substituting these values into equation (2):

$$5x + 3y = 5 \times (2) + 3 \times (5) = 10 + 15 = 25 \qquad \text{as required.}$$

Take care when subtracting a quantity that is negative to begin with; a double minus generates a plus in this case.

EXAMPLE

Solve the simultaneous equations:

$$x + 2y = 4$$
$$5x - 7y = 3$$

SOLUTION

$$x + 2y = 4 \qquad (1)$$
$$5x - 7y = 4 \qquad (2)$$

(1) ×5: $\qquad 5x + 10y = 20 \qquad (3)$

(2): $\qquad \underline{5x - 7y = 3} \qquad (4)$

Subtracting: $\qquad 17y = 17 \quad \longleftarrow \quad \boxed{10y - -7y \text{ gives } 10y + 7y = 17y}$
$$y = 1$$

Now substitute this value into equation (1) to obtain:

$$x + 2 \times (1) = 4$$
$$x + 2 = 4$$
$$x = 4 - 2$$
$$x = 2$$

So the solution is $x = 2$ and $y = 1$

Check by substituting these values into equation (2):

$$5x - 7y = 5 \times (2) - 7 \times (1) = 10 - 7 = 3 \qquad \text{as required.}$$

This final example shows the subtraction method applied again, this time when both the matching terms are negative.

EXAMPLE

Solve the simultaneous equations:

$$5x - 2y = 25$$
$$4x - 3y = 13$$

SOLUTION

$$5x - 2y = 25 \quad (1)$$
$$4x - 3y = 13 \quad (2)$$

Look at the *y* terms. If you multiply equation (1) by 3 and equation (2) by 2 then they will both contain $-6y$.

(1) ×3: $\quad 15x - 6y = 75 \quad (3)$

(2) ×2: $\quad\quad 8x - 6y = 26 \quad (4)$

The matching parts are $-6y$ and $-6y$.
These are both negative, so you use the subtraction method.
When you subtract $-6y$ from $-6y$ there are no *y*'s left at all.

Subtracting: $\quad 7x \quad\quad = 49$
$$x = 49 \div 7$$
$$x = 7$$

Now substitute this value into equation (1) to obtain

$$5 \times (7) - 2y = 25$$
$$35 - 2y = 25$$
$$10 - 2y = 0$$
$$2y = 10$$
$$y = 5$$

So the solution is $\underline{x = 7}$ and $\underline{y = 5}$

Check by substituting these values into equation (2):

$$4x - 3y = 4 \times (7) - 3 \times (5) = 28 - 15 = 13 \quad\quad \text{as required.}$$

EXERCISE 7.2

Solve questions **1** to **8** using the algebraic *addition* method. Write out all the stages clearly, as in the worked examples above.

1 $4x + 2y = 22$
$\quad 3x - 2y = 6$

2 $x - 3y = 4$
$\quad 4x + 3y = 1$

3 $5x - y = 9$
$\quad 3x + 2y = 8$

4 $x + y = 1$
$\quad 4x - 3y = 11$

5 $2x + 5y = 20$
$\quad x - 2y = 1$

6 $3x + 2y = 5$
$\quad 5x - 4y = 1$

7 $x - 2y = 9$
$\quad 2x + 3y = 4$

8 $3x + 4y = -8$
$\quad 11x - 5y = 10$

Solve questions **9** to **16** using the algebraic *subtraction* method, showing all your working clearly.

9 $2x + y = 6$
$\quad x + 3y = 13$

10 $2x + 3y = 13$
$\quad x + 2y = 8$

11 $9x + 2y = 5$
$\quad 3x + y = 1$

12 $4x - 3y = 5$
$\quad x - y = 1$

13 $3x - 2y = 2$
$\quad 5x - 3y = 3$

14 $x + 4y = 2$
$\quad 2x + 5y = 10$

15 $6x - y = 4$
$\quad 2x - 3y = 28$

16 $x - 4y = 10$
$\quad 2x - 7y = 18$

Solve questions **17** to **32** using algebra. For each question you will have to decide whether the addition method or the subtraction method is appropriate. Remember to show all the stages of your working.

17 $2x + 3y = 9$
$x - y = 2$

18 $x - y = 5$
$4x - 3y = 19$

19 $6x + y = 18$
$7x - 2y = 2$

20 $4x - y = -1$
$3x - 4y = 9$

21 $x + 2y = 5$
$3x - 4y = 10$

22 $3x - 6y = 9$
$x + 2y = 9$

23 $x + y = 0$
$x - y = 6$

24 $2x + y = 10$
$x + 11y = 5$

25 $14x + 3y = 7$
$5x - 2y = 24$

26 $3x - 4y = 3$
$x + 6y = 12$

27 $5x + 3y = 1$
$7x + 5y = 1$

28 $3x - 8y = 22$
$2x - 12y = 23$

29 $3x - 2y = 33$
$2x + 3y = -4$

30 $x - y = 4$
$4x - 6y = 21$

31 $5x - 3y = 34$
$7x - 4y = 47$

32 $x - 4y = 18$
$2x - 5y = 21$

7.3 Solving simultaneous equations by a graphical method

This method is quick and simple – it is particularly effective if the answers are whole numbers. When they are decimals, however, it becomes less accurate than the algebraic method.

EXAMPLE

Solve, graphically, the simultaneous equations:

$$4x + y = 6$$
$$5x - 4y = 18$$

SOLUTION

Consider, first, the equation $4x + y = 6$.

When $x = 0$ then $4x + y = 6$, giving $y = 6$.

Thus the graph passes through $(0, 6)$.

When $y = 0$ then $4x + y = 6$, giving $x = 1.5$.

Thus the graph passes through $(1.5, 0)$.

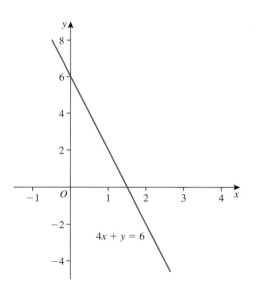

123

Next, consider the second equation, $5x - 4y = 18$.

When $x = 0$ then $5x - 4y = 18$, giving $y = -4.5$.

Thus the graph passes through $(0, -4.5)$.

When $y = 0$ then $5x - 4y = 18$, giving $x = 3.6$.

Thus the graph passes through $(3.6, 0)$.

Adding this line to the previous graph, we obtain this graph:

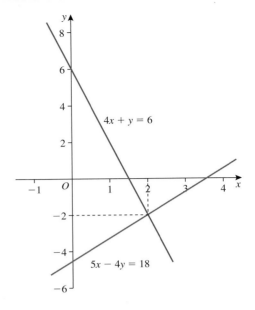

The solution occurs where these two lines cross.

From the graph, this can be read off as $x = 2, y = -2$

EXERCISE 7.3

For each of these questions draw a set of coordinate axes on squared paper (or graph paper). Draw the lines corresponding to each equation, and hence solve the simultaneous equations graphically.

1 $3x + y = 6$
   $x + y = 4$

2 $x + y = 10$
   $y = 2x - 2$

3 $x + 2y = 10$
   $2x + y = 14$

4 $x - y = 6$
   $2x + y = 12$

5 $2x + 3y = 18$
   $x + y = 7$

6 $y = x + 2$
   $x + y = 10$

7 $= x - 1$
   $x + y = 7$

8 $x + 4y = 14$
   $3x - y = 3$

7.4 Setting up and solving problems using simultaneous equations

Although many exam questions on simultaneous equations will already be set up for you, it is important that you learn how to set them up when needed. This section shows you how to formulate such problems, which can then be solved by the algebraic method.

EXAMPLE

A theatre has two different ticket prices, one for adults and another for children.
A party of 6 adults and 10 children costs £38, while for 5 adults and
12 children the cost is £39.

a) Write this information as two simultaneous equations.

b) Solve your equations to find the cost of an adult ticket and the cost of a
child ticket.

SOLUTION

a) Let the cost of an adult ticket be £x, and that of a child's ticket, £y.

$$6x + 10y = 38$$
$$5x + 12y = 39$$

> Remember to define the symbols you are going to use …

> …. then use them to represent the given information.

b) Multiplying the first equation by 6
and the second by 5, we obtain:

$$36x + 60y = 228$$
$$25x + 60y = 195$$

Subtracting:

$$11x = 33$$
$$x = 33 \div 11$$
$$x = 3$$

Substituting back into the first equation, we have:

$$6 \times (3) + 10y = 38$$
$$18 + 10y = 38$$
$$10y = 38 - 18$$
$$10y = 20$$
$$y = 2$$

Thus an adult ticket costs £3 and a child's ticket costs £2.

EXERCISE 7.4

Use simultaneous equations to help you solve the following problems. Remember to show all your working
carefully.

1 A clothes shop is having a sale. All the shirts are reduced to one price. All the jackets are reduced to a
single price as well, though they remain more expensive than the shirts. Arthur buys 10 shirts and
3 jackets, and pays £104. Alan buys 4 shirts and one jacket, and pays £38.
  a) Write two simultaneous equations to express this information.
  b) Solve your equations, to find the price of a shirt and the price of a jacket.

2 A hire company has a fleet of coaches and minibuses. Three coaches and four minibuses can carry
180 passengers, while five coaches and two minibuses can carry 230 passengers.
  a) Write two simultaneous equations to express this information.
  b) How many passengers can one coach carry?

3 A mathematics teacher buys some books for her A-level and IGCSE students. A-level books cost £10
each, and IGCSE books £15 each. She spends a total of £1800, buying a total of 160 books in all.
  a) Write two simultaneous equations to express this information, defining your symbols clearly.
  b) Solve your equations to find how many of each type of book she buys.

4 A shop sells tins of paint in 2 litre and 5 litre cans. The manager checks the amount of paint he has in stock, and finds that there are 500 cans altogether. These cans hold a total of 1420 litres of paint.
  a) Write two simultaneous equations to express this information. Explain the meaning of the symbols you use.
  b) Solve your equations to find the number of each size of can in stock.

5 A plant stall at a school fete sells tomato plants and pepper plants. Martin buys two tomato plants and four pepper plants for £2.50, while Suzy buys five tomato plants and three pepper plants for £3.10. Work out the cost of each type of plant.

REVIEW EXERCISE 7

Solve these simultaneous equations by inspection.

1 $3x + 4y = 24$
  $3x + 5y = 27$

2 $x + 5y = 16$
  $x - 2y = 16$

3 $5x + 2y = 5$
  $3x + 2y = 7$

4 $8x - 3y = 11$
  $8x - 7y = 15$

Solve these by the elimination (addition or subtraction) method.

5 $x + 3y = 7$
  $4x + y = 17$

6 $6x + y = 11$
  $4x + 5y = 3$

7 $3x - 2y = 13$
  $4x + 3y = 6$

8 $3x - 4y = 5$
  $2x - 5y = 8$

9 $x + 2y = 6$
  $x - 2y = 4$

10 $3x - 4y = 8$
  $5x - 6y = 13$

11 $2x - 3y = 13$
  $10x + y = 1$

12 $5x + 4y = 4$
  $x - 2y = 5$

13 The diagram shows part of the graph of $2x + 3y = 18$.

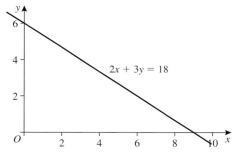

  a) Make a copy of this graph on squared paper or graph paper.
  b) On the same diagram, plot the graph of the line $y = x + 1$.
  c) Hence solve the simultaneous equations $y = x + 1$, $2x + 3y = 18$.

14 Use a graphical method to solve the simultaneous equations:

$$5x + 3y = 30$$
$$y = x - 2$$

15 At a seaside drinks stall you can buy cans of cola and cans or orange drink. Five cans of cola and one can of orange cost £2.07. Two cans of cola and three cans of orange cost £1.66.
  a) Using x to represent the cost of a can of cola and y to represent the cost of a can of orange, in pence, write this information as two simultaneous equations.
  b) Solve your equations to find the cost of each type of drink.

16 A potter is making cups and saucers. Each cup takes c minutes to produce, and each saucer takes s minutes. The potter can produce three cups and two saucers in 19 minutes, while it would take exactly half an hour to produce four cups and five saucers.

a) Write this information as two simultaneous equations.

b) Solve your equations, to find the values of c and s.

c) How long would it take to produce a set of 6 cups and 6 saucers?

17 A phone network charges x pence per minute for telephone calls, and y pence for each text message sent. 100 minutes and 50 texts cost £4, while 150 minutes and 100 texts cost £6.50.

a) Write this information as two simultaneous equations.

b) Solve your equations to find the values of x and y.

c) How much would it cost for 300 minutes and 50 texts?

18 Solve the simultaneous equations:

$$4x + y = 8$$
$$2x - 3y = 11$$

[Edexcel]

19 Solve:

$$2x - 3y = 11$$
$$5x + 2y = 18$$

[Edexcel]

Key points

1 When you solve simultaneous equations you must ensure you show full algebraic working. You will lose marks if you show no working at all or incomplete working.

2 In practice, the most frequently used method is that of algebraic elimination. Multiply one or both of the equations by a suitable scaling factor, so the x (or y) coefficients are numerically the same in both equations. If the matching coefficients are one positive and one negative then you add the two equations to achieve the elimination. If they are both positive, or both negative, then you must subtract one equation from the other instead. Remember DASS:

Different sign	**S**ame sign
Add	**S**ubtract

3 The graphical method of solution can be quite neat, but it is not reliable if the solutions are not whole numbers or simple decimals.

Internet Challenge 7

Magic squares

In a magic square each row, column and diagonal adds up to the same total, known as the square constant.

Here is a 3 by 3 magic square, with a square constant of 15.

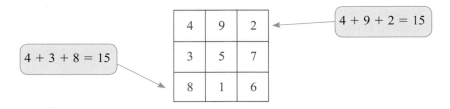

1 Try to make a 4 by 4 magic square using the numbers 1 to 16. The square constant will be 34. (This is quite difficult!)

2 Use the internet to find a picture of Albrecht Dürer's engraving *Melancholia*. What do you find in the top right corner of the picture?

3 Magic squares with an odd number of rows/columns are much easier to make than those with an even number of rows. Use the internet to find a procedure for making odd-sized magic squares. Then use the procedure to make:
   a) a 5 by 5 magic square
   b) an 11 by 11 magic square.

4 An 8 by 8 magic square was constructed by Benjamin Franklin in the nineteenth century.
   a) Use the internet to find a copy of Franklin's 8 by 8 square, and print it out.
   b) Using a red pen, join the numbers 1, 2, 3, …, 16 using a set of straight lines. Now do the same for the numbers 17, 18, 19, …, 32. What do you notice?
   c) Using a blue pen, join the numbers 33, 34, 35, …, 48 using a set of straight lines. Now do the same for the numbers 49, 50, 51, …, 64. What do you notice?
   d) Try to find out some other interesting properties of Franklin's square.
   e) Find out a little about the life and achievements of Benjamin Franklin.

5 The image to the right shows the world's oldest known magic square.
   a) By what name is this square known?
   b) Approximately when does it date from?

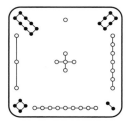

CHAPTER 8

Inequalities

In this chapter you will **learn how to**:

- solve simple linear inequalities in one variable
- represent solution sets on a number line
- solve linear inequalities in two variables and find the solution set.

You will also be **challenged to**:

- investigate mathematical symbols.

Starter: **Treasure hunt**

Here is a map of Treasure Island. The pirates have buried their treasure at a place where the x and y coordinates are whole numbers. Use the clues to work out where the treasure is buried.

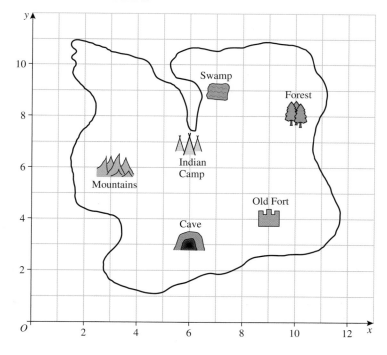

Clue 1: x is greater than 5.
Clue 2: y is greater than 6.
Clue 3: One of x and y is prime, and the other is not.
Clue 4: x and y add up to 16.

8.1 Whole-number solutions to inequalities

Inequalities are similar, in many ways, to equations. Simple inequality problems with whole number (integer) solutions may be solved at sight, but you will need to use algebraic and graphical methods for dealing with harder problems where the solutions need not be integers.

In this chapter you will be working with these four symbols:

$<$ less than $>$ greater than
$\leqslant$ less than or equal to $\geqslant$ greater than or equal to

For example, $x < 3$ would be read as 'x is less than 3'. $y \geqslant 6$ would be read as 'y is greater than or equal to 6'. The narrow end of the symbol points to the smaller quantity.

EXAMPLE

Write down whole-number solutions to these inequalities:
a) $x > 5$ b) $y \leqslant 6$ c) $3 \leqslant z < 10$

> This means that z can equal 3 but not equal 10.
> z lies from 3 up to but not including 10.

SOLUTION

a) 6, 7, 8, 9, 10, …
b) …, 2, 3, 4, 5, 6
c) 3, 4, 5, 6, 7, 8, 9

> Dots indicate that the number pattern continues beyond those written down.

EXAMPLE

Find the whole-number solutions to these inequalities:
a) $2x > 5$ b) $3y \leqslant 6$ c) $3 \leqslant 2z < 10$ d) $5 < 3x - 1 \leqslant 14$

Solution

a) $2x > 5$
    Dividing both sides by 2, we obtain $x > 2\frac{1}{2}$.
    Since x has to be a whole number, the possible values are 3, 4, 5, 6, 7, …

b) $3y \leqslant 6$
    Dividing both sides by 3, we obtain $y \leqslant 2$.
    The whole-number solutions are …, −2, −1, 0, 1, 2

c) $3 \leqslant 2z < 10$
    Dividing through by 2, we obtain $1\frac{1}{2} \leqslant z < 5$.
    The integer solutions are 2, 3 and 4

d) $5 < 3x - 1 \leqslant 14$
    Adding 1 throughout, we obtain $6 < 3x \leqslant 15$
    Dividing through by 3, we obtain $2 < x \leqslant 5$.
    The integer solutions are 3, 4 and 5.

EXERCISE 8.1

Find the whole-number (integer) solutions to each of these inequalities.

1 $x > 3$ **2** $2x > 5$ **3** $y < 4$ **4** $3y < 7$

5 $0 < y \leqslant 6$ **6** $0 \leqslant 2p \leqslant 13$ **7** $1 < x < 3$ **8** $1 \leqslant 2w \leqslant 3$

9 $1 < z - 1 < 5$ **10** $1 < 2z < 5$ **11** $1 \leqslant z \leqslant 5$ **12** $1 \leqslant 2z \leqslant 5$

13 $2x \leqslant 11$ **14** $3x > 10$ **15** $7 < 2y + 3 \leqslant 13$ **16** $5 > g \geqslant 1$

17 $14 \geqslant 3x > 0$ **18** $10 < 2x \leqslant 20$ **19** $98 \leqslant t + 1 \leqslant 99$ **20** $-3 < u < 3$

8.2 Using algebra to solve linear inequalities

In the previous section you solved inequalities using whole numbers. The solution could be listed as a set of whole numbers, for example, 2, 3, 4, 5, 6.

If you are not told that the solution is a whole number, then you must leave your solution as an inequality, covering a range of possible values. When this inequality is in its simplest form, you are said to have solved the inequality.

To solve an inequality, you can use similar methods to those used for solving equations, such as:

- You can add (or subtract) the same number to both sides.
- You can multiply (or divide) both sides by the same positive number.

You should *not*, however, multiply or divide both sides by a *negative* number, as this would cause the direction of the inequality to reverse, and may well introduce a mistake.

EXAMPLE

Solve, algebraically, the inequality $3x - 8 \leqslant 30 + x$.

SOLUTION

$$3x - 8 \leqslant 30 + x$$
$$3x - 8 - x \leqslant 30$$ First, subtract x from both sides.
$$2x - 8 \leqslant 30$$
$$2x \leqslant 30 + 8$$ Next, add 8 to both sides.
$$2x \leqslant 38$$
$$x \leqslant 19$$ Finally, divide both sides by 2.

Your work will be more accurate, and easier to follow, if you process only a small step at each stage. Also, the lines of working should be aligned at the inequality signs.

Note: Make sure you do **not** use any equals signs ($=$) anywhere.

EXAMPLE

Solve, algebraically, the inequality $10 - 3x < 30 + 2x$.

SOLUTION

Note the alignment here.

$$10 - 3x < 30 + 2x$$
$$10 < 30 + 2x + 3x$$
$$10 < 30 + 5x$$
$$10 - 30 < 5x$$
$$-20 < 5x$$
$$-4 < x$$

It would be a *bad idea* to take $2x$ from both sides to begin with, because $10 - 5x < 30$ is leading towards a solution that will require you to divide by a negative number: this is best avoided.

EXERCISE 8.2

Solve, algebraically, these inequalities.

1 $x + 5 \geqslant 13$

2 $3x - 1 > 14$

3 $10x + 43 < 13$

4 $6 + x < 10 - x$

5 $2x - 5 < x + 1$

6 $3x + 11 \leqslant 17 + x$

7 $5x \geqslant x + 20$

8 $32 + x < 12 + 6x$

9 $16 + x \leqslant 10 + 4x$

10 $3(x + 2) > x + 4$

11 $2x + 13 \geqslant 41 - 5x$

12 $6x + 1 < 28$

13 $13 - x > 5 + 3x$

14 $x + 15 \leqslant 7 - x$

15 $3x + 7 \leqslant x + 7$

16 $16 - x < 2x + 31$

17 $144 \geqslant 360 - 6x$

18 $4 + 3x < 4 - x$

19 $6(x + 2) \geqslant x + 7$

20 $3(2x + 3) < 4(x + 4)$

21 $5 < 2x + 1 \leqslant 9$

22 $8 < 5x + 3 < 13$

23 $6 \leqslant 2(x - 3) \leqslant 12$

24 $1 \leqslant 4x + 5 < 17$

25 $\dfrac{x}{2} - 3 \leqslant 4$

26 $\dfrac{3x - 1}{2} > 7$

27 $\dfrac{x}{3} + 1 < 10$

28 $\dfrac{x + 1}{3} \geqslant 4$

8.3 Illustrating inequalities on a number line

In the first section of this chapter you learned how to list the solution set as a list of whole numbers, but this is not usually possible in more general problems where x is not restricted to being an integer. The solutions may, however, be illustrated graphically by means of a 'thermometer diagram' drawn alongside a number line. The end of the line is left as an open 'bulb' if it is not to be included, or filled in if it is included.

This symbol shows a region not including the end-points:

○────────────────○ , for example $-2 < x < 5$

This region includes the left-hand end but not the right-hand end:

 , for example $-2 \leqslant x < 5$

This region includes both end-points:

 , for example $-2 \leqslant x \leqslant 5$

Note: We never combine $<$ and $>$ in one single inequality.

EXAMPLE

Solve the inequality $6 + x < 13$ and illustrate the solution with a number line diagram.

The solid line shows that x can take any value below 7 …

SOLUTION

$6 + x < 13$
$x < 13 - 6$
$x < 7$

… while the open circle shows that 7 itself is not part of the solution set.

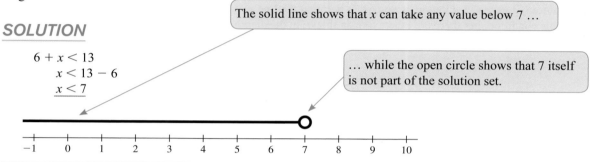

EXAMPLE

Solve the inequality $x - 6 \leqslant 2x - 5$ and illustrate the solution with a diagram.

SOLUTION

$x - 6 \leqslant 2x - 5$
$-6 \leqslant 2x - 5 - x$
$-6 \leqslant x - 5$
$-6 + 5 \leqslant x$
$-1 \leqslant x$

The filled circle shows that -1 is part of the solution.

The solid line shows that all values above -1 are part of the solution.

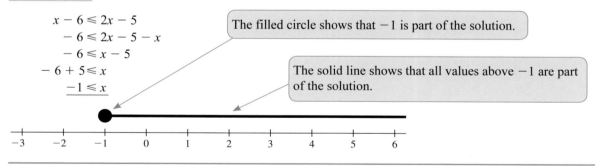

EXERCISE 8.3

In questions **1** to **10** you are given the solution to an inequality. Draw a suitable number line diagram to illustrate the solution in each case.

1 $x < 5$ **2** $x \geqslant 2$ **3** $1 \leqslant x \leqslant 5$ **4** $-2 < x < 2$

5 $x \leqslant 7$ **6** $x < -1$ **7** $-3 < x < 0$ **8** $-4 < x \leqslant -1$

9 $-2 \leqslant x \leqslant 2$ **10** $x > 2.5$

In questions **11** to **20**, solve each inequality and then illustrate it with a line diagram.

11 $5 + x < 19$

12 $3x - 2 \leqslant 13$

13 $20 - x > 10$

14 $5x + 4 \geqslant 19$

15 $10 < 2x < 17$

16 $15 - x < 2x - 3$

17 $2x + 1 \leqslant 5x - 11$

18 $3x - 1 < 11 - x$

19 $20 - x < 6x - 1$

20 $9x + 1 \leqslant 2(x - 3)$

8.4 Graphs of linear inequalities in two variables

Suppose two variables x and y are restricted by a rule such as $x + y \leqslant 5$. The solution set is the set of all possible combinations of values for x and y so that $x + y$ does not exceed 5. Obviously any (x, y) point on the line $x + y = 5$ will meet this condition, but there are others too, as shown in the diagram below.

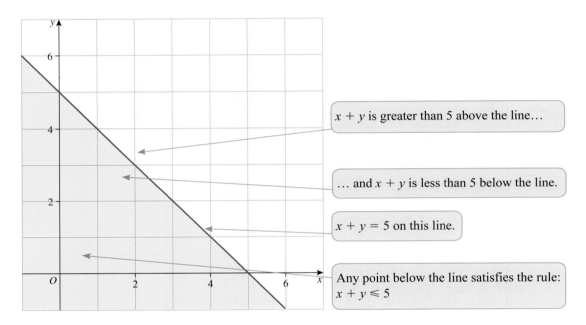

$x + y$ is greater than 5 above the line…

… and $x + y$ is less than 5 below the line.

$x + y = 5$ on this line.

Any point below the line satisfies the rule: $x + y \leqslant 5$

Examination questions are likely to include more than one inequality, so the required region is bounded by several straight lines, as in this example below.

EXAMPLE

The point (x, y) satisfies the following inequalities:

$\quad x > 1, x < 3, y < 4, y > x$

a) Illustrate these inequalities on a graph, shading the region that satisfies all four inequalities. Label your region R.

b) The point P lies in the region R. The coordinates of P are integers. Write down the coordinates of P.

SOLUTION

a) First, draw the four straight lines
$x = 1$, $x = 3$, $y = 4$ and $y = x$.

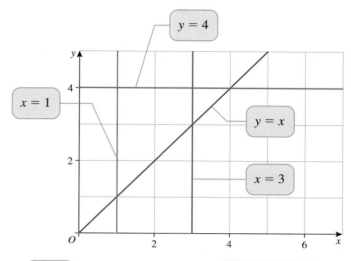

Next, shade the region
corresponding to the
inequalities.

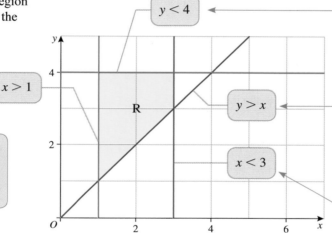

$x > 1$ means the
x coordinates lie
to the *right* of
the line $x = 1$

$y < 4$ means the
y coordinates lie
below the line
$y = 4$

$y > x$ means the
y coordinates are
greater than if
they were on the
line $y = x$, so the
region is *above*
the line.

$x < 3$ means the x coordinates
lie to the *left* of the line $x = 3$

b) P lies within the region R and
has integer coordinates:

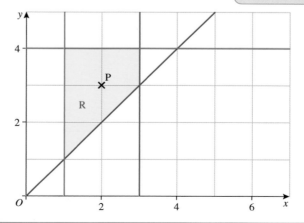

Thus P is at (2, 3)

EXERCISE 8.4

For each of questions **1** to **5**, draw a coordinate grid in which x and y can range from 0 to 10.

1 Draw the graphs of these straight lines:

$x = 2, x = 5, y = 1, y = x$

Hence shade the region R corresponding to the inequalities:

$x \geqslant 2, x < 5, y > 1, y \leqslant x$

2 Draw the graphs of these straight lines:

$x = 2, y = 2, y = 7, y = x - 1$

Hence shade the region R corresponding to the inequalities:

$x > 2, y > 2, y < 7, y > x - 1$

3 Draw the graphs of these straight lines:

$x = 9, y = 7, y = x$

Hence shade the region R corresponding to the inequalities:

$x < 9, y > 7, y < x$

4 Shade the region R corresponding to the inequalities:

$x > 3, x \leqslant 7, y > 1, x + y \leqslant 10$

5 Shade the region R corresponding to the inequalities:

$x > 0, y > 1, y < x + 4, x + y < 8$

6 The diagram shows a region R bounded by three straight lines, L_1, L_2 and L_3.

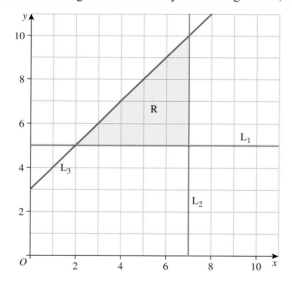

a) Write down the equations of the three straight lines, L_1, L_2 and L_3. Show clearly which equation applies to which line.

b) Write down three inequalities that define the region R.

8.5 Quadratic inequalities

Sometimes you may need to simplify a quadratic inequality, such as $x^2 \leqslant 9$.
By taking the square root of both sides, the solution might seem to be $x \leqslant 3$.
This is fine for positive values of x, and also for negative values of x provided
they are not too far below zero.

However, if x is substantially negative, the original inequality can fail. For
example, if $x = -4$, then $x^2 = 16$ which is now greater than 9. Clearly we need
to restrict the negative values of the solution so that x does not become any
lower than -3.

The result is that the solution consists of $0 \leqslant x \leqslant 3$ for positive values of x and
$-3 \leqslant x \leqslant 0$ for negative values of x. These can be combined into the single
statement $-3 \leqslant x \leqslant 3$

EXAMPLE

You are given that x satisfies the inequality $x^2 \leqslant 16$
a) Solve the inequality.
b) Draw a diagram to illustrate the solution on a number line.

SOLUTION

a) Since the square root of 16 is 4, the solution must be $-4 \leqslant x \leqslant 4$
b) Illustrating this on a number line:

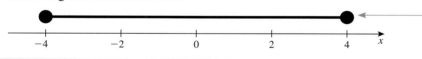

> Remember that the filled bubble shows that 4 **is** included in the solution.

The above example shows you how to deal with inequalities of the form $x \leqslant k$.
A similar method holds for $x < k$, the only difference being that the end points
are now excluded.

If, however, the inequality is in the other direction, then the solution consists of
two disconnected regions. The next example illustrates this type of question.

EXAMPLE

a) Solve the inequality $x^2 + 3 > 12$
b) Draw a diagram to illustrate the solution on a number line.

137

SOLUTION

a) Since $x^2 + 3 > 12$, then $x^2 > 9$
Taking square roots, we obtain $x > 3$ for positive values of x and $x < -3$
for negative values of x.
So the solution is $\underline{x < -3 \text{ or } x > 3}$

b) On a number line:

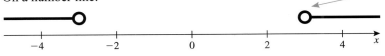

> Remember that the open bubble shows that 3 is **not** included in the solution.

EXERCISE 8.5

Solve, algebraically, these inequalities.

1 $x^2 < 25$ **2** $x^2 \leqslant 81$ **3** $y^2 > 16$ **4** $x^2 \geqslant 4$

5 $x^2 + 7 \leqslant 71$ **6** $x^2 < 49$ **7** $y^2 \geqslant 144$ **8** $4x^2 < 49$

Solve these inequalities. Illustrate each one on a number line.

9 $x^2 + 4 \leqslant 8$ **10** $x^2 + 1 > 50$ **11** $x^2 \geqslant 25$ **12** $x^2 \leqslant 9$

REVIEW EXERCISE 8

Find whole-number solutions to these inequalities:

1 $0 \leqslant 2x \leqslant 7$ **2** $1 < 3y < 27$ **3** $8 \leqslant 2n < 18$

4 $5 < 2x + 1 < 20$ **5** $-3 \leqslant 2t - 1 \leqslant 12$ **6** $2 < \frac{1}{2}x < 4$

7 $4 \leqslant 3m + 2 < 9$ **8** $5 < 2(x + 1) \leqslant 12$ **9** $6 \leqslant 3(x - 2) \leqslant 12$

10 $8 < 5t < 11$

Solve these linear inequalities. Illustrate each one on a number line.

11 $7 \leqslant 2x + 5$ **12** $6 < 4 - x$ **13** $5x + 3 < 18$

14 $3x - 7 \leqslant 11$ **15** $2 \leqslant 2x + 4 < 7$ **16** $3 < 4x + 3 \leqslant 11$

17 $5x + 1 < 2x + 7$ **18** $2x + 17 < 7x + 2$ **19** $3 \leqslant x + 5 < 19 - x$

20 $15 - x < 2x - 3 \leqslant x + 6$

21 The diagram shows the graphs of the lines $y = \frac{1}{2}x + 4$, $y = x - 1$ and $x = 3$

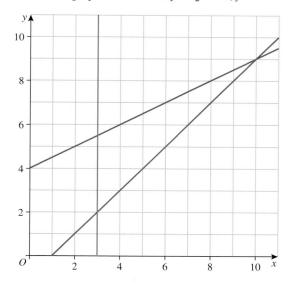

a) Make a copy of this diagram.

The point P(x, y) has integer coordinates. P satisfies the inequalities:

$$y < \tfrac{1}{2}x + 4, \ y > x - 1, \ x > 3$$

b) Mark on your diagram, with a cross, each of the points where P could lie.

22 a) $-2 < x \le 1$ and x is an integer. Write down all the possible values of x.
b) $-2 < x \le 1$, $y > -2$ and $y < x + 1$ and x and y are integers. On the grid, mark with a cross ($\times$) each of the six points which satisfies *all* these three inequalities.

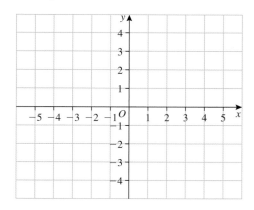

[Edexcel]

23 n is a whole number such that $6 < 2n < 13$. List all the possible values of n. [Edexcel]

24 a) Solve the inequality $4y + 3 \ge 1$
b) Write down the smallest **integer** value of y which satisfies the inequality

$$4y + 3 \ge 1$$

[Edexcel]

25 The line with equation $6y + 5x = 15$ is drawn on the grid below.

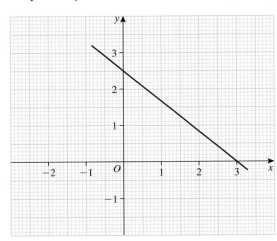

a) Rearrange the equation $6y + 5x = 15$ to make y the subject.
b) The point $(-21, k)$ lies on the line. Find the value of k.
c) (i) On a copy of the grid, shade the region of points whose coordinates satisfy the four inequalities:

$$y > 0, x > 0, 2x < 3, 6y + 5x < 15$$

Label this region R.
P is a point in the region R. The coordinates of P are both integers.
(ii) Write down the coordinates of P. [Edexcel]

26 a) (i) Solve the inequality $5x - 7 < 2x - 1$
(ii) Copy this number line, and use it to represent the solution set to part (i).

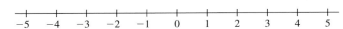

n is an integer such that $-4 \le 2n < 3$
b) Write down the possible values of n. [Edexcel]

27 a) Solve the inequality $x^2 > 36$
b) Copy this number line, and use it to represent the solution set of the inequality $x^2 > 36$

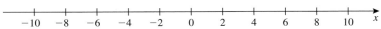

28 Solve the inequality $x^2 \ge 100$

29 Solve the inequality $y^2 < 9$ [Edexcel]

30 a) Solve the inequality $x^2 \le 4$
b) Copy this number line, and use it to represent the solution set of the inequality $x^2 \le 4$

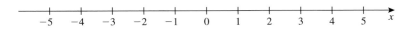

[Edexcel]

Key points

1. Inequalities may be manipulated using many of the methods applicable to ordinary equations.

2. The solution to an inequality is usually a range of values, rather than just a single value.

   For example, the solution to $5x + 1 < 16$ is $x < 3$

3. It is best to avoid multiplying or dividing an inequality by a negative number, since this causes the direction of the inequality to reverse.

4. Solutions to inequalities in one variable may be shown on a number line, using a 'thermometer diagram'. For example, to represent $-1 < x \leqslant 3$ we have:

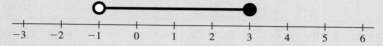

5. The open bulb shows that -1 is not to be included. The filled bulb shows that 3 is included.

6. Inequalities in two variables are usually represented on a coordinate grid. Turn the inequalities into equations first, and plot the lines. Then decide which side of the line represents the required solution set.

7. In the examination you will usually have to plot several lines, and find the intersection of the corresponding regions. Integer solutions will often be asked for in both one and two variable inequalities.

8. Take special care when taking square roots to solve simple quadratic inequalities. For example, the solution to $x^2 \leqslant 9$ is not simply $x \leqslant 3$; the lower end has to be closed off to give $-3 \leqslant x \leqslant 3$. Similarly $y^2 > 100$ does not just lead to $y > 10$; at the lower end we also have $y < -10$ so the solution set now consists of two separate (but symmetric) regions. You should revisit Section 8.5 carefully if you are not entirely sure about this.

Internet Challenge 8

Investigating mathematical symbols

The mathematical symbols we use nowadays have evolved over a long period of time, from many different and diverse cultures. In this investigation you will try to uncover the origins of some of the more widely used symbols – some of these symbols will be quite familiar to you, but there may be others that you have not yet encountered.

Here are some mathematical symbols, and some facts about them. Unfortunately the symbols and the facts have become jumbled up. Match the symbols to the corresponding fact.

∞	The Golden Ratio.
<	The eighteenth letter of the Greek alphabet, denotes 'the sum of'.
√	This 17th century symbol was formerly used in Europe to indicate subtraction.
φ	A sculpture of this symbol, by Marta Pan, stands on the A6 roadside in France.
θ	The (not real) square root of minus one.
÷	First used in Harriot's *Artis Analyticae Praxis* in 1631.
=	This originated from Hindu mathematics, where it was known as *sunya*.
i	Invented by Robert Recorde in 1557.
Σ	The eighth letter of the Greek alphabet, used to denote an unknown angle.
0	This 16th century symbol may be a corrupted abbreviation for *radix*.

Try to find out some more facts about each symbol.

CHAPTER 9

Sequences and series

In this chapter you will learn how to:

- recognise and use common number sequences
- use rules to generate number sequences
- find a general formula for the nth term of an arithmetic sequence
- find the sum of an arithmetic series.

You will also be challenged to:

- investigate Fibonacci numbers.

> Starter: Circles, lines and regions

Look at the sequence of circles below.

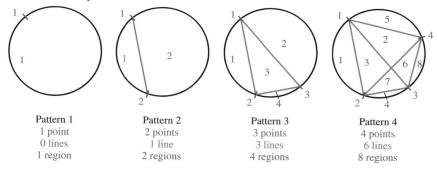

Pattern 1	Pattern 2	Pattern 3	Pattern 4
1 point	2 points	3 points	4 points
0 lines	1 line	3 lines	6 lines
1 region	2 regions	4 regions	8 regions

The diagram shows a sequence of circles. Each circle has some points marked around its circumference. Each point is joined to every other point by a line.

The lines and regions are then counted. The lines and regions are not all the same size.

Task 1

Describe a rule for how the number of points increases in this sequence.

Task 2

Describe a rule for how the number of lines increases.

Task 3

Describe a rule for how the number of regions increases.

Task 4

Now draw pattern 5 and pattern 6, and see if your rules seem correct. You should space out the points so that no triple intersections can occur, otherwise you lose a region, for example:

No   Yes

9.1 Number sequences

Here are some number sequences that occur often in mathematics.

Name of sequence	First six terms	Formula for the nth term
Positive integers	1, 2, 3, 4, 5, 6, …	n
Even numbers	2, 4, 6, 8, 10, 12, …	$2n$
Odd numbers	1, 3, 5, 7, 9, 11, …	$2n - 1$
Square numbers	1, 4, 9, 16, 25, 36, …	n^2
Cube numbers	1, 8, 27, 64, 125, 216, …	n^3
Powers of 2	2, 4, 8, 16, 32, 64 …	2^n
Powers of 10	10, 100, 1000, 10 000, 100 000, 1 000 000, …	10^n

You may encounter these number patterns when solving mathematical problems based on counting patterns.

EXAMPLE

Look at this pattern of squares.

Pattern 1   Pattern 2   Pattern 3   Pattern 4

a) How many squares would there be in pattern 5?
b) Find a formula for the number of squares in pattern n.
c) Use your formula to find the number of squares in pattern 100.

SOLUTION

The number of squares forms a pattern 2, 4, 6, 8, that is, the even numbers.

a) Pattern 5 contains $2 \times 5 = \underline{10 \text{ squares.}}$

b) Pattern n contains $\underline{2n \text{ squares.}}$

c) Pattern 100 contains $2 \times 100 = \underline{200 \text{ squares.}}$

Some number sequences are disguised versions of the common ones, perhaps with a constant number added or multiplied.

EXAMPLE

Find the next three terms in this number sequence. Find also a formula for the nth term.

101, 104, 109, 116, 125, …

SOLUTION

101, 104, 109, 116, 125, … are all 100 more than the square numbers.

The next three terms are $100 + 36$, $100 + 49$ and $100 + 64$, that is, $\underline{136, 149, 164}$

The nth term is $\underline{100 + n^2}$

EXERCISE 9.1

Write down the next two terms in each of these number sequences, and explain how each term is worked out. Give an expression for the nth term in each case.

They are all related to the list of common sequences in the table on the previous page.

1 10, 20, 30, 40, 50, 60, …

2 5, 7, 9, 11, 13, 15, …

3 51, 53, 55, 57, 59, 61, …

4 4, 8, 12, 16, 20, 24, …

5 2, 8, 26, 80, 242, …

6 0.1, 0.01, 0.001, 0.0001, …

7 10, 30, 60, 100, 150, 210, …

8 2, 8, 18, 32, 50, 72, …

9 Look at this pattern of triangles.

  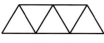 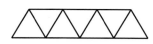

Pattern 1   Pattern 2   Pattern 3   Pattern 4

a) How many triangles would there be in pattern 7?

b) Find a formula for the number of triangles in pattern n.

10 Look at this pattern of spots.

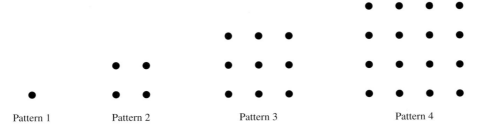

Pattern 1   Pattern 2   Pattern 3   Pattern 4

a) Find an expression for the number of spots in pattern n.

b) How many spots would there be in pattern 30?

9.2 Describing number sequences with rules

It can be very useful to be able to describe number sequences using rules. One way of doing this is to say how each term is connected to the next one in the sequence. (This is sometimes called a **term-to-term rule**, because it explains the link between one term and the next.)

EXAMPLE

A number sequence is defined as follows:

- The first term is 3.
- Each new term is double the previous one.

Use this rule to generate the first five terms of the number sequence.

SOLUTION

Start with 3:

$3 \times 2 = 6$
$6 \times 2 = 12$
etc.

The first five terms of the sequence are 3, 6, 12, 24, 48, ….

EXAMPLE

A number sequence is defined as follows:

- The first term is 7.
- Each new term is 3 more than the previous one.

Use this rule to generate the first six terms of the number sequence.

SOLUTION

Start with 7:

$$7 + 3 = 10$$
$$10 + 3 = 13$$
etc.

The first six terms of the sequence are 7, 10, 13, 16, 19, 22, ….

If you wanted to work out the 100th number in a sequence, it would be very tedious to have to write out all 100 numbers, one at a time. In this case it is better if you can use an algebraic expression for the nth term. (This is sometimes called a **position-to-term rule**, since you can work out the value of any term as long as you know its position in the sequence.)

EXAMPLE

The nth term of a number sequence is given by the expression $2n^2 + 1$.

a) Write down the first four terms of the sequence.
b) Find the value of the 20th term.

SOLUTION

a) $n = 1$ gives $2 \times 1^2 + 1 = 2 + 1 = 3$
$n = 2$ gives $2 \times 2^2 + 1 = 8 + 1 = 9$
$n = 3$ gives $2 \times 3^2 + 1 = 18 + 1 = 19$
$n = 4$ gives $2 \times 4^2 + 1 = 32 + 1 = 33$

The first four terms are 3, 9, 19, 33

b) When $n = 20$, $2 \times 20^2 + 1 = 800 + 1 = 801$.

EXERCISE 9.2

1 A number sequence is defined as follows:
- The first term is 5.
- Each new term is 2 more than the previous one.

Use this rule to generate the first five terms of the number sequence.

2 A number sequence is defined as follows:
- The first term is 1.
- To find each new term, add 1 to the previous term, and double this total.

Use this rule to generate the first four terms of the number sequence.

3 The nth term of a number sequence is given by the expression $8n - 1$.
a) Write down the values of the first five terms.
b) Work out the value of the 20th term.

4 The nth term of a number sequence is given by the expression $\dfrac{3n + 1}{2}$.
   a) Write down the values of the first six terms.
   b) Work out the value of the 23rd term.

5 Andy has been doing a mathematical investigation. He gets this sequence of numbers:

      12, 15, 18, 21, 24, …

   a) Describe Andy's pattern in words.
   b) Find the tenth term in Andy's number sequence.

6 The nth term of a number sequence is given by the expression $100 - n$.
   a) Write down the values of the first five terms.
   b) Work out the value of the 50th term.

7 In a certain number sequence, the first term is 3. Each new term is found by multiplying the previous term by 3.
   a) Write down the first five terms of the number sequence.
   b) What name is given to this particular number sequence?

8 The nth term of a number sequence is given by the formula $7n + 3$.
   a) Work out the first three terms.
   b) Find the value of the 10th term.
   c) One of the numbers in the sequence is 1053. Which term is this?

9 The nth term of a number sequence is given by the expression $\dfrac{n(n + 1)}{2}$.
   a) Write down the values of the first four terms.
   b) Work out the value of the 30th term.
   c) Explain why all the terms in this sequence are integers.
   d) What name is often given to the number sequence generated by this rule?

10 David is working with a number sequence. The nth term of his sequence is given by the expression $6n + 7$. He gets the number 2770 as one of his terms. Show that David must have made a mistake.

9.3 Arithmetic sequences

A number sequence in which the terms go up or down in equal steps is called an **arithmetic sequence**. The size of the step is called the **common difference**.

- The first term of an arithmetic sequence is a.
- The common difference is d. — When d is negative, each term is less than the preceding term.

EXAMPLE

For each sequence, say whether it is arithmetic or not.
For each arithmetic sequence state the value of the first term, a, and the common difference, d.

a) 2, 3, 5, 8, 12, …
b) 2, 5, 8, 11, 14, …
c) 1, 2, 4, 8, 16, …
d) 40, 36, 32, 28, 24, …

SOLUTION

a) 2, 3, 5, 8, 12, … is not an arithmetic sequence.

> The terms go up by 1, then 2, then 3 and so on.

b) 2, 5, 8, 11, 14, … is an arithmetic sequence.

First term $a = 2$ and common difference $d = 3$

c) 1, 2, 4, 8, 16, … is not an arithmetic sequence.

> The terms go up by 1, then 2, then 4 and so on.

d) 40, 36, 32, 28, 24, … is an arithmetic sequence.

First term $a = 40$ and common difference $d = -4$

You can find the rule to give the nth term of an arithmetic sequence using the first term, a, and the common difference, d.

Look at the arithmetic sequence with first term $a = 3$ and common difference $d = 4$.

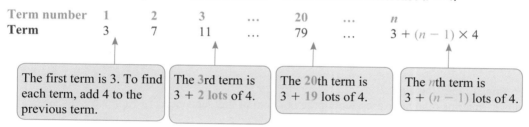

Term number	1	2	3	…	20	…	n
Term	3	7	11	…	79	…	$3 + (n - 1) \times 4$

> The first term is 3. To find each term, add 4 to the previous term.

> The 3rd term is $3 + 2$ lots of 4.

> The 20th term is $3 + 19$ lots of 4.

> The nth term is $3 + (n - 1)$ lots of 4.

Now look at the general arithmetic sequence with first term a and common difference d.

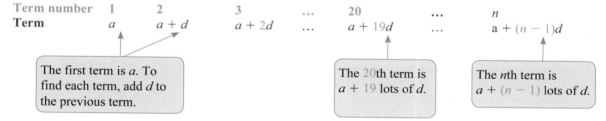

Term number	1	2	3	…	20	…	n
Term	a	$a + d$	$a + 2d$	…	$a + 19d$	…	$a + (n - 1)d$

> The first term is a. To find each term, add d to the previous term.

> The 20th term is $a + 19$ lots of d.

> The nth term is $a + (n - 1)$ lots of d.

The nth term of an arithmetic sequence is

$$a + (n - 1)d$$

where a is the first term and d is the common difference.

EXAMPLE

a) Find a formula for the nth term of the arithmetic sequence:
   7, 10, 13, 16, 19, …
b) Find the 50th term of the sequence.
c) The nth term of the sequence is 205.
   Find the value of n.

SOLUTION

a) $a = 7$ and $d = 3$

nth term is $a + (n - 1)d$

$$\begin{aligned}
n\text{th term} &= 7 + (n - 1) \times 3 \\
&= 7 + 3(n - 1) \\
&= 7 + 3n - 3 \\
&= \underline{3n + 4}
\end{aligned}$$

b) Substitute $n = 50$ into the formula for the nth term.

So the 50th term $= 3 \times 50 + 4 = \underline{154}$

c) $\quad 3n + 4 = 205$

$$\begin{aligned}
3n &= 201 \\
n &= 67
\end{aligned}$$

So the <u>67th term is 205</u>.

EXERCISE 9.3

1 The first five terms in an arithmetic sequence are:

$$12, 17, 22, 27, 32, \ldots$$

a) Find the value of the 10th term.

b) Write down, in terms of n, an expression for the nth term of this sequence.

2 The first four terms in an arithmetic sequence are:

$$58, 50, 42, 34, \ldots$$

a) Find the value of the first negative term.

b) Write down, in terms of n, an expression for the nth term of this sequence.

Here are some arithmetic sequences. For each one, find, in terms of n, an expression for the nth term of the sequence.

3 $8, 11, 14, 17, 20, \ldots$ **4** $2, 7, 12, 17, 22, \ldots$ **5** $10, 9, 8, 7, 6, \ldots$

6 $4, 9, 14, 19, 24, \ldots$ **7** $21, 24, 27, 30, 33, \ldots$ **8** $12, 10, 8, 6, 4, \ldots$

9 Nina has been making patterns with sticks. Here are her first three patterns.

Pattern 1
4 sticks

Pattern 2
7 sticks

Pattern 3
10 sticks

a) Work out the number of sticks in pattern 6.

b) Write down, in terms of n, an expression for the nth term of this sequence.

c) Explain how the coefficients in your formula are related to the way the sticks fit together.

10 The tenth term of an arithmetic sequence is 68 and the eleventh term is 75.
   a) Write down value of the common difference for this sequence.
   b) Work out the value of the first term.
   c) Write down, in terms of n, an expression for the nth term of this sequence.

   Check that your formula works when $n = 10$ and $n = 11$.

9.4 Arithmetic series

When you add terms of a sequence it is called a **series**.
   So 1, 4, 7, ...   is an arithmetic sequence.
   $1 + 4 + 7 + \ldots$ is an arithmetic series.

ite the
es out
wards

There is a neat method you can use to add the terms in an arithmetic series.
Look at this method of adding all the numbers from 1 to 100.

then
ckwards.

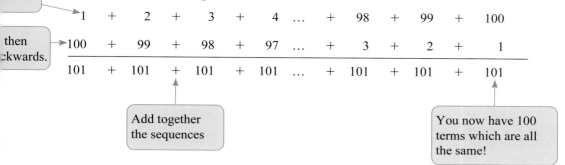

$$1 \quad + \quad 2 \quad + \quad 3 \quad + \quad 4 \quad \ldots \quad + \quad 98 \quad + \quad 99 \quad + \quad 100$$
$$100 \quad + \quad 99 \quad + \quad 98 \quad + \quad 97 \quad \ldots \quad + \quad 3 \quad + \quad 2 \quad + \quad 1$$
$$\overline{101 \quad + \quad 101 \quad + \quad 101 \quad + \quad 101 \quad \ldots \quad + \quad 101 \quad + \quad 101 \quad + \quad 101}$$

Add together
the sequences

You now have 100
terms which are all
the same!

The sum of **both** series is $100 \times 101 = 10\,100$
So the sum of the numbers from 1 to 100 is $\frac{1}{2} \times 10\,100 = 5050$

You can use the same method to find the sum of the terms, S_n, of any arithmetic series with first term a and common difference d.

Write the series out
forwards ...

$$S_n \quad = \quad a \qquad\qquad\quad + \quad (a + d) \qquad + \quad (a + 2d) \qquad\quad + \quad \ldots \quad + \quad (a + (n - 1)d)$$
$$S_n \quad = \quad (a + (n - 1)d) \quad + \quad (a + (n - 2)d) \quad + \quad (a + (n - 3)d) \quad + \quad \ldots \quad + \quad a$$
$$\overline{2 \times S_n \quad = \quad 2a + (n - 1)d \quad + \quad 2a + (n - 1)d \quad + \quad 2a + (n - 1)d \quad + \quad \ldots \quad + \quad 2a + (n - 1)d}$$

... then backwards.

So you have n terms which are all the same; the sum of these terms is:

$$2 \times S_n = n \times (2a + (n - 1)d)$$

This is the sum of two identical series, so you need to halve it:

$$S_n = \frac{n}{2}(2a + (n - 1)d)$$

You will be given this
formula in the exam.

EXAMPLE

The first term of an arithmetic series is 40 and the common difference is $-\frac{1}{2}$.
Find the sum of the first 21 terms of the arithmetic series.

SOLUTION

Substitute $a = 40$, $d = -\frac{1}{2}$ and $n = 21$ into $S_n = \frac{n}{2}(2a + (n-1)d)$

$$S_{21} = \frac{21}{2} \times \left(2 \times 40 + (21 - 1) \times \left(-\frac{1}{2}\right)\right)$$

$$= 10.5 \times \left(80 + 20 \times \left(-\frac{1}{2}\right)\right)$$

$$= 10.5 \times (80 - 10)$$

$$= \underline{735}$$

Some problems are more complicated – be prepared to solve simultaneous equations when tackling questions on arithmetic sequences and series.

EXAMPLE

The 6th term of an arithmetic series is 20.
The 11th term of the same arithmetic series is 35.
Find the sum of the first 100 terms of this arithmetic series.

SOLUTION

The nth term of an arithmetic series is $a + (n-1)d$.
The 6th term is 20, so $a + 5d = 20$
and the 11th term is 35, so $a + 10d = 35$
You now have two equations and two unknowns, so you can solve them simultaneously to find a and d.

$$
\begin{array}{r}
a + 10d = 35 \\
-\ a +\ \ 5d = 20 \\
\hline
5d = 15
\end{array}
$$
so $\underline{d = 3}$

Subtract to eliminate a.

Now substitute $d = 3$ into $a + 5d = 20$

$a + 5 \times 3 = 20$ so $\underline{a = 5}$

You can use the formula for the sum of a series, substitute $a = 5$, $d = 3$ and $n = 100$ into

$$S_n = \frac{n}{2}(2a + (n-1)d)$$

$$S_{100} = \frac{100}{2} \times (2 \times 5 + (100 - 1) \times 3)$$

$$= 50 \times (10 + 99 \times 3)$$

$$= 50 \times 307$$

$$= \underline{15\,350}$$

EXERCISE 9.4

1 Find the sum of the first 20 terms of each of the following series.
   a) First term is 5, common difference is 3
   b) First term is 3, common difference is 5
   c) First term is 3, common difference is -5
   d) First term is 5, common difference is -3

2 The first three terms in an arithmetic sequence are 7, 9, 11.
   Find:
   a) an expression for the nth term of the sequence
   b) the 20th term
   c) the sum of the first 50 terms.

3 Find the sum of the whole numbers from 1 to 1000.

4 Find the sum of the first 100 multiples of 3.

5 The first term in an arithmetic series is 3.
   The sum of the first three terms is 21.
   a) Find the common difference.
   The last term is 99.
   b) How many terms are in the series?
   c) Find the sum of all the terms of the series.

6 The 12th term of an arithmetic series is 62.
   The 20th term of the series is 102.
   Find the sum of the first 20 terms of this arithmetic series.

7 The 6th term of an arithmetic series is 20.
   The 11th term of the series is 35.
   Find the sum of the first 100 terms of this arithmetic series.

8 The second term of an arithmetic series is 95 and the fourth term is 91.
   a) Find the first term and the common difference.
   b) Find the sum of the first 100 terms.
   c) S_n is the sum of the first n terms.
      What is the maximum value of S_n?

REVIEW EXERCISE 9

Find the next three terms in each of these number sequences. For those
that form arithmetic sequences, write down, in terms of n, an expression
for the nth term of this sequence.

1 11, 22, 33, 44, 55, …

2 2, 4, 8, 16, 32, …

3 2, 5, 8, 11, …

4 1, 4, 9, 16, 25, …

5 10, 9, 8, 7, 6, …

6 100, 99, 97, 94, 90, …

7 A number sequence is defined as follows:
   • The first term is 7.
   • To get each new term, multiply the previous one by 3 and subtract 15.
   Work out the first four terms of this sequence.

8 The nth term of a number sequence is given by the expression $\dfrac{n^2 + 3n}{2}$.

   a) Work out the first five terms of this sequence.

   b) Do the first five terms form an arithmetic sequence?

9 Timothy has been drawing patterns. Here are his first three patterns.

      Pattern 1   Pattern 2   Pattern 3
       6 sticks   11 sticks   16 sticks

   a) Write down the number of sticks in pattern 5.

   b) Work out the number of sticks in pattern 12.

   c) Write down, in terms of n, an expression for the nth term of this sequence.

10 Here are the first five terms of a sequence.

        30, 29, 27, 24, 20, …

   a) Write down the next two terms in the sequence.

   Here are the first five terms of a different sequence.

        1, 5, 9, 13, 17, …

   b) Find, in terms of n, an expression for the nth term of the sequence.   **[Edexcel]**

11 Here are the first five numbers of a simple sequence.

        1, 5, 9, 13, 17

   a) Write down the next two numbers of the sequence.

   b) Write down, in terms of n, an expression for the nth term of this sequence.   **[Edexcel]**

12 Here are the first five terms of an arithmetic sequence.

        6, 11, 16, 21, 26

   Find an expression, in terms of n, for the nth term of this sequence.   **[Edexcel]**

13 Find **i)** the nth term and **ii)** the sum of the first 20 terms of each of the following arithmetic series.

   a) First term is 9, common difference is 4

   b) First term is 4, common difference is 9

   c) First term is 90, common difference is –4

   d) First term is 50, common difference is –9

14 The first three terms in an arithmetic sequence are 15, 19, 23.
   Find:

   a) an expression for the nth term of the sequence

   b) the 20th term

   c) the sum of the first 50 terms.

15 Find the sum of the first 50 multiples of 5.

16 The first term in an arithmetic series is 3.
The sum of the fifth term and the sixth term is 42.

a) Find the common difference.

The last term is 499.

b) How many terms are in the series?

c) Find the sum of all the terms of the series.

17 The 11th term of an arithmetic series is 55.
The 21st term of the series is 50.
Find the first term and the sum of the first 100 terms of this arithmetic series.

Key points

1 Common number sequences include the positive integers, the even numbers and the odd numbers. Others you should learn to recognise are:

Square numbers	1, 4, 9, 16, 25, 36, …
Cube numbers	1, 8, 27, 64, 125, 216, …
Powers of 2	1, 2, 4, 8, 16, 32, …
Powers of 10	1, 10, 100, 1000, 10 000, 100 000, …
Triangular numbers	1, 3, 6, 10, 15, 21, …

2 A **term-to-term rule** is a rule which explains how to find the next term in a sequence using the term before.

3 A **position-to-term rule** is a rule that gives an expression for the nth term of a sequence.

4 In an **arithmetic sequence** there is a constant difference between successive terms.

5 The nth term of an arithmetic sequence is

$a + (n - 1)d$

where a is the first term and d is the common difference.

6 A **series** is the sum of a sequence.

5, 7, 9, …   is an arithmetic sequence.

5 + 7 + 9 …   is an arithmetic series.

7 The sum S_n of the first n terms of an arithmetic series is

$$S_n = \frac{n}{2}(2a + (n - 1)d)$$

where a is the first term and d is the common difference.

Internet Challenge 9

Fibonacci numbers

Fibonacci numbers are used to model the behaviour of living systems. Fibonacci numbers also lead to the Golden Ratio, widely used in classical art and architecture. In this challenge you will need to use a spreadsheet at first, before looking on the internet to complete your work.

Here is the Fibonacci number sequence:

1, 1, 2, 3, 5, 8, 13, 21, ….

1. Type these numbers into a computer spreadsheet, such as Excel. (It is a good idea to enter them in a vertical list, rather than a horizontal one.)

2. Each term (apart from the first two) is found by adding together the two previous ones, for example, $13 = 8 + 5$. Use your spreadsheet replicating functions to automatically generate a list of the first 50 Fibonacci numbers.

3. Divide each Fibonacci number by the one before it, for example $8 \div 5 = 1.6$. Set up a column on your spreadsheet to do this up to the 50th Fibonacci number. What do you notice?

The quantities you found in question **3** approach a limit called the Golden Ratio, ϕ.

4. Using your spreadsheet value for ϕ, calculate $1 - \phi$ and $\dfrac{1}{\phi}$. What do you notice?

Now use the internet to help answer the following questions. Find pictures where appropriate.

5. How was the Golden Ratio used by the builders of the Parthenon in Athens?

6. Whose painting of 'The Last Supper' was based on Golden Ratio constructions?

7. Which painter was said to have 'attacked every canvas by the golden section'?

8. When was Fibonacci born? When did he die?

9. Is there a position-to-term rule for Fibonacci numbers, that is, is there a formula for finding the nth number?

10. What sea creature has a spiral shell that is often (mistakenly) said to be based on a Golden Ratio spiral?

CHAPTER 10

Travel and other graphs

In this chapter you will learn how to:

- use straight-line graphs to model real-life situations
- draw graphs to represent rates of change, such as in containers being filled with water
- solve problems using travel graphs.

You will also be challenged to:

- investigate speeds of artificial objects.

Starter: Animal races

Here is some information about the speeds of various animals.

Giant tortoise

0.27 km/h

Ostrich

64 km/h

Human

45 km/h

Cheetah

112 km/h

Black mamba snake

32 km/h

Garden snail

0.05 km/h

1 How many times faster is the ostrich, compared with the giant tortoise?

2 How many times faster is the cheetah, compared with the garden snail?

3 How long would it take a black mamba to travel 100 metres?

4 How far can a garden snail travel during one night of 12 hours?

10.1 Distance–time graphs

Linear graphs are often used to illustrate the movement of an object away from a given point, or back towards it. Distance is plotted up the vertical axis, against time along the horizontal axis, and the result is called a distance–time graph.

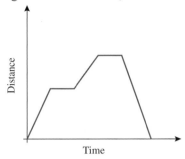

On a distance–time graph:

- *Straight lines* correspond to motion with a *constant speed*.
- The *gradient* of the line indicates the value of the *speed*.
- The *steeper* the gradient, the *faster* the speed.
- Lines with *positive gradient* indicate movement *away* from the starting point.
- Lines with *negative gradient* indicate movement back *towards* the starting point.
- *Horizontal lines* indicate no movement at all, which means that the object is *stationary*.

EXAMPLE

Lance walks from home to the bicycle shop. He spends 20 minutes choosing a new bicycle. He then rides it home, at a constant speed of 9.6 km/h. The distance–time graph below shows part of his journey.

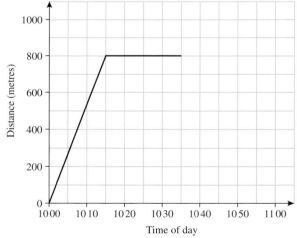

a) Find the distance from Lance's home to the bicycle shop.
b) Find the speed at which Lance walks to the shop.
   Give your answer in km/h.
c) At what time does he arrive back home?
d) Complete the graph, to show his return journey.

SOLUTION

a) From the vertical scale, the distance from home to the shop is <u>800 metres</u>

b) Lance walks 800 m in 15 minutes …
   … which is 1600 m in 30 minutes …
   … which is 3200 m in 60 minutes …
   … which is 3.2 km in 1 hour.

> This type of simple proportion calculation is usually a very good way of solving distance–time graph calculations.

His speed is <u>3.2 km per hour</u>

c) His return speed on the bike is 9.6 km per hour …
   … which is 9600 m in 60 minutes …
   … which is 4800 m in 30 minutes …
   … which is 1600 m in 10 minutes …
   … which is 800 m in 5 minutes.

It takes 5 minutes for the return journey, starting at 1035.

Therefore Lance arrives back home at <u>1040</u>.

d)

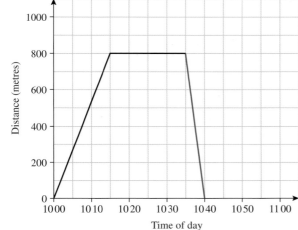

EXERCISE 10.1

1 Tim cycles from home to his grandmother's for tea. He has a puncture on the way. He fixes the puncture, and is able to complete his journey. After tea, he cycles back home again. The travel graph shows his journey.

  a) How long did Tim spend at his grandma's house?

  b) Work out his speed, in km per hour, for the journey *home* from his grandma.

  c) Did he cycle at the same speed as this on the *outward* journey?

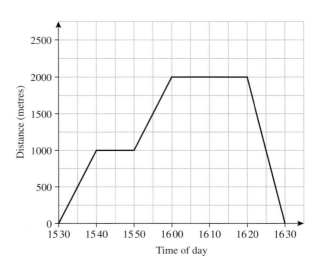

2 Tom is a polar explorer. He is pulling a sledge across the Antarctic icecap. Tom had planned a schedule to travel 16 kilometres every day, but because of poor weather conditions he managed only 6 kilometres on Day 1. Then the weather improved, and he managed to travel 18 kilometres per day from Day 2 onwards.

a) Copy and complete this table to show the distance travelled by the end of each of the first four days.

Day number (n)	Total distance travelled (D km)
1	6
2	24
3	
4	

b) Construct a graph to show the total distance travelled over the first 10 days.

c) Add a second line to your graph to show his progress if he had travelled at the planned rate of 16 kilometres per day. On which day does Tom manage to get back on schedule?

d) Write down a formula for D in terms of n.

Tom has a resupply depot located 240 kilometres from the start point.

e) Use your formula to work out on which day he arrives at the depot. How does this compare with his original schedule?

3 Some teenagers are doing an outdoor walk. They set off from their base at 0900.
They walk for two hours at 4 km per hour.
Then they rest for one hour.
After their rest, they walk on for a further two hours at 5 km per hour.
Then they rest for one hour again.
Finally, they walk for another two hours at 5 km per hour.

a) On a copy of the grid, complete the travel graph.

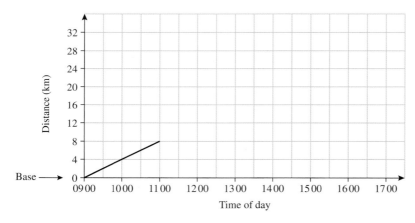

A teacher is camped 20 km from the start point. At 1200 he starts walking towards the group at 4 km per hour. He keeps walking until he meets the group.

b) Add a line on your graph to show the teacher's journey.

c) At what time does the teacher meet the group?

10.2 Modelling with graphs

Many real-life situations can be described, or modelled, by linear graphs. The price charged by a carpet shop increases steadily as the length of the carpet increases, for example. Similarly the amount of fuel in a car fuel tank decreases at a steady rate as the car cruises along a motorway at constant speed.

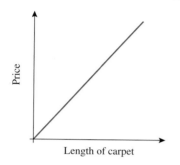

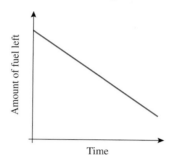

EXAMPLE

Jenny runs a bath. The water from the taps fills the bath at a rate of 12 litres per minute. She runs the bath water for 10 minutes.

a) Work out the amount of water in the bath when Jenny has finished running it.

Jenny stays in the bath for 20 minutes. She then empties it. The bath drains at a rate of 15 litres per minute.

b) Work out how long it takes the bath to empty.

c) Draw a graph to show how the amount of water in the bath changes.

SOLUTION

a) The amount of water is $12 \times 10 = \underline{120\ litres}$

b) Draining time $= 120 \div 15 = \underline{8\ minutes}$

c)

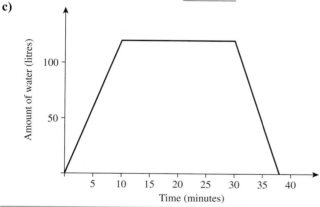

Some graphs can be built up by constructing a table of values first.

When quantities change at a uniform rate, they can be modelled with straight line graphs. Other quantities may change at a varying rate; the corresponding graphs then become curves.

EXAMPLE

Water is poured at a steady rate into four different containers A, B, C and D.

The graphs P, Q, R and S show how the depth of water in each container changes over time. Match the shapes to their corresponding graphs. Explain your reasoning.

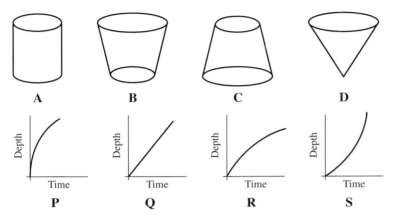

SOLUTION

Shape A has constant cross-section as you move upwards, so its depth increases at a uniform rate. Thus shape A must correspond to graph Q.

Shape C becomes narrower near the top, so its depth will rise more quickly as time goes on. Thus shape C must correspond to graph S.

Shapes B and D are both wider at the top, so the rate at which their depth rises will tail off in both cases. Because shape D has a point at the bottom, however, its initial rate of increase of depth is very high, as in graph P. So shape B matches graph P, leaving shape B to match graph R.

The matchings are: A – Q, B – R, C – S and D – P.

EXAMPLE

Victoria takes part in a cycle race. The speed–time graph shows her speed in metres per second during the race.

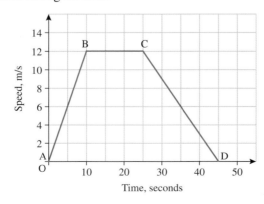

a) Describe what is happening during each of the parts AB, BC and CD on the graph.

b) Write down the highest speed that Victoria achieves during the race.

SOLUTION

a) AB: Victoria's <u>speed is increasing</u>, i.e. she is accelerating.

BC: Victoria is cycling at <u>constant speed.</u>

CD: Victoria's <u>speed is decreasing</u>, i.e. she is decelerating.

b) The highest speed Victoria achieves is <u>12 m/s.</u>

EXERCISE 10.2

1 John climbs a mountain. He gains height at a rate of 10 metres per minute, and it takes him 45 minutes to reach the top. He then stops for 20 minutes to have lunch. Then he descends at 15 metres per minute.
 a) How high is the mountain?
 b) How long does John's descent take?
 c) Copy and complete the graph below, labelling the scales on both axes.

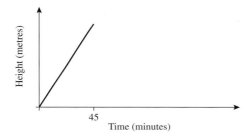

2 The diagram below shows a bowl. It is in the shape of a hemisphere (half a sphere). Water is poured into the bowl at a steady rate.

 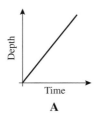 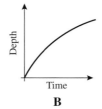 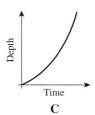

A B C

Say which of graphs A, B or C best describes how the depth of water in the bowl varies over time. Explain your reasoning.

3 Jeremy has a full tank of petrol. It holds 50 litres of fuel. He then drives for 3 hours at a steady speed, during which time the car consumes 1 litre of fuel every 5 minutes. At the end of the 3 hours, Jeremy stops and refills the tank at a service station, which takes 5 minutes. He rests for a further 25 minutes. He completes his journey by travelling at the same steady speed for a further one hour.

a) Work out how much petrol remains in the tank after 3 hours.

b) Copy and complete the graph below.

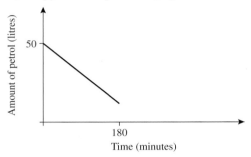

c) Use your graph to find how much fuel is in the tank at the end of the journey.

4 Water is poured into a container at a constant rate. The container is in the shape of a cone and a cylinder joined together as shown in the diagram. Sketch a set of depth/time axes, and complete the diagram to show how the depth of water in the container changes over time.

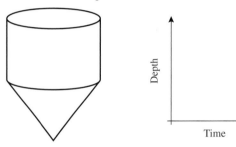

5 A car is travelling at a constant speed of 25 m/s when it passes a police car P. It continues at this speed, for 8 seconds until it reaches a point Q. Then it decelerates uniformly over the next five seconds, when it passes a road sign R at 15 m/s. It continues at this new speed for 12 seconds, to reach the point S.

a) Illustrate this information on a speed–time graph. Indicate the points P, Q, R and S on your graph.

b) Work out how far the car travels between points P and Q.

c) Work out how far the car travels between points R and S.

REVIEW EXERCISE 10

1 Anil cycled from his home to the park. Anil waited in the park. Then he cycled back home. Here is a distance–time graph for Anil's complete journey.

a) At what time did Anil leave home?

b) What is the distance from Anil's home to the park?

c) How many minutes did Anil wait in the park?

d) Work out Anil's average speed on his journey home. Give your answer in kilometres per hour. **[Edexcel]**

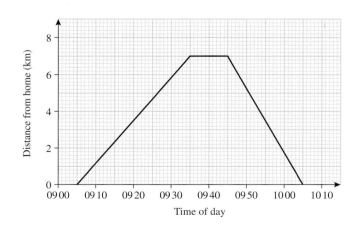

2 Here is a part of a travel graph of Siân's journey from her house to the shops and back.

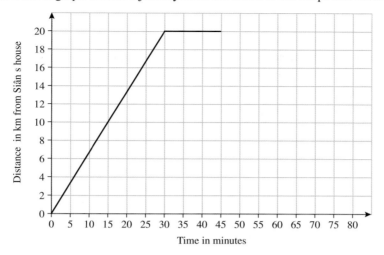

a) Work out Siân's speed for the first 30 minutes of her journey. Give your answer in km/h.
   Siân spends 15 minutes at the shops. She then travels back to her house at 60 km/h.

b) Complete the travel graph.   [Edexcel]

3 The diagram shows a water tank. The tank is a hollow cylinder joined to a hollow hemisphere at the top. The tank has a circular base.

The empty tank is slowly filled with water.

On a copy of the axes, sketch a graph to show the relation between the volume V cm^3, of water in the tank and the depth, d cm, of water in the tank.

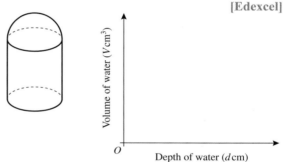

[Edexcel]

4 The diagram shows four empty containers. Water is poured at a constant rate into each of these containers. Each sketch graph shows the relationship between the height of water in a container and the time as the water is poured in.

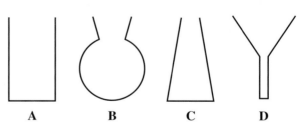

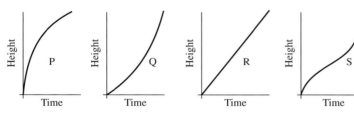

Copy this table and write the letter of each graph in the correct place.

Container	Graph
A	
B	
C	
D	

[Edexcel]

5 Ken and Wendy go from home to their caravan site. The caravan site is 50 km from their home. Ken goes on his bike. Wendy drives in her car. The diagram shows information about the journeys they made.

a) At what time did Wendy pass Ken?
b) Between which two times was Ken cycling at his greatest speed?
c) Work out Wendy's average speed for her journey.   [Edexcel]

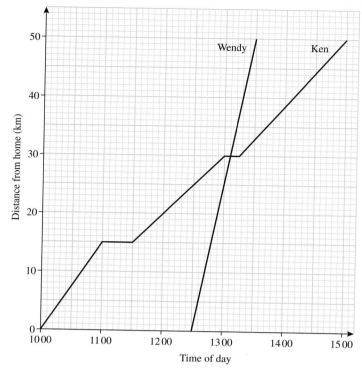

6 Linford runs in a 100 metres race. The graph shows his speed, in metres per second, during the race.

a) Write down Linford's speed, after he has covered a distance of 10 m.
b) Write down Linford's greatest speed.
c) Write down the distance Linford has covered when his speed is 7.4 m/s.   [Edexcel]

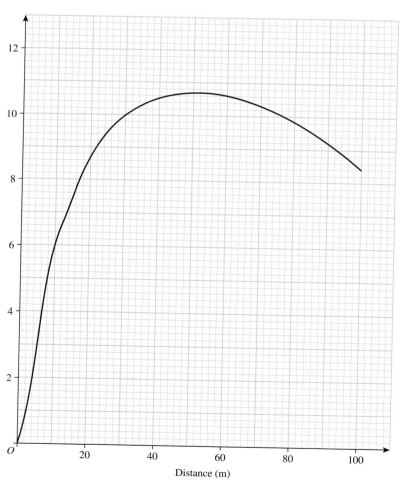

Key points

1 Linear graphs are used to model a variety of real-life situations. They can be constructed from a table, or by looking at start and finish points.

2 If containers such as cylinders and cuboids are filled at a uniform rate, then the graph of depth against time is a straight line. Other shaped containers, such as cones and spheres, will generate curved graphs, and you need to be able to recognise how these are formed.

3 Distance–time graphs show how the position of an object (or person) changes over time. On a distance–time graph:

- Straight lines correspond to motion with a constant speed.
- Lines with positive gradient indicate movement away from the start point.
- Lines with negative gradient indicate movement back towards the start point.
- The steeper the gradient, the faster the speed.
- Horizontal lines indicate no movement at all (the object is stationary).

4 The gradient of the line corresponds to its speed, but care must be taken with units. For example, some distance–time graphs show distances in metres and times in minutes, but expect speeds to be computed in kilometres per hour. In this case, it is better to work out speeds or distances by a proportional method, rather than using formulae such as speed = distance ÷ time.

Internet Challenge 10

Faster and faster

Some artificial objects are capable of travelling at very high speeds.

Use your judgement to arrange these in order of speed, slowest to fastest. Then use the internet to check if your order was correct.

- Intercity 225 train
- Porsche 911 GT3 RS car
- Apollo 11 spacecraft
- Speed of sound (in air)
- Eurofighter *Typhoon* jet aircraft
- Challenger 2 tank
- Disney's *Space Mountain* roller coaster (Paris)
- Orbiting Space Shuttle
- Boeing 747-400 passenger jet aircraft
- The tea clipper *Cutty Sark*

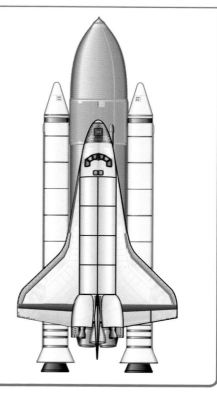

CHAPTER 11

Working with shape and space

In this chapter you will revise earlier work on:

- basic angle properties including vertically opposite angles.

You will learn how to:

- use corresponding and alternate angles
- work with angles in triangles and quadrilaterals
- calculate interior and exterior angles of polygons
- find areas of triangles and quadrilaterals
- calculate surface areas and volumes of solids
- convert between systems of units.

You will also be challenged to:

- investigate the four-colour theorem.

Starter: Alphabet soup

Work out the values of the angles at the letters. Explain your reasoning clearly.

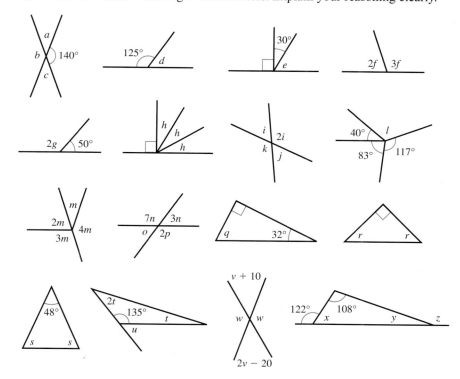

11.1 Corresponding and alternate angles

Imagine an infinitely long railway track, made up of two rails and a set of sleepers.

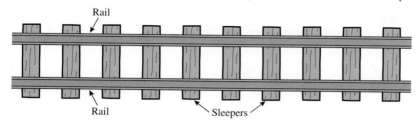

The rails and the sleepers are made up of straight lines, but there is a subtle mathematical distinction between them.

The rails are infinitely long straight lines. The rails are **lines**.

The sleepers are pieces of straight line, with definite start and finish points, so they are finite in length. The sleepers are **line segments**.

In this chapter you will be revising and practising your knowledge of geometry with straight lines and line segments.

The diagram shows two parallel lines and two line segments, or **transversals**, that cross the parallel lines at an angle.

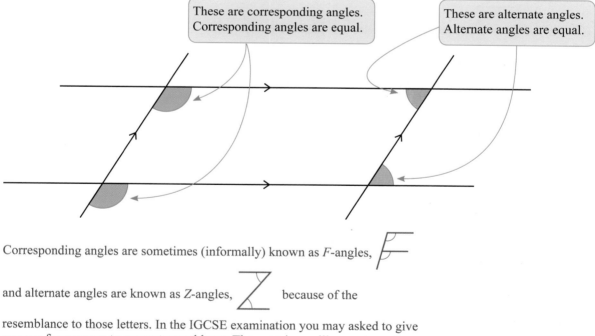

Corresponding angles are sometimes (informally) known as F-angles,

and alternate angles are known as Z-angles, because of the

resemblance to those letters. In the IGCSE examination you may asked to give reasons for answers to geometry problems. The examiner will expect you to use the correct mathematical names (alternate, corresponding and so on) and you may lose marks if you just call them Z-angles or F-angles.

Remember that angles on a straight line add up to 180° and angles around a point add up 360°.

EXAMPLE

Find the angles represented by letters in the diagram below. Give a reason in each case.

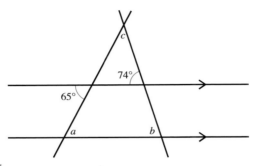

SOLUTION

<u>$a = 65°$</u> (alternate to marked 65° angle)

<u>$b = 74°$</u> (corresponding to marked 74° angle)

Angles a, b, c are at the three vertices of a triangle, so they add up to 180°.

> Always explain your reasons.

Therefore
$$c = 180° - (65° + 74°)$$
$$c = 180° - 139°$$
$$\underline{c = 41°}$$

Here are two other results about equal angles that you have probably met before.

> Angles on opposite sides of a vertex are equal; they are called **vertically opposite**.
>
> $a = b$

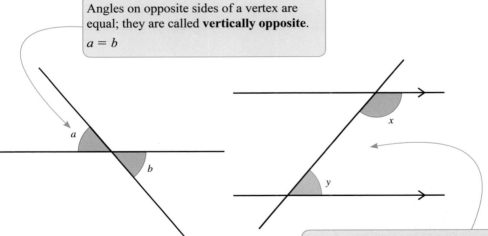

> Angles inside two parallels add up to 180°; they are called **co-interior angles**.
>
> $x + y = 180°$

In this next exercise you may use any angle properties you know, including those about vertically opposite angles, alternate angles and corresponding angles.

170

EXERCISE 11.1

Find the values of the angles represented by the letters in each question. Explain
your reasoning clearly.

1

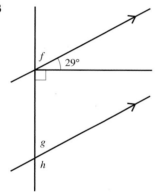

2

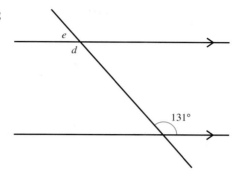

3

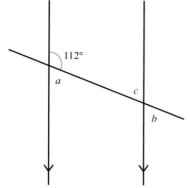

4

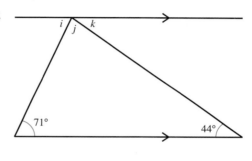

5

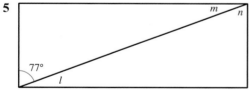

6

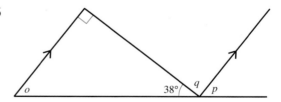

7

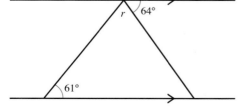

8

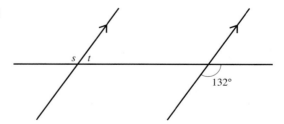

11.2 Angles in triangles and quadrilaterals

Here is a reminder of some definitions to do with angles and triangles.

Acute angle

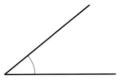

An **acute angle** is less than 90°.

Right-angle

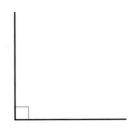

A **right angle** is 90°.

Obtuse angle

An **obtuse angle** is greater than 90° and less than 180°.

Reflex angle

A **reflex angle** is greater than 180° and less than 360°.

Equilateral triangle

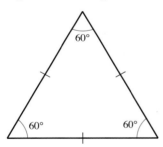

An **equilateral triangle** has 3 equal sides and 3 angles of 60°.

Isosceles triangle

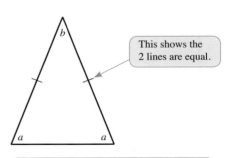

This shows the 2 lines are equal.

An **isosceles triangle** has 2 equal sides and 2 equal angles.

$2a + b = 180°$

Here are three important results about angles in triangles and quadrilaterals.
You could be asked to prove them in an exam.

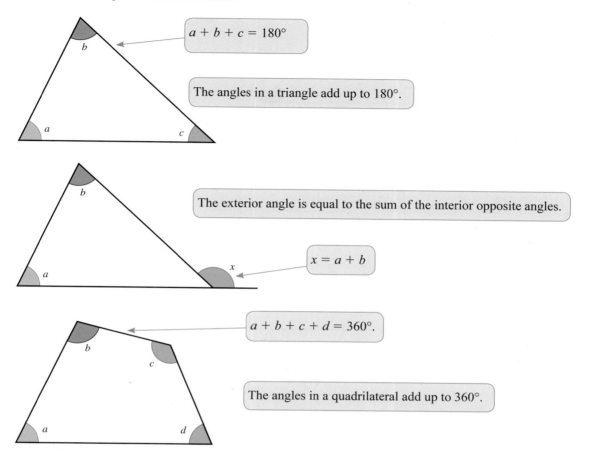

$a + b + c = 180°$

The angles in a triangle add up to 180°.

The exterior angle is equal to the sum of the interior opposite angles.

$x = a + b$

$a + b + c + d = 360°$.

The angles in a quadrilateral add up to 360°.

Here are proofs of these results.

THEOREM

The angles in a triangle add up to 180°.

PROOF

Construct a line through one vertex, parallel to the opposite side.
Angles x and a are alternate, so $x = a$. Likewise, angles y
and c are alternate, so $y = c$.

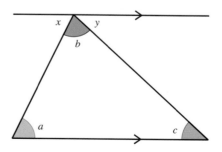

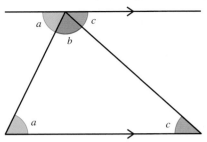

Now angles a, b and c form a straight line at the top of the diagram.

Therefore $a + b + c = 180°$.

So the angles in the triangle add up to 180°.

THEOREM

The angles in a quadrilateral add up to 360°.

PROOF

Consider any quadrilateral PQRS, and draw the diagonal PR, so as to divide it into two triangles.

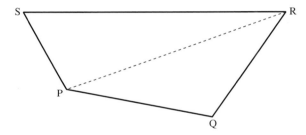

Now consider the angles inside each of the two triangles.

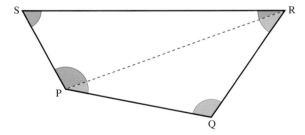

Clearly, the total of the angles inside quadrilateral PQRS is equal to the sum of the angles in triangle PSR plus the sum of the angles in triangle PQR. Hence:

Sum of angles in quadrilateral PQRS $= 180° + 180°$
$$= 360°$$

EXAMPLE

The diagram shows a quadrilateral PQRS. Find the value of the angle x.

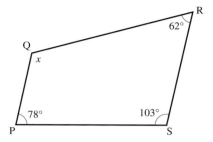

SOLUTION

The angles in a quadrilateral add up to $360°$, so

$$x + 78° + 103° + 62° = 360°$$
$$x + 243° = 360°$$
$$x = 360° - 243°$$
$$\underline{x = 117°}$$

Some examination questions will set problems on quadrilaterals that lead to simple equations, as in this example.

EXAMPLE

The diagram shows a quadrilateral PQRS. Find the value of x. Hence find the values of the angles.

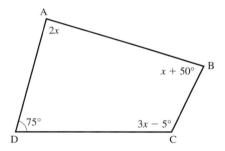

SOLUTION

Since the angles in a quadrilateral add up to $360°$:

$$2x + x + 50° + 3x - 5° + 75° = 360°$$
$$6x + 120° = 360°$$
$$6x = 360° - 120°$$
$$6x = 240°$$
$$x = \frac{240°}{6°}$$
$$\underline{x = 40°}$$

Then angle DAB $= 2x = \underline{80°}$, angle ABC $= x + 50° = \underline{90°}$ and angle BCD $= 3x - 5 = \underline{115°}$

EXERCISE 11.2

1 Find the value of a.

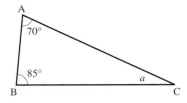

2 Find the value of x.

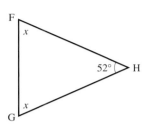

3 Find the value of x.
Hence work out the size of the largest angle.

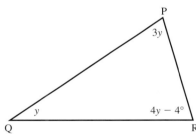

4 Form and solve an equation in y.
Hence find the sizes of the angles in the triangle.

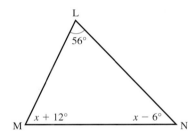

5 A triangle has angles $x + 8°$, $2x - 8°$ and $90°$.
  a) Set up an equation in x.
  b) Solve your equation, to find the value of x.
  c) Work out the sizes of the angles in the triangle.

6 The angles in a triangle are $4c + 4°$, $5c - 7°$ and $7c + 7°$.
  a) Set up an equation in c.
  b) Solve your equation, to find the value of c.
  c) Work out the sizes of the angles in the triangle.
  d) What kind of triangle is this?

7 The diagram shows a quadrilateral. Work out the size
of the angle m.

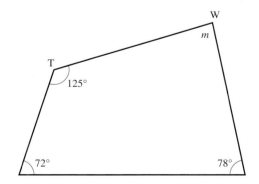

8 The diagram shows a quadrilateral.
Work out the size of the angle y.

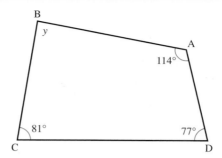

9 Find the size of each angle marked z.

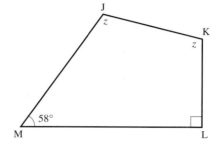

10

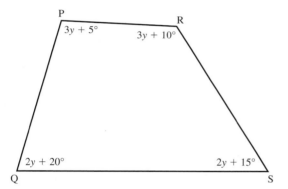

a) Form and solve an equation in y.

b) Hence find the sizes of the angles in the quadrilateral.

c) What do your answers tell you about the line segments PR and QS?

11 Form and solve an equation in k. Hence find the values of the angles in the quadrilateral.

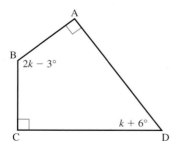

12 The angles in a quadrilateral are $x + 16°$, $2x - 2°$, $3x + 9°$ and $5x - 15°$.

    a) Set up an equation in x.

    b) Solve your equation, to find the value of x.

    c) Work out the sizes of the angles in the quadrilateral.

    d) Check that your four answers add up to 360°.

11.3 Angles in polygons

The angles in a triangle add up to 180° and those in a quadrilateral add up to 360°. For polygons with more sides, another 180° is added for each extra side. For example, the angles in a pentagon must add up to $360° + 180° = 540°$.

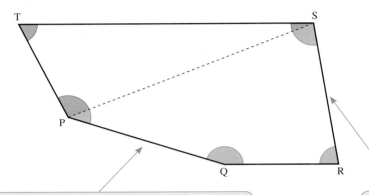

The angles in quadrilateral PQRS add up to 360° … and the angles in triangle PST, 180°…

… so the total for pentagon PQRST is $360° + 180° = 540°$.

This can also be expressed as a mathematical formula:

    Sum of interior angles of a polygon with n sides $= 180° \times (n - 2)$

EXAMPLE

Find the sum of all the interior angles of:

a) a hexagon   **b)** a ten-sided polygon.

SOLUTION

a) Since a pentagon has a sum of 540°, a hexagon must have a sum of $540° + 180° = 720°$.

b) Using the formula with $n = 10$:

$$\text{Angle sum} = 180° \times (n - 2)$$
$$= 180° \times (10 - 2)$$
$$= 180° \times 8$$
$$= \underline{1440°}$$

EXAMPLE

Six of the angles in a seven-sided polygon are 100°, 110°, 130°, 145°, 145° and 150°. Find the value of the seventh angle.

SOLUTION

The angle sum for a seven-sided polygon is $180° \times (7 - 2) = 900°$.

The given angles have a sum of:

$100° + 110° + 130° + 145° + 145° + 150° = 780°$

Thus the remaining angle is $900° - 780° = \underline{120°}$

If you were to travel all the way around the perimeter of a polygon, you would need to change direction at each corner, or vertex. The angle by which you change direction is called the **exterior angle** at that vertex. To indicate an exterior angle on a diagram, you need to **produce** (extend) each of the sides slightly, in the same sense (clockwise or anti-clockwise) each time. The diagram shows the exterior angles for a pentagon, with each side produced in a clockwise direction.

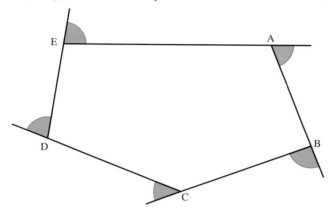

The sum of the exterior angles is simply the total angle you would turn through by travelling all the way around the perimeter. This must be a complete turn, or 360°.

Sum of exterior angles of a polygon with n sides $= 360°$

Some polygons have all their sides the same length and all their angles equal; these are called **regular polygons**. If a regular polygon has n sides, then each exterior angle must be $360° \div n$.

EXAMPLE

A regular polygon has 12 sides.
a) Calculate the size of each exterior angle. ◄———
b) Hence find the size of each interior angle.

> This kind of question works only for a regular polygon – all the interior (or exterior) angles are the same size.

SOLUTION

a) Exterior angle $= 360° \div 12 = \underline{30°}$

b) Interior angle $= 180° - 30° = \underline{150°}$

EXAMPLE

The diagram shows one vertex of a regular polygon with *n* sides.

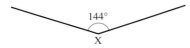

Calculate the value of *n*.

SOLUTION

Each exterior angle is 180° − 144° = 36°.

Number of sides is 360° ÷ 36° = 10 sides. So *n* = 10.

EXERCISE 11.3

1 Find the sum of the interior angles of:
   a) an octagon
   b) a 20-sided polygon.

2 Five of the angles in a hexagon are 102°, 103°, 118°, 125° and 130°. Find the sixth one.

3 Work out the value of the exterior angle at each vertex of:
   a) a regular hexagon
   b) a regular 15-sided polygon.

4 a) The diagram shows part of a regular polygon.

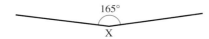

   Work out how many sides the polygon has.

   b) Albert draws this diagram. He says it shows part of a regular polygon.

   Explain how you can tell that Albert must have made a mistake.

5 The diagram shows an irregular pentagon. Work out the value of the angle marked *y*.

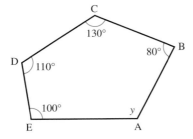

6 Work out the value of *x*. Hence find the size of each angle.

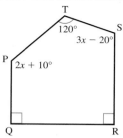

7 The diagram shows a hexagon. All the angles marked *a* are equal. Calculate the value of *a*.

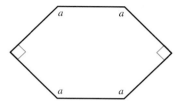

8 The diagram shows a hexagon. It has a vertical line of symmetry.

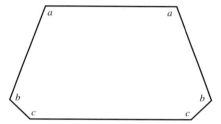

a) Explain carefully why $a + b + c = 360°$.

Angle *a* is 20° smaller than angle *c*.

Angle *b* is 10° smaller than angle *c*.

b) Use this information to rewrite the equation $a + b + c = 360°$ so that it does not contain either *a* or *b*. Solve this equation, to find the value of *c*.

c) Hence find the other angles in the hexagon.

d) Justin says the hexagon is regular. Is he right? Explain your answer.

9 Follow these instructions to make an accurate drawing of a regular hexagon.

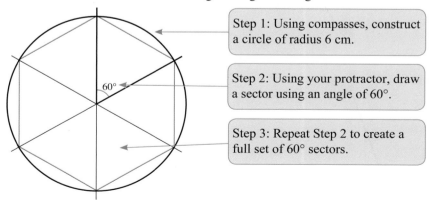

Step 1: Using compasses, construct a circle of radius 6 cm.

Step 2: Using your protractor, draw a sector using an angle of 60°.

Step 3: Repeat Step 2 to create a full set of 60° sectors.

Complete the construction by joining the six points around the circumference of the circle.

10 Adapt the instructions from question **9** to make:
  a) a regular octagon
  b) a regular nine-sided polygon
  c) nine-pointed star.

11.4 Areas and perimeters of simple shapes

A quadrilateral shape drawn at random will have four unequal sides, and the angles will all be different.

In practice, quadrilaterals with some sides (or angles) the same tend to be more useful, so these have special names. You will have met them before, but here is a reminder of the special quadrilaterals and their geometric properties.

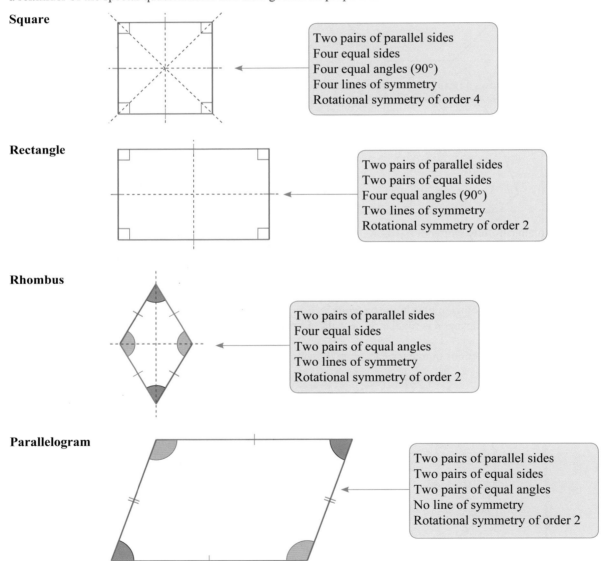

Square

Two pairs of parallel sides
Four equal sides
Four equal angles (90°)
Four lines of symmetry
Rotational symmetry of order 4

Rectangle

Two pairs of parallel sides
Two pairs of equal sides
Four equal angles (90°)
Two lines of symmetry
Rotational symmetry of order 2

Rhombus

Two pairs of parallel sides
Four equal sides
Two pairs of equal angles
Two lines of symmetry
Rotational symmetry of order 2

Parallelogram

Two pairs of parallel sides
Two pairs of equal sides
Two pairs of equal angles
No line of symmetry
Rotational symmetry of order 2

Kite

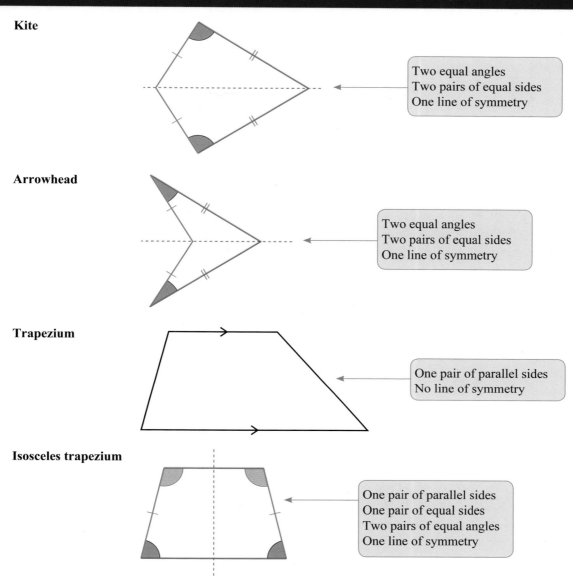

Two equal angles
Two pairs of equal sides
One line of symmetry

Arrowhead

Two equal angles
Two pairs of equal sides
One line of symmetry

Trapezium

One pair of parallel sides
No line of symmetry

Isosceles trapezium

One pair of parallel sides
One pair of equal sides
Two pairs of equal angles
One line of symmetry

In the IGCSE examination you will need to know how to find the area of a square, rectangle, triangle, parallelogram and trapezium.

You may already be familiar with some of these results for calculating areas:

Rectangle

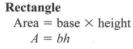

Area = base × height
$A = bh$

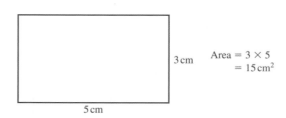

3 cm

Area = 3 × 5
= 15 cm²

Triangle

Area = half × base × height

$A = \frac{1}{2}bh$

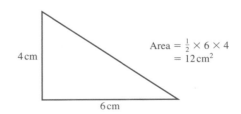

4 cm

6 cm

Area $= \frac{1}{2} \times 6 \times 4$
$= 12\,\text{cm}^2$

Parallelogram

The area of a parallelogram can be found by using the triangle formula twice:

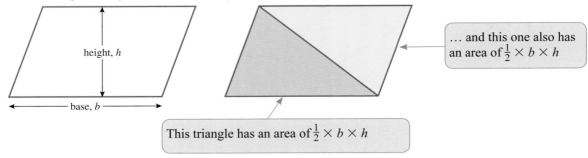

height, h

base, b

… and this one also has an area of $\frac{1}{2} \times b \times h$

This triangle has an area of $\frac{1}{2} \times b \times h$

Area of **parallelogram** $= \frac{1}{2} \times b \times h + \frac{1}{2} \times b \times h$

$A = b \times h$

EXAMPLE

Find the area of this parallelogram.

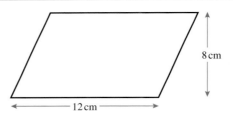

8 cm

12 cm

SOLUTION

The parallelogram has a base $b = 12$ cm and a height $h = 8$ cm.

Therefore area $= b \times h$

$= 12 \times 8$

$= \underline{96\ \text{cm}^2}$

Remember to include the units in your answer

Finally, there is a formula for finding the area of a trapezium. The trapezium requires some surgery:

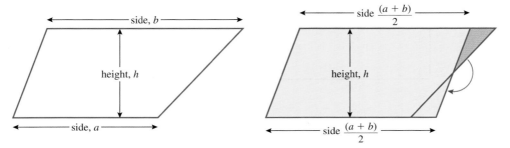

side, b

height, h

side, a

side $\dfrac{(a + b)}{2}$

height, h

side $\dfrac{(a + b)}{2}$

By slicing off and swivelling a triangle, as shown, the trapezium can be converted into a parallelogram with the equivalent area. The original trapezium had two different lengths of a and b. The length of the new parallelogram is found by taking the average of these, namely $\dfrac{a + b}{2}$

Thus the area of the parallelogram is base $\times$ height $= \dfrac{a + b}{2} \times h$, and the trapezium must have the same area. This formula is often written using bracket notation:

$$\boxed{\text{Area of trapezium} = \tfrac{1}{2}(a + b)h}$$

EXAMPLE

Find the area of this trapezium.

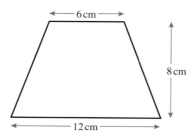

SOLUTION

The trapezium has parallel sides $a = 6$ cm and $b = 12$ cm, and a height $h = 8$ cm.

$$\begin{aligned}
\text{Therefore area} &= \tfrac{1}{2}(a + b)h \\
&= \tfrac{1}{2} \times (6 + 12) \times 8 \\
&= \tfrac{1}{2} \times 18 \times 8 \\
&= 9 \times 8 \\
&= \underline{72 \text{ cm}^2}
\end{aligned}$$

Areas of simple shapes such as triangles, rectangles, parallelograms and trapeziums may be found directly, by using the appropriate formulae. Make sure you learn them! You may also need to work out the areas of compound shapes, by breaking them down to two or more simpler pieces.

EXAMPLE

The diagram shows a shape made from two rectangles and a triangle.

a) Calculate the perimeter of the shape.
b) Calculate the area of the shape.

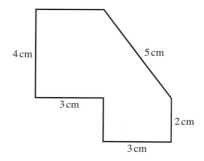

185

SOLUTION

a) Marking the missing lengths:

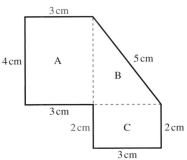

Perimeter = 3 + 4 + 3 + 2 + 3 + 2 + 5

= <u>22 cm</u>

b) Denoting the three parts as A, B, C (see diagram above), then:

Area of A = 4 × 3 = 12 cm^2

Area of B = $\frac{1}{2}$ × 4 × 3 = 6 cm^2

Area of C = 3 × 2 = 6 cm^2

Total area = 12 + 6 + 6 = <u>24 cm^2</u>

Some questions on perimeters may be suitable for solving with algebra.

EXAMPLE

This rectangle has a perimeter of 58 cm.
Work out the value of x.

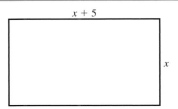

SOLUTION

The rectangle has sides of length x, $x + 5$, x and $x + 5$.

Its perimeter is $x + x + 5 + x + x + 5 = 4x + 10$.

Therefore $4x + 10 = 58$

$4x = 58 - 10$

$4x = 48$

<u>$x = 12$ cm</u>

EXERCISE 11.4

Find the perimeter and the area of each shape. You may use standard formulae
to help.

1

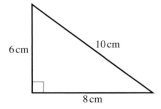

6 cm   10 cm   8 cm

2

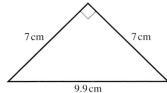

7 cm   7 cm   9.9 cm

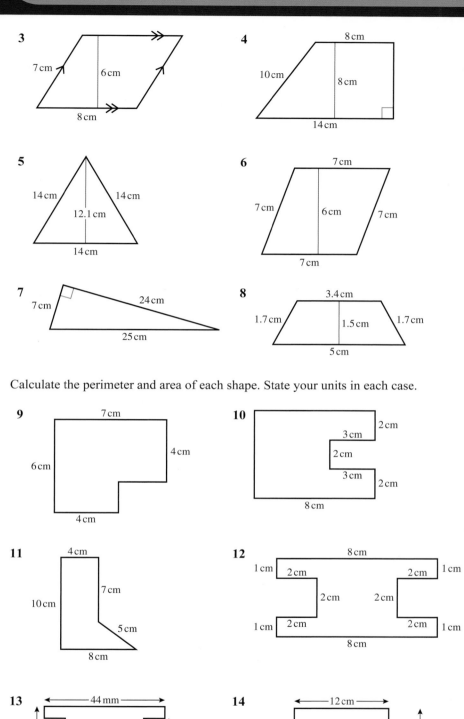

3

7 cm 6 cm 8 cm

4

8 cm 10 cm 8 cm 14 cm

5

14 cm 14 cm 12.1 cm 14 cm

6

7 cm 7 cm 6 cm 7 cm 7 cm 7 cm

7

7 cm 24 cm 25 cm

8

3.4 cm 1.7 cm 1.5 cm 1.7 cm 5 cm

Calculate the perimeter and area of each shape. State your units in each case.

9

7 cm 4 cm 6 cm 4 cm

10

2 cm 3 cm 2 cm 3 cm 2 cm 8 cm

11

4 cm 7 cm 10 cm 5 cm 8 cm

12

8 cm 1 cm 2 cm 2 cm 1 cm 2 cm 2 cm 1 cm 2 cm 2 cm 1 cm 8 cm

13

44 mm 30 mm 28 mm 40 mm

14

12 cm 8 cm 10 cm 14 cm

187

15 A quadrilateral has exactly one set of parallel sides. What type of quadrilateral is it?

16 A quadrilateral has all four sides the same length. Tina says: 'It must be a square'.
Is Tina right? Explain your answer.

17 The diagram shows a rectangle.
All lengths are in centimetres.
The perimeter of the rectangle is 32 cm.
Work out the value of x.

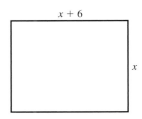

18 The diagram shows an isosceles triangle. AB = CB. Lengths are in centimetres.

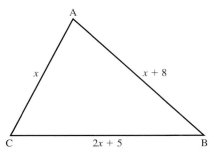

a) Work out the value of x.
b) Hence work out the perimeter of the triangle.

19 The diagram shows a triangle.
All lengths are in centimetres.
a) What type of triangle is this?
b) Set up, and solve, an equation in x.
c) Hence work out the perimeter of the triangle.

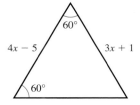

20 The diagram shows a parallelogram shape.
All lengths are in centimetres.

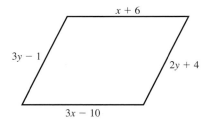

a) Set up, and solve, an equation in x.
b) Set up, and solve, an equation in y.
c) Hence work out the lengths of the sides of the parallelogram.
d) Suggest a better name for this shape.

11.5 Surface area and volume

The **surface area** of a cuboid is found by calculating the areas of its six separate faces, and then adding them together. The **volume of a cuboid** is found by multiplying the three dimensions of the cuboid together.

Here is a reminder of some definitions to do with 3-D shapes.

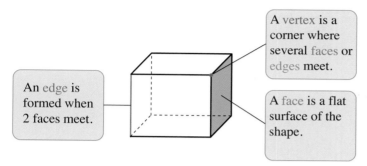

A vertex is a corner where several faces or edges meet.

An edge is formed when 2 faces meet.

A face is a flat surface of the shape.

So a cube has 12 edges, 8 vertices and 6 faces.

EXAMPLE

Find **a)** the surface area and **b)** the volume of this solid cuboid.

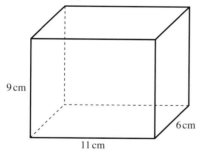

9 cm

11 cm

6 cm

SOLUTION

a) Consider the left and right ends:

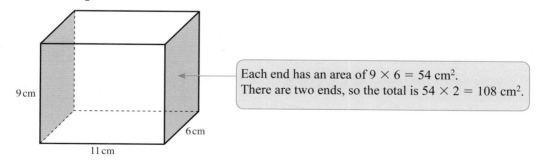

9 cm

11 cm

6 cm

Each end has an area of $9 \times 6 = 54$ cm^2.
There are two ends, so the total is $54 \times 2 = 108$ cm^2.

Similarly for the top and bottom:

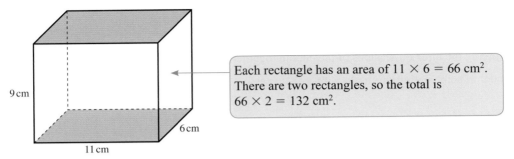

Each rectangle has an area of $11 \times 6 = 66$ cm^2.
There are two rectangles, so the total is
$66 \times 2 = 132$ cm^2.

Finally, look at the front and back:

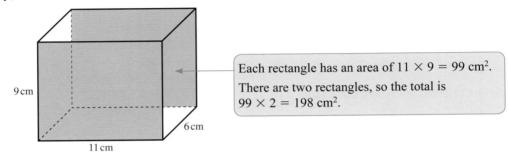

Each rectangle has an area of $11 \times 9 = 99$ cm^2.
There are two rectangles, so the total is
$99 \times 2 = 198$ cm^2.

So the total area is $(9 \times 6) \times 2 + (11 \times 6) \times 2 + (11 \times 9) \times 2$
$$= 108 + 132 + 198$$
$$= \underline{438 \text{ cm}^2}$$

b) The volume is $11 \times 6 \times 9 = \underline{594 \text{ cm}^3}$

A cuboid is a simple example of a **prism**. Prisms are three-dimensional solids with a constant cross-section. To find the volume of a prism, multiply its cross-sectional area by its length.

Volume of prism = area of cross section $\times$ length

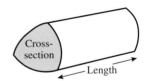

Sometimes you will be told the cross-sectional area, and you can then simply multiply it by the length.

EXAMPLE

The diagram shows a prism of length 10 cm and cross-sectional area 8 cm^2.
Calculate its volume.

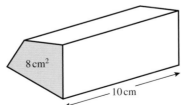

SOLUTION

Volume = area of cross section × length
$\quad$ = 8 × 10
$\quad$ = 80 cm³

$cm^2 × cm = cm^3$

If the cross section is a simple shape, such as a triangle, then you might be asked to work its area out first.

EXAMPLE

The diagram shows a prism. The cross section of the prism is a right-angled triangle.

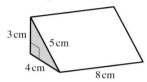

3 cm, 5 cm, 4 cm, 8 cm

a) Calculate the area of the cross section.
b) Find the volume of the prism.
c) Work out the surface area of the prism.

SOLUTION

a) Area of cross section = $\frac{1}{2}$ × 4 × 3
$\qquad\qquad\qquad$ = 6 cm²

b) Volume of prism = 6 × 8
$\qquad\qquad\quad$ = 48 cm³

c) The two triangular ends have areas of 6 cm² each.
Top rectangle has area 5 × 8 = 40 cm²
Base has area 4 × 8 = 32 cm²
Back has area 3 × 8 = 24 cm²
Total surface area = 6 + 6 + 40 + 32 + 24 = 108 cm²

EXERCISE 11.5

1 The diagram shows a cube of side 10 cm.
Calculate its surface area and also its volume.
State the units in your answers.

10 cm, 10 cm, 10 cm

2 The diagram shows a cuboid, with dimensions 8 cm, 12 cm and 15 cm.
a) Work out the surface of the cuboid, in cm².
b) Work out the volume of the cuboid, in cm³.

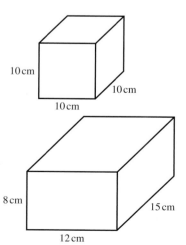

8 cm, 15 cm, 12 cm

3 The diagram shows a prism. Its cross section is formed by
a right-angled triangle of sides 5 cm, 12 cm, 13 cm.
The prism has a length of 6 cm.

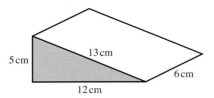

   a) Calculate the area of the cross section, shaded in the diagram.
   b) Work out the volume of the prism.
   c) Calculate the surface area of the prism.

4 The cross section of a steel girder is in the shape of a letter L. The cross section is shown in the
diagram below.

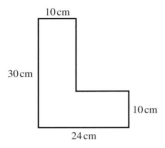

   a) Work out the area of the L-shaped cross section.

   The girder is 80 cm long.
   b) Work out the volume of the girder.

5 A cube measures 12 cm along each side.
   a) Work out the volume of the cube.
   b) Work out the surface area of the cube.

6 A cuboid measures 15 cm by 20 cm by 30 cm.
   a) Work out the volume of the cuboid.
   b) Work out the surface area of the cuboid.

7 The diagram shows a water tank. It is in the shape of a cuboid. It has no lid.

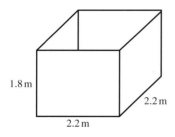

   a) Work out the volume of the tank, correct to 3 significant figures.
   b) Work out the total surface area of the inside of the tank.

8 The diagram shows a sketch of a swimming pool.

The pool is 1.2 m deep at the shallow end, and 2.4 m deep at the deep end.
The pool is 25 m long, and is 10 m wide.

a) Work out the volume of the pool.

1 cubic metre = 1000 litres.

b) Work out the number of litres of water in the pool when it is full.

9 A cube has a volume of 10 648 cm³.
  a) Work out the dimensions of the cube.
  b) Calculate the surface area of the cube.

10 A cuboid has a volume of 455 cm³. Its dimensions are all different.
  Each dimension is a whole number of centimetres. Each dimension is greater than 1 cm.
  a) Work out the dimensions of the cuboid.
  b) Calculate the surface area of the cuboid.

11.6 Converting between units

Sometimes you may want to convert an area or a volume from one set of units to another. This needs to be done carefully!

There are, for example, 10 millimetres in 1 centimetre, but there are **not** 10 square millimetres in 1 square centimetre. The diagram shows why:

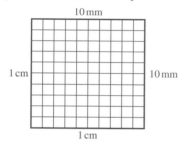

Area = 1 cm × 1 cm = 1 cm²

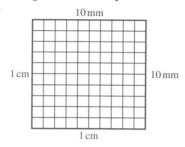

Area = 10 mm × 10 mm = 100 mm²

Thus we have:

  1 cm = 10 mm
  1 cm² = 10 × 10 = 100 mm²
  1 cm³ = 10 × 10 × 10 = 1000 mm³

These results illustrate a general principle, that **areas** must be multiplied by the **square** of the conversion factor, and **volumes** by its **cube**.

EXAMPLE

a) Convert 2 m² into cm².
b) Convert 5000 cm³ into m³.

SOLUTION

a) 2 m² = 2 × 100² cm²
        = 20 000 cm²

b) 5000 cm³ = 5000 ÷ 100³ m³
          = 5000 ÷ 1 000 000 m³
          = 0.005 m³

EXERCISE 11.6

1 Convert 2 m³ into cm³.

2 Convert 5000 cm² into m².

3 Convert 3 000 000 m² into km².

4 Convert 660 mm² into cm².

5 Convert 1 m³ into mm³.

6 A sphere has a volume of 35 000 cm³. Express its volume in m³.

7 A cone has a surface area of 2.4 m². Express its area in cm².

For the next three questions, you may use the information that 1 litre = 1000 cm³.

8 A bucket contains 20 litres of water.
   a) Convert 20 litres into cm³.
   b) Hence find the amount of water in the bucket in m³.

9 A water tank in the shape of a cuboid measures 1.4 m by 1.5 m by 2 m.
   a) Find the volume of the tank, in m³.
   b) Convert this answer into cm³.
   c) How many litres of water can the tank hold?

10 A tank in the shape of a cube has a capacity of 512 litres.
   a) Express this capacity in cm³.
   b) Convert your answer to **a)** into m³.
   c) Find the dimensions of the tank, in metres.

REVIEW EXERCISE 11

In some of the questions that follow, you may find three capital letters being used to describe an angle, e.g. angle PQR. This means the angle formed by the line segment PQ joining the line segment QR, i.e. the angle Q.

1 Find the size of the angles marked a and b on the diagram.

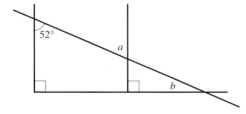

2 The angles in a quadrilateral are $4y - 10°$, $y + 40°$, $3y + 20°$ and $2y + 10°$ in order as you go around the quadrilateral.
   a) Set up and solve an equation, to find the value of y.
   b) Hence work out the value of each angle.
   c) What type of quadrilateral is this?

3 PQ is a straight line.

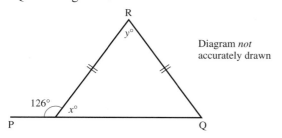

Diagram *not* accurately drawn

a) Work out the size of the angle marked $x°$.
b) Work out the size of the angle marked $y°$. Give reasons for your answer.

[Edexcel]

4 Work out the size of exterior angle of a regular 12-sided polygon. Hence find the size of each interior angle.

5 The angles inside a certain polygon add up to $1980°$. How many sides has it?

6 A regular polygon has interior angles of size $176°$. How many sides has it?

7 A certain quadrilateral has all its angles equal, but its sides are not all the same length.
a) Is it regular?
b) What type of quadrilateral is this?

8 The diagram shows a rectangle.
All lengths are in centimetres.
a) Work out the value of x.
b) Calculate the perimeter of the rectangle.
c) Work out the area of the rectangle.

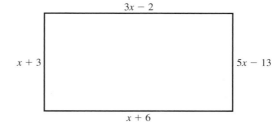

9 a) Convert 3500 mm² into cm².

b) Convert 2.5 m³ into cm³.

10 Triangle ABC is isosceles, with AC = BC. Angle ACD = $62°$. BCD is a straight line.

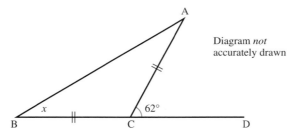

Diagram *not* accurately drawn

a) Work out the size of angle x.
b) The diagram shows part of a **regular** octagon. Work out the size of angle x.

Diagram *not* accurately drawn

[Edexcel]

195

11 A cuboid has a volume of 175 cm³. Two of its dimensions are 2.5 cm and 3.5 cm.
Work out the remaining dimension.

12 Three different rectangles each have an area of 28 cm². The lengths of all the sides are whole
numbers of centimetres. For each rectangle work out the lengths of the two sides.   [Edexcel]

13 PQRS is a parallelogram. Angle QSP = 47°. Angle QSR = 24°. PST is a straight line.

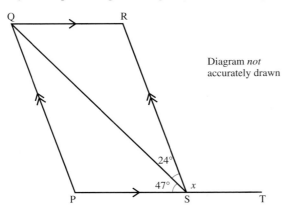

Diagram *not*
accurately drawn

a) (i) Find the size of the angle marked *x*.
   (ii) Give a reason for your answer.
b) (i) Work out the size of angle PQS.
   (ii) Give a reason for your answer.   [Edexcel]

14 The diagram shows a prism.

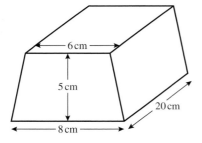

Diagram *not*
accurately drawn

The cross section of the prism is a trapezium.
The lengths of the parallel sides of the trapezium are 8 cm and 6 cm.
The distance between the parallel sides of the trapezium is 5 cm.
The length of the prism is 20 cm.
a) Work out the volume of the prism.

The prism is made out of gold. Gold has a density of 19.3 grams per cm³.
b) Work out the mass of the prism. Give your answer in kilograms.   [Edexcel]

15 ABC and EBD are straight lines. BD = BC. Angle CBD = 42°.

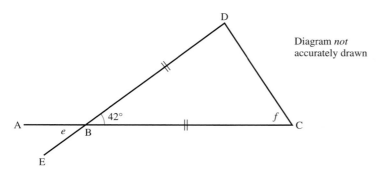

Diagram *not*
accurately drawn

a) Write down the size of the angle marked *e*.
b) Work out the size of the angle marked *f*. [Edexcel]

16 In this diagram, the lines AB and CD are parallel.

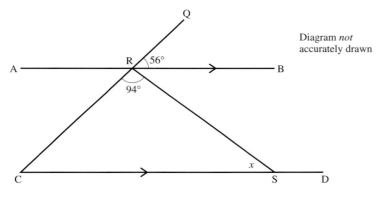

Diagram *not*
accurately drawn

CRQ is a straight line. Angle CRS = 94°. Angle QRB = 56°. Angle RSC = $x°$.
Find the value of x. [Edexcel]

17 The diagram shows a regular hexagon.

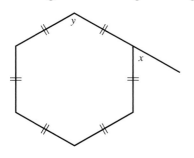

Diagram *not*
accurately drawn

a) Work out the value of x.
b) Work out the value of y. [Edexcel]

18 The diagram shows a prism.

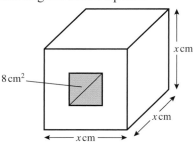

Diagram *not* accurately drawn

8 cm²

x cm

x cm

x cm

The prism is made from a cube of side *x* cm. A hole of uniform cross-sectional area 8 cm² is cut through the cube. Find, in terms of *x*, an expression for the volume of the prism. **[Edexcel]**

19 The diagram shows a trapezium ABCD.

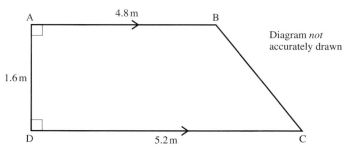

4.8 m

A   B

Diagram *not* accurately drawn

1.6 m

D   5.2 m   C

AB is parallel to DC. AB = 4.8 m, DC = 5.2 m, AD = 1.6 m.
Angle BAD = 90°, angle ADC = 90°.
Calculate the area of the trapezium. **[Edexcel]**

20 The diagram shows a water tank in the shape of a cuboid. The measurements of the cuboid are 20 cm by 50 cm by 20 cm.

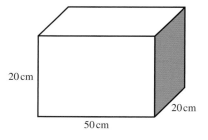

Diagram *not* accurately drawn

20 cm

20 cm

50 cm

a) Work out the volume of the water tank.

Water is poured into the tank at a rate of 5 litres per minute. 1 litre = 1000 cm³.
b) Work out the time it takes to fill the water tank completely. Give your answer in minutes. **[Edexcel]**

21 The lengths, in cm, of the sides of the triangle are $x + 1$, $2x + 5$ and $3x + 2$.

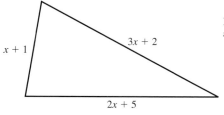

Diagram *not* accurately drawn

$3x + 2$

$x + 1$

$2x + 5$

a) Write down, in terms of x, an expression for the perimeter of the triangle. Give your expression in its simplest form.

The perimeter of the triangle is 50 cm.

b) Work out the value of x. **[Edexcel]**

22 The diagram shows a pentagon. AB = AE, and BC = CD = DE.

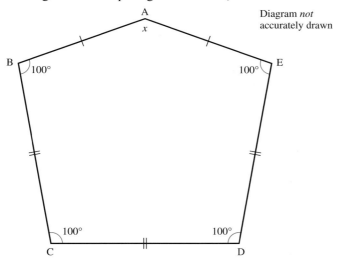

Diagram *not* accurately drawn

Find the size of the angle marked $x°$. **[Edexcel]**

23 The perimeter of this rectangle has to be more than 11 cm and less than 20 cm.

Diagram *not* accurately drawn

a) Show that $5 < 2x < 14$
b) x is an **integer**. List all the possible values of x. **[Edexcel]**

24 The diagrams show a paperweight.

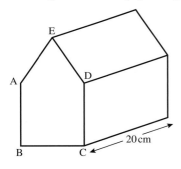 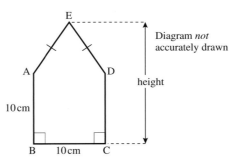

Diagram *not* accurately drawn

ABCDE is a cross-section of the paperweight.
AB, BC and CD are three sides of a square of side 10 cm. AE = DE.
The area of the cross-section is 130 cm².
a) Work out the height.
The paperweight is a prism of length 20 cm.
b) Work out the volume of the paperweight. Give the units with your answer. **[Edexcel]**

25 The diagram shows a shape.

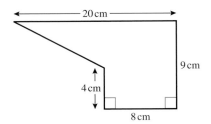

Diagram *not* accurately drawn

Work out the area of the shape.

26 ABCD is a quadrilateral.

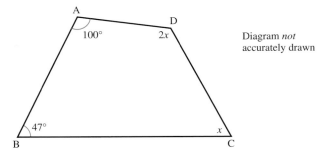

Diagram *not* accurately drawn

Work out the size of the largest angle in the quadrilateral.

27 This is part of the design of a pattern found at the theatre of Diana at Alexandria. It is made up of a regular hexagon, squares and equilateral triangles.

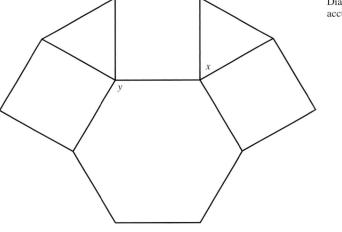

Diagram *not* accurately drawn

a) Write down the size of the angle marked x.

b) Work out the size of the angle marked y.

The area of each equilateral triangle is 2 cm^2.

c) Work out the area of the regular hexagon.

28 The diagram represents a large tank in the shape of a cuboid.

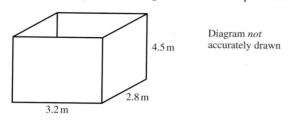

4.5 m   Diagram *not* accurately drawn

2.8 m

3.2 m

The tank has a base. It does not have a top.
The width of the tank is 2.8 metres, the length is 3.2 metres, the height is 4.5 metres.
The outside of the tank is going to be painted. 1 litre of paint will cover 2.5 m² of the tank.
The cost of the paint is £2.99 per litre.
Calculate the total cost of the paint needed to paint the outside of the tank.   [Edexcel]

29 The width of a rectangle is x centimetres. The length of the rectangle is $(x + 4)$ centimetres.

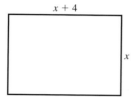

$x + 4$

x

a) Find an expression, in terms of x, for the perimeter of the rectangle. Give your expression in its simplest form.

The perimeter of the rectangle is 54 centimetres.
b) Work out the length of the rectangle.   [Edexcel]

30

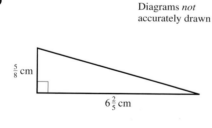

Diagrams *not* accurately drawn

$\frac{5}{8}$ cm

$6\frac{2}{5}$ cm

The area of the square is 18 times the area of the triangle.
Work out the **perimeter** of the square.   [Edexcel]

Key points

1 Corresponding (F) angles are equal.   Alternate (Z) angles are equal.

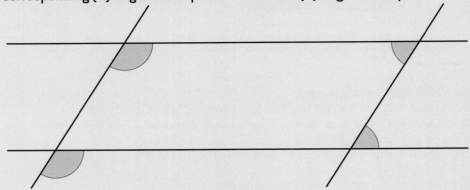

2 The angles in any triangle add up to 180°.

3 The angles in any quadrilateral add up to 360°.

4 For a polygon with n sides, the interior angles will sum to $180 \times (n - 2)°$

5 The exterior angles of any polygon add up to 360°.

6 If the polygon has equal sides and equal angles, it is regular.

7 Each exterior angle of an n-sided regular polygon is 360° ÷ n.

8 Areas:

Area of a rectangle $= bh$

Area of a triangle $= \frac{1}{2}bh$

Area of a parallelogram $= bh$

Area of a trapezium $= \frac{1}{2}(a + b)h$

9 Remember to include units in your answers to numerical problems.

10 The volume of a cuboid is found by multiplying its three dimensions, so $V = abc$

11 A prism has constant cross-sectional area.

12 Volume of a prism = cross-sectional area × length

13 To find the surface area of a solid, work out the area of each separate flat surface, then add them up.

14 When converting between different units of area, remember to square the ordinary linear conversion factor. For example, 100 cm = 1 m, but 10 000 cm² = 1 m². In a similar way, for volumes the factor must be cubed: 1 000 000 cm³ = 1 m³.

Internet Challenge 11

The four-colour theorem

The four-colour theorem claims that four colours are sufficient to colour in a map, in such a way that no two regions share the same colour along a boundary (except at a point).

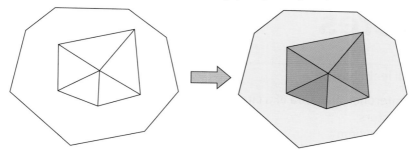

Try drawing some maps of your own, and colouring them in. Does the four-colour theorem appear to be true?

Now use the internet to help answer these questions.

1 Who first proposed this theorem, in 1852/3?

2 Who presented a flawed proof, in 1879?

3 Who presented another flawed proof, in 1880?

4 When was the four-colour theorem first successfully proved?

5 Who achieved this first proof?

6 What major innovation was used to support the proof?

7 Supposing a map is drawn on a sphere instead of a plane? How many colours are sufficient now?

8 Mathematicians refer to a three-dimensional ring doughnut shape as a *torus*. How many colours are sufficient to colour in any map on a torus?

9 What April Fool's joke concerning the four-colour theorem was perpetrated by the mathematician and mathematical games writer Martin Gardner in 1975?

10 Why might a real map-maker need to use more than four colours?
Clue: What is unusual about Alaska?

CHAPTER 12

Circles, cylinders, cones and spheres

In this chapter you will learn how to:

- calculate the circumference and area of a circle
- calculate areas of sectors
- use circle formulae in reverse
- find the surface area and volume of a cylinder
- find surface area and volume of cones and spheres
- obtain exact expressions for results in terms of π.

You will also be challenged to:

- investigate measuring the Earth.

Starter: **Three and a bit ...**

Use your calculator to work out the value of each of these expressions. Write down all the figures on your calculator display. Each answer should be a little over 3.

$$3 + \frac{1}{8}$$

$$\frac{22}{7}$$

$$\sqrt{10}$$

$$3 + \frac{8}{60} + \frac{30}{60^2}$$

$$\frac{333}{106}$$

$$\left(\frac{2143}{22}\right)^{\frac{1}{4}}$$

$$\left(\frac{4}{3}\right)^4$$

$$\frac{88}{\sqrt{785}}$$

$$\sqrt{2} + \sqrt{3}$$

$$\frac{355}{113}$$

These are all approximations to an important mathematical quantity called pi. This is stored on your calculator as a key marked with a π symbol. Use this key to obtain the value of pi correct to as many significant figures as possible, and write it down.

Which one of these approximations is the closest?

12.1 Circumference and area of a circle

The distance all the way around the perimeter of a circle is known as its **circumference**. The circumference of any circle is just over three times its diameter. More precisely, this ratio is 3.141 592 6 … and is known as **pi** (the Greek letter p), written π. The value of pi will be stored in your calculator, and can be called up at the press of a key.

> **Circumference of a circle = pi × diameter**
> $$C = \pi d$$

Sometimes it is more convenient to work with a circle's radius instead. The radius is exactly half the diameter, so it must be doubled in order for this method to work.

> **Circumference of a circle = two × pi × the radius**
> $$C = 2\pi r$$

EXAMPLE

Find the circumference of these two circles:

a)

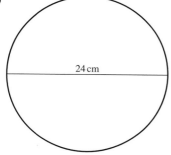

24 cm

b)

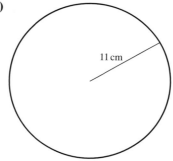

11 cm

SOLUTION

a) This circle has diameter $d = 24$.

$C = \pi d$
$\quad = \pi \times 24$
$\quad = 75.398\ 223\ 69$
$\quad = \underline{75.4\ \text{cm}}$ (3 s.f.)

> It is usual to round the answer to 3 significant figures in this kind of question. Make sure you show your unrounded answer too.

b) This circle has radius $r = 11$.

$C = 2\pi r$
$\quad = 2 \times \pi \times 11$
$\quad = 69.115\ 038\ 38$
$\quad = \underline{69.1\ \text{cm}}$ (3 s.f.)

Your calculator makes easy work of finding the area of a circle, too. The area of a circle is found by using this formula.

> **Area of a circle = pi × the square of the radius**
> $$A = \pi r^2$$

EXAMPLE

Find the areas of these two circles:

a)

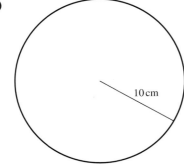

10 cm

b)

18 cm

SOLUTION

a) This circle has radius $r = 10$.

$$A = \pi r^2$$
$$= \pi \times 10^2$$
$$= \pi \times 100$$
$$= 314.159\ 265\ 4$$
$$= \underline{314\ \text{cm}^2}\ (3\ \text{s.f.})$$

Square the radius *first*, then multiply the result by pi.

b) This circle has radius $r = 18 \div 2 = 9$.

$$A = \pi r^2$$
$$= \pi \times 9^2$$
$$= \pi \times 81$$
$$= 254.469\ 004\ 9$$
$$= \underline{254\ \text{cm}^2}\ (3\ \text{s.f.})$$

You can type this expression straight in your calculator – it knows that it must work out the square first.

Take care to choose the right formula when working with circles. It might help to remember that the formula containing a **squared** term – πr^2 – is used for finding **area**, which is measured in **square units**.

Some questions may require you to use these circle formulae in combination with other area or perimeter calculations.

EXAMPLE

The diagram shows an ornamental flowerbed. It is in the shape of a rectangle, with semicircles at each end. The rectangle is of length 2.8 metres. Each semicircle has a radius of 1.1 metre.

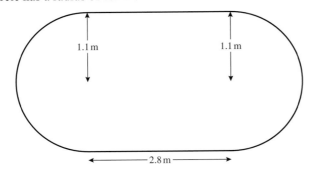

1.1 m

1.1 m

2.8 m

a) Calculate the perimeter of the flowerbed.
b) The gardener is going to put edging around the flowerbed. Edging costs £2.50 per metre. Work out how much the edging will cost.
c) Calculate the area of the flowerbed.
d) The gardener plans to add fertiliser to the flowerbed. One bag of fertiliser will be sufficient for 0.8 square metres of flowerbed. How many bags of fertiliser will the gardener need to buy?

SOLUTION

a) The two semicircles are equivalent to a single circle with radius $r = 1.1$.

$$C = 2\pi r$$
$$= 2 \times \pi \times 1.1$$
$$= 6.911\ 503\ 838$$

The two rectangular edges are 2.8 metres each, so the total perimeter is:

$$P = 2.8 + 2.8 + 6.911\ 503\ 838$$
$$= 12.511\ 503\ 838$$
$$= \underline{12.5\ \text{m}}\ (3\ \text{s.f.})$$

b) Cost of edging $= 12.5 \times £2.50$
$$= \underline{£31.25}$$

c) The area of the semicircles is equivalent to the area of a single circle with $r = 1.1$.

Area of circle $= \pi r^2$
$$= \pi \times 1.1^2$$
$$= 3.801\ 327\ 111$$
$$= 3.80\ \text{m}^2\ (3\ \text{s.f.})$$

The rectangular part measures 2.8 m by 2.2 m, so:

Area of rectangle $= 2.8 \times 2.2$
$$= 6.16\ \text{m}^2$$

The total area is $3.80 + 6.16 = \underline{9.96\ \text{m}^2}$

d) 1 bag of fertiliser is sufficient for 0.8 m².
Therefore the gardener needs $9.96 \div 0.8 = 12.45$ bags.
The gardener needs to buy $\underline{13\ \text{bags}}$.

EXERCISE 12.1

1 A circle has radius 12 mm. Find its circumference, correct to 3 significant figures.

2 A circle has diameter 22 cm. Find its circumference, correct to 3 significant figures.

3 A circle has radius 18 cm. Find its area, correct to 3 significant figures.

4 A circle has diameter 11.5 cm. Find its area, correct to 3 significant figures.

5 Find, correct to 4 significant figures, the circumference of a circle with radius 21.25 cm.

6 Find, correct to 4 significant figures, the area of a circle with diameter 66.25 mm.

7 Find, correct to 4 significant figures, the circumference of a circle with diameter 1.25 cm.

8 Find, correct to 4 significant figures, the area of a circle with radius 0.455 cm.

9 A circle has diameter 11 cm. Find its area, correct to 3 significant figures.

10 A circle has radius 0.85 mm. Find its area, correct to 3 significant figures.

11 A circle has diameter 250 cm. Find its circumference, correct to 3 significant figures.

12 A circle has radius 1.06 m. Find its circumference, correct to 3 significant figures.

13 A tennis ball has a diameter of 66 mm. Calculate its circumference, correct to 2 significant figures.

14 A face of a one euro coin is a circle of diameter of 23.25 mm. Calculate its area, correct to 3 significant figures.

15 Emma decides to run around a circular race track. The radius of the track is 25 metres.
   a) Work out the length of one lap of the track, correct to 3 significant figures.

   Emma wants to run at least 5000 metres. She wants to run a whole number of laps.
   b) Work out the minimum number of laps that Emma must run.

16 A circular CD disc is cut from a plastic square of side 12 cm. A hole of diameter 1.5 cm is then cut from the centre. Calculate the area of the CD disc, correct to 3 significant figures.

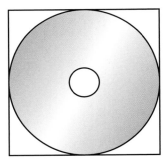

17 The diagram shows a simple 'eclipse viewer' observing aid. The frame is made of cardboard, and comprises a rectangle with two circular holes cut into it. The holes are then filled with a reflective safety film that blocks harmful radiation from the Sun.

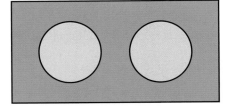

The holes are each of diameter 4 cm. The rectangle measures 15 cm by 6 cm.
   a) Calculate the area of one of the circular holes, correct to the nearest 0.1 cm^2.
   b) Work out the area of the cardboard frame, correct to the nearest 0.1 cm^2.

18 The diagram shows an ornamental stained glass window.

The window is a circle, of radius 30 cm. The rectangle measures 48 cm by 36 cm. The glass inside the rectangle is stained blue; the glass outside the rectangle is stained yellow. There is a boundary, made of lead, indicated by the heavy black line. (The lead is of negligible thickness.)

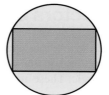

a) Work out the length of the boundary, correct to the nearest centimetre.

b) Work out the area of the blue glass, correct to 3 significant figures.

c) Work out the area of the yellow glass, correct to 3 significant figures.

19 The diagram shows a running track. It is made up of two straight sections, and two semicircular ends. The dimensions are marked on the diagram.

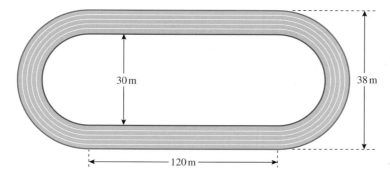

Steve runs around the outside boundary of the track, marked with a red line. Seb runs around the inside boundary, marked in blue. They each run one lap of the track.

a) Work out the length of the outside boundary of the track.

b) Work out the length of the inside boundary of the track.

c) How much longer is the outside boundary, compared with the inside one? Give your answer as a percentage.

20 The diagram shows a metal washer. It is made from a circular sheet of radius 6 mm. A smaller circle of radius 3 mm is then removed from the centre and discarded.

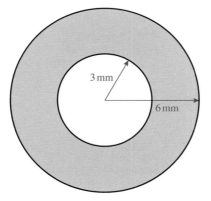

a) Calculate the area of the large circle.

b) Calculate the area of the smaller circle.

c) Hence find the area of the washer.

Fred says 'Since 3 mm is half of 6 mm, then 50% of the metal is wasted by discarding the smaller circle.' Fred is incorrect.

d) Calculate the correct percentage of metal wasted.

12.2 Sectors of a circle

In some of the previous examples you have worked with **semicircles**. Circles can be sliced into quarters, called **quadrants**, or other sized fractions of a complete circle. These are called **sectors**.

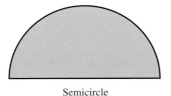

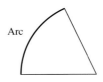

Semicircle   Quadrant   Sector   Arc

The curved boundary along the edge of a sector is called an **arc**. You can find the length of an arc by calculating the corresponding fraction of a circumference of a circle. The area of a sector can be found in a similar manner.

$$\text{arc length} = \frac{\text{sector angle}}{360°} \times 2\pi r \qquad \text{sector area} = \frac{\text{sector angle}}{360°} \times \pi r^2$$

EXAMPLE

Calculate the perimeter and area of each of these sectors of a circle.

a)

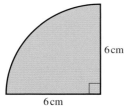

6 cm

6 cm

b)

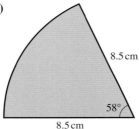

8.5 cm

8.5 cm

58°

SOLUTION

a) The sector is a quadrant; one-quarter of a circle.

The circumference of the full circle would be $2 \times \pi \times 6 = 37.699\ 111\ 84$.

The arc length of the quadrant is $\frac{1}{4} \times 37.699\ 111\ 84 = 9.42$ cm.

Thus the perimeter is $9.42 + 6 + 6 = \underline{21.42 \text{ cm}}$ (4 s.f.)

The area of the full circle would be $\pi \times 6^2 = 113.097\ 335\ 5$ cm².

The area of the quadrant is $\frac{1}{4} \times 113.097\ 335\ 5 = \underline{28.27 \text{ cm}^2}$ (4 s.f.)

b) This sector has a central angle of 58°, so it represents $\frac{58}{360}$ of a circle.

The arc length of the sector is $\frac{58}{360} \times 2 \times \pi \times 8.5 = 8.60$ cm.

Thus the perimeter is $8.60 + 8.5 + 8.5 = \underline{25.60 \text{ cm}}$ (4 s.f.)

The area of the sector is $\frac{58}{360} \times \pi \times 8.5^2 = \underline{36.57 \text{ cm}^2}$ (4 s.f.)

EXERCISE 12.2

Calculate the perimeter and area of each sector. Give your answer correct to 3 significant figures.

1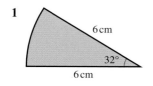
6 cm
32°
6 cm

2
9 cm
83°
9 cm
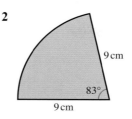

3
7.5 cm
120°
7.5 cm

4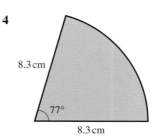
8.3 cm
77°
8.3 cm

5
44 mm
10°
44 mm

6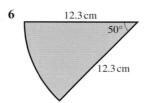
12.3 cm
50°
12.3 cm

7
170°
2.5 cm
2.5 cm

8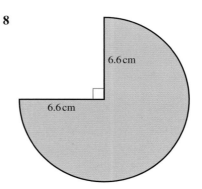
6.6 cm
6.6 cm

9 Bob wants to display some statistical data in a pie chart. The three sectors of the pie chart are to have angles of 160°, 140° and 60°. The radius of the pie chart is to be 10 cm.
  a) Work out the area of each sector of Bob's pie chart, correct to the nearest cm².
  b) Work out the perimeter of the smallest sector, correct to 3 significant figures.

10 A pizza of diameter 12 inches is to be shared between five people. It is cut into five equal segments.
  a) Work out the angles at the centre of each segment of pizza.
  b) Work out the area of one segment, correct to 3 significant figures.

12.3 Circumference and area in reverse

Supposing you want to find the dimensions of a circle in order for it to have a given circumference or area. Then it is necessary to apply the circle formulae in reverse.

EXAMPLE

A circular hula-hoop has a circumference of 2.4 metres. Find its diameter in centimetres, correct to the nearest centimetre.

SOLUTION

Let the diameter be d metres.

$$\pi \times d = 2.4$$

So $d = 2.4 \div \pi$

$$= 0.763\,943\,726\,8 \text{ m}$$

$$= \underline{76 \text{ cm}} \text{ (nearest cm)}$$

Reverse area problems require a little more care. It is best to find r^2 first, then square root at the end to find r. If the question asks for the diameter, just double the final r value.

EXAMPLE

A coin has an area of 4 cm². Work out its diameter, in millimetres, correct to 3 significant figures.

SOLUTION

Let the radius of the coin be r cm.

Then $\pi r^2 = 4$

$$r^2 = 4 \div \pi$$

$$= 1.273\,239\,545$$

$$r = \sqrt{1.273\,239\,545}$$

$$= 1.128\,379\,167$$

The diameter is $2r = 2.256\,758\,334$ cm

$$= \underline{22.6 \text{ mm}} \text{ (3 s.f.)}$$

EXERCISE 12.3

Give the answers to each of these problems correct to 3 significant figures.

1 A circle has circumference 15.5 cm.

Find its radius.

2 A circle has circumference 12.8 cm.

Find its diameter.

3 A circle has area 120 cm².

Find its radius.

4 A circle has area 44 cm².

Find its diameter.

5 A circle has circumference 1.45 cm.

Find its radius.

6 A circle has area 850 cm².

Find its diameter.

7 A circle has circumference 6.25 cm.

Find its diameter.

8 A circle has area 225 cm².

Find its radius.

9 The diagram shows a sector of a circle. The angle at the centre of the sector is 40°.

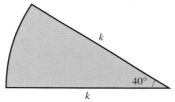

The radius of the segment is k centimetres. The area of the segment is 427.6 cm².
a) Work out the area of the corresponding complete circle.
b) Hence find the value of k.
c) Calculate the perimeter of the segment.

10 The diagram shows the boundary of a running track. The ends are semicircles of radius x metres. The straights are of length 35 metres each. The total distance around the outside of the track is 100 metres.

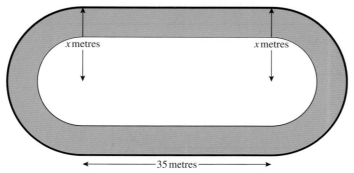

Calculate the value of x.

12.4 Surface area and volume of a cylinder

You can make a hollow cylinder by rolling up a rectangular sheet of paper.

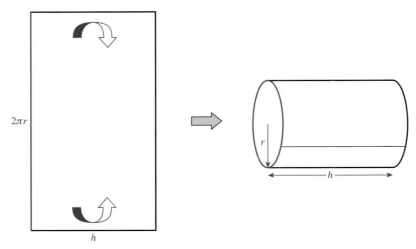

Suppose the cylinder has height h and radius r.

Then the distance marked in red on the diagram is the circumference of the end of the cylinder, which is $2\pi r$.

The cylinder forms a **curved surface area**, which must be equal in area to the original rectangle.

This is $2\pi r$ times h, giving the formula:

Curved surface area of a cylinder $= 2\pi rh$

A cylinder can be thought of as a prism with a circular base. Then its volume is found by multiplying the cross-sectional area (πr^2) by the length (h) to obtain this formula:

Volume of a cylinder $= \pi r^2 h$

EXAMPLE

A metal pipe is in the form of a cylinder, 1.5 metres long and 22 centimetres in diameter.

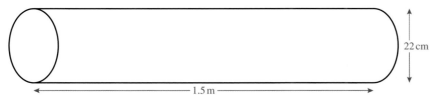

a) Calculate the curved surface area of the pipe, in square centimetres.
b) Work out the volume of the cylindrical pipe, in cubic centimetres.

SOLUTION

Using centimetres as a standard unit, $r = 11$ and $h = 150$.

a) Curved surface area $= 2\pi rh$

$$= 2 \times \pi \times 11 \times 150$$
$$= 10\,367.255\,76$$
$$= \underline{10\,400\text{ cm}^2}\ (3\text{ s.f.})$$

b) Volume $= \pi r^2 h$

$$= \pi \times 11^2 \times 150$$
$$= 57\,019.906\,66$$
$$= \underline{57\,000\text{ cm}^3}\ (3\text{ s.f.})$$

EXERCISE 12.4

1 A cylinder has radius 12 cm and height 19 cm. Find its volume, correct to 3 significant figures.

2 A cylinder has radius 5 cm and height 2 cm. Find its curved surface area, correct to 3 significant figures.

3 A cylinder has diameter 22 cm and height 8 cm.
   a) Find its volume, correct to 3 significant figures.
   b) Find its curved surface area, correct to 3 significant figures.

4 A cylinder has radius 6 cm and height 4.5 cm. Find its volume, correct to 3 significant figures.

5 The diagram shows a hollow cylinder.

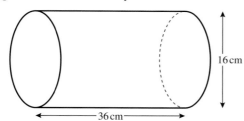

Work out the curved surface area of the cylinder, correct to 3 significant figures.

6 A hollow cylindrical pipeline has an internal diameter of 15 cm. The pipeline is 120 metres in length.
   a) Work out the volume of the pipeline. Give your answer in cm³, correct to 3 significant figures.
   b) 1000 cm³ = 1 litre. Express the volume of the pipeline in litres.

7 A biscuit tin is in the shape of a cylinder. It has radius 9 cm and height 14 cm. Work out the volume of the cylinder. Give your answer to the nearest cm³.

8 A cylinder of radius 8.5 cm has a volume of 3178 cm³, correct to 4 significant figures. Work out the height of the cylinder.

9 A sweet packet is in the shape of a hollow cardboard cylinder. The inside diameter of the cylinder is 2.5 cm and it has a height of 15 cm.
   a) Work out the volume of the cylinder, correct to 3 significant figures.
   b) The sweets have a volume of 1.5 cm³ each. Show that the packet cannot contain as many as 50 sweets.

10 Nick and Alan have been working on an exercise about cylinders. They have to find the volume of a cylinder with diameter 14 cm and height 24 cm.

Nick says 'The volume is 3690 cm³ correct to 3 significant figures.'

Alan says 'The volume is 14800 cm³ correct to 3 significant figures.'

a) Work out who was right.
b) Suggest what mistake has been made by the person who was wrong.

12.5 Exact calculations using pi

So far in this chapter you have used the pi key on your calculator. Although this is very convenient, it does introduce a slight inaccuracy in your work.

You can make exact statements about areas and volumes of circles and cylinders, by leaving π in your working and final answer. Some exam questions will instruct you to do this.

EXAMPLE

The diagram shows a circular washer. It is made from a circular sheet of radius 4 mm, with a circular hole of radius 2 mm removed. Work out the area of the washer. Leave your answer in terms of π.

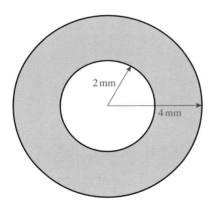

SOLUTION

The area of the larger circle is $\pi r^2 = \pi \times 4^2$
$$= \pi \times 16$$
$$= 16\pi \text{ mm}^2$$

The area of the smaller circle is $\pi r^2 = \pi \times 2^2$
$$= \pi \times 4$$
$$= 4\pi \text{ mm}^2$$

Thus the area of the washer $= 16\pi - 4\pi$
$$= \underline{12\pi \text{ mm}^2}$$

EXAMPLE

A cylinder has height 8.5 cm and radius 4 cm. Work out its curved surface area and volume. Leave your answer in terms of π.

SOLUTION

For this cylinder, $h = 8.5$ and $r = 4$.

Curved surface area $= 2\pi rh$
$= 2 \times \pi \times 4 \times 8.5$
$= 2 \times \pi \times 34$
$= \pi \times 68$
$= \underline{68\pi \text{ cm}^2}$

Volume $= \pi r^2h$
$= \pi \times 4^2 \times 8.5$
$= \pi \times 16 \times 8.5$
$= \pi \times 136$
$= \underline{136\pi \text{ cm}^3}$

EXERCISE 12.5

1 A circle has diameter 24 cm. Work out its circumference and area. Leave your answer as an exact multiple of π.

2 A circle has radius 11 cm. Work out its circumference and area. Leave your answer in terms of π.

3 A cylinder has radius 12 cm and height 8 cm.
   a) Find the curved surface area of the cylinder.
   b) Find the volume of the cylinder.
   Give your answers as exact multiples of π.

4 A circle has circumference 24π centimetres.
   a) Find the exact radius of the circle.
   b) Find the exact area of the circle. Leave your answer in terms of π.

5 A circle has area 121π square centimetres.
   a) Find the exact radius of the circle.
   b) Find the exact circumference of the circle.

6 A cylinder has volume 300π cm^3. It has radius 10 cm. Work out its height.

7 A cylinder has volume 480π cm^3. It has diameter 8 cm. Work out its height.

8 The diagram shows a quadrant of a circle. The radius is 16 cm.

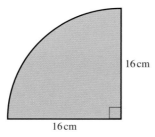

16 cm

16 cm

   a) Find the area of the quadrant, in terms of π.
   b) Find an exact expression for the perimeter of the quadrant.

9 The diagram shows an ornamental design. It is in the shape of a square, with semicircles on each of the four sides. The square is of side 12 cm.

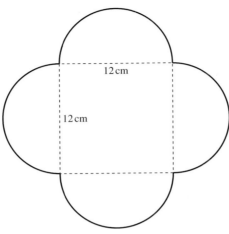

a) Find the area of one of the semicircles, leaving your answer in terms of π.

b) Hence find an exact expression for the area of the ornamental design.

10 The diagram shows two cylinders. Cylinder A has diameter 6 cm and height 8 cm. Cylinder B has diameter 8 cm and height 6 cm.

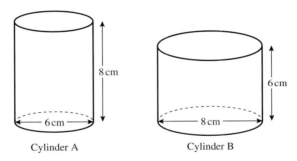

Cylinder A   Cylinder B

a) Show that both cylinders have exactly the same curved surface area.

b) Work out the volume of each cylinder, leaving your answers in terms of π. Which cylinder has the larger volume?

12.6 Volume and surface area of cones and spheres

A **pyramid** has a **base** and an **apex**, or point. Rays drawn from the edge of the base converge at the apex. Pyramids may have square or triangular bases, but other polygons can be bases as well. A **cone** is like a pyramid with a circular base.

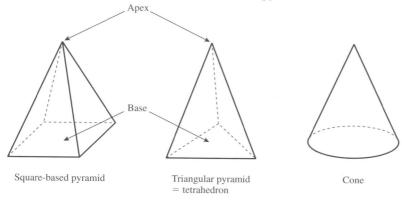

Square-based pyramid   Triangular pyramid = tetrahedron   Cone

A remarkable property of a cone is that it fills exactly one-third of the enveloping cylinder that would just contain it.

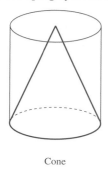

Cone

So the volume of a cone is

$$V = \tfrac{1}{3}\pi r^2 h$$

EXAMPLE

Find the volume of this cone.

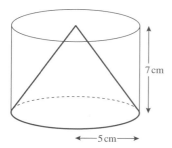

7 cm

5 cm

SOLUTION

For the cone:

radius $r = 5$ and height $h = 7$

So the volume V is:

$V = \frac{1}{3}\pi r^2 h$

$= \frac{1}{3} \times \pi \times 5^2 \times 7$

$= 183.259\ 571\ 5$

$= \underline{183\ \text{cm}^3}$ (3 s.f.)

> Remember to include units with the answers, e.g. mm³ or cm³ for volumes and mm² or cm² for areas.

To find the **volume of a sphere**, radius r, you may use this formula:

$V = \frac{4}{3}\pi r^3$

EXAMPLE

A spherical ball bearing has a diameter of 6 mm. Find its volume.

SOLUTION

Since $d = 6$ we have $r = 3$ mm, so the volume V is:

$V = \frac{4}{3}\pi r^3$

$= \frac{4}{3} \times \pi \times 3^3$

$= 113.097\ 335\ 5$

$= \underline{113\ \text{mm}^3}$ (3 s.f.)

Finally, you may need to calculate the surface area of one of these shapes. For a cone or a sphere, these formulae may be used:

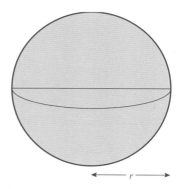

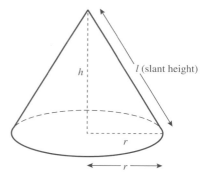

Surface area of a sphere
$A = 4\pi r^2$

(Curved) Surface area of a cone
$A = \pi r l$

EXAMPLE

Find the total surface area of each of these shapes, giving your answers exactly in terms of π.

a)

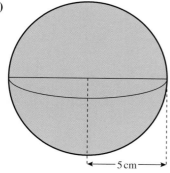

b)

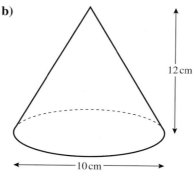

SOLUTION

a) For the sphere, $r = 5$ cm, so:

$$A = 4\pi r^2$$
$$= 4 \times \pi \times 5^2$$
$$= 4 \times \pi \times 25$$
$$= \underline{100\pi \text{ cm}^2}$$

b) For the cone, $r = 5$ cm and $h = 12$ cm, but the slant height h is needed.

By Pythagoras' theorem:

$$l^2 = r^2 + h^2$$
$$= 25 + 144$$
$$= 169$$
$$l = 13 \text{ cm}$$

> Pythagoras' theorem states that the square of the longest side of a right-angled triangle equals the sum of the squares of the two shorter sides.

> You will meet Pythagoras' theorem more formally in chapter 16.

Then the curved surface area:

$$\text{CSA} = \pi r l$$
$$= \pi \times 5 \times 13$$
$$= 65\pi \text{ cm}^2$$

The circular base has area:

$$\pi r^2 = \pi \times 5^2$$
$$= 25\pi \text{ cm}^2$$

Thus the total surface area is:

$$65\pi + 25\pi = \underline{90\pi \text{ cm}^2}$$

EXERCISE 12.6

Find the volume and total surface area of each of these solids. Give your answers correct to 3 significant figures.

1
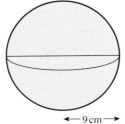
←9 cm→

2
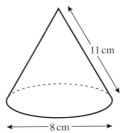
11 cm
←8 cm→

3
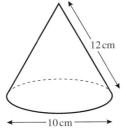
12 cm
←10 cm→

Find the volume and surface area of each of these solids, in terms of π.

4
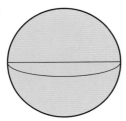
←12 cm→

5
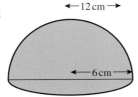
←6 cm→

6
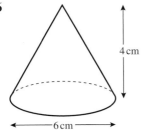
4 cm
←6 cm→

7 A sphere has surface area 900π cm². Find its volume in terms of π.

8 A cone of slant height 7 cm has curved surface area of 28π cm². Find the area of its base in terms of π.

REVIEW EXERCISE 12

1 A circle has radius 28 cm. Work out its area, correct to 3 significant figures.

2 A circle has diameter 90 mm. Work out its circumference, correct to 3 significant figures.

3 A circle has radius 1.9 cm. Work out its circumference, correct to 4 significant figures.

4 A circle has diameter 64 mm. Work out its area, correct to 4 significant figures.

5 A circle has an area of 64π cm².
   a) Write down its radius.
   b) Find its circumference. Give your answer in terms of π.

6 A closed cylinder has a radius of 10 cm and a height of 15 cm.
   a) Calculate the curved surface area of the cylinder.
   b) Calculate the area of one of its circular ends.
   c) Hence find the total surface area of the cylinder.

7 A circle has a circumference of 15.71 cm, correct to 4 significant figures.
   a) Calculate the radius of this circle.
   b) Hence find the area of the circle. Give your answer to 3 significant figures.

8 The diagram shows a rectangle inscribed in a circle.
AB = 5 cm, BC = 12 cm, AC = 13 cm.
The line segment AC is a **diameter** of the circle.
Work out the size of the shaded area.
Give your answer to 3 significant figures.

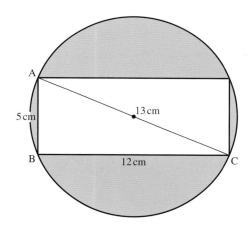

9 The radius of a circle is 5.1 m.

Diagram *not* accurately drawn

Work out the area of the circle. State the units of your answer.   [Edexcel]

10 A circle has a radius of 3 cm.

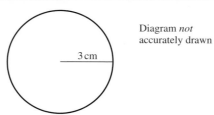

Diagram *not*
accurately drawn

3 cm

a) Work out the area of the circle. Give your answer correct to 3 significant figures.

A semicircle has a diameter of 9 cm.

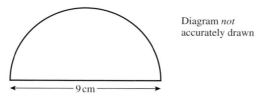

Diagram *not*
accurately drawn

9 cm

b) Work out the perimeter of the semicircle. Give your answer correct to
3 significant figures. [Edexcel]

11 The diagram shows a semicircle.
The diameter of the semicircle is 15 cm.

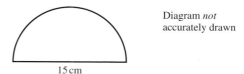

Diagram *not*
accurately drawn

15 cm

Calculate the area of the semicircle. Give your answer correct to 3 significant figures. [Edexcel]

12 A circle has a radius of 32 cm. Work out the circumference of the circle. Give your answer
correct to the nearest centimetre. [Edexcel]

13 The diagram shows a right-angled triangle ABC and a circle.

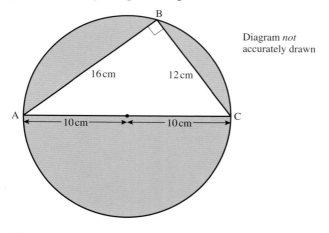

Diagram *not*
accurately drawn

A, B and C are points on the circumference of the circle. AC is a diameter of the circle.
The radius of the circle is 10 cm. AB = 16 cm and BC = 12 cm.
Work out the area of the shaded part of the circle. Give your answer correct to the nearest cm². [Edexcel]

14 A can of drink is in the shape of a cylinder. The can has a radius of 4 cm and a height of 15 cm.

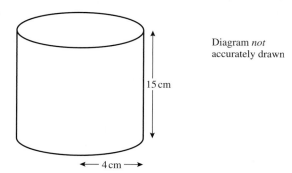

Diagram *not* accurately drawn

15 cm

4 cm

Calculate the volume of the cylinder. Give your answer correct to 3 significant figures.   [Edexcel]

15 A ten pence coin has a diameter of 2.45 cm.

2.45 cm

Work out the circumference of the coin. Give your answer in cm correct to 1 decimal place.   [Edexcel]

16 The diagram shows a shape, made from a semicircle and a rectangle.

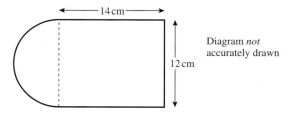

14 cm

Diagram *not* accurately drawn

12 cm

The diameter of the semicircle is 12 cm. The length of the rectangle is 14 cm.
Calculate the **perimeter** of the shape. Give your answer correct to 3 significant figures.   [Edexcel]

17 The diagram shows a sector of a circle, centre O.

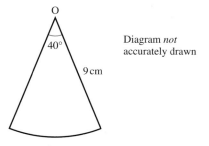

O

40°

Diagram *not* accurately drawn

9 cm

The radius of the circle is 9 cm. The angle at the centre of the circle is 40°.
Find the perimeter of the sector. Leave your answer in terms of π.   [Edexcel]

18 The diagram shows the shape PQRST.

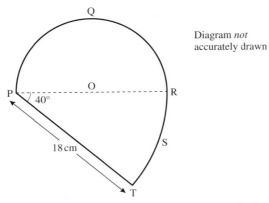

Diagram *not* accurately drawn

RST is a circular arc with centre P and radius 18 cm. Angle RPT = 40°.

a) Calculate the length of the circular arc RST. Give your answer correct to 3 significant figures.

PQR is a semicircle with centre O.

b) Calculate the total area of the shape PQRST. Give your answer correct to 3 significant figures.

[Edexcel]

19 The diagram shows a water tank. The tank is a hollow cylinder joined to a hollow hemisphere at the top. The tank has a circular base.

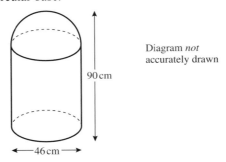

Diagram *not* accurately drawn

Both the cylinder and the hemisphere have a diameter of 46 cm. The height of the tank is 90 cm. Work out the volume of water which the tank holds when it is full. Give your answer, in cm³, correct to 3 significant figures.

[Edexcel]

20 A sphere has a radius of 5.4 cm. A cone has a height of 8 cm. The volume of the sphere is equal to the volume of the cone. Calculate the radius of the base of the cone. Give your answer, in centimetres, correct to 2 significant figures.

[Edexcel]

21 A cone fits exactly on top of a hemisphere to form a solid toy. The radius, CA, of the base of the cone is 3 cm. AB = 5 cm. Show that the total surface area of the toy is 33π cm².

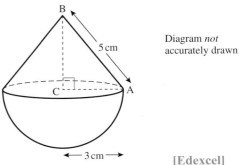

Diagram *not* accurately drawn

[Edexcel]

22

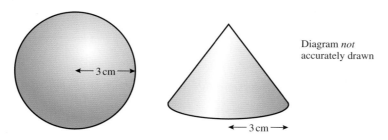

Diagram *not* accurately drawn

The radius of a sphere is 3 cm. The radius of the base of a cone is also 3 cm. The volume of the sphere is three times the volume of the cone.

Work out the curved surface area of the cone. Give your answer as a multiple of π.

[Edexcel]

Key points

1   The circumference of a circle of radius r is found from the formula $C = 2\pi r$

   Writing the diameter as d, you could also use $C = \pi d$

2   The area of a circle is found from the formula $A = \pi r^2$

3   When reading exam questions, take care to check whether you have been told the radius or the diameter of the circle.

4   A sector of a circle is a slice formed by two radii. To compute the area of a sector, begin by working out the area for the full circle. Then compute the corresponding fraction of this. If the sector forms an angle of x degrees, then its area is $\dfrac{x}{360}$ of the full circle.

5   A sector is bounded by a curved arc and two radii. The arc length of a sector can be found in a similar way, by taking a fraction of the full circumference. If a question asks you to find the perimeter of a sector, remember to include the two radii as well as the curved arc.

6   The volume of a cylinder of radius r and height h is $V = \pi r^2 h$

7   The curved surface area of the cylinder is $A = 2\pi r h$

8   Volumes and surface areas of cones and spheres may be found using these formulae:

   Volume of a cone, radius r, height h: $V = \frac{1}{3}\pi r^2 h$

   Volume of a sphere, radius r: $V = \frac{4}{3}\pi r^3$

   Surface area of a sphere, radius r: $A = 4\pi r^2$

   Curved surface area of a cone, radius r, slant height l: $A = \pi r l$

9   Make sure that you know how to use your calculator's π key correctly. Answers to calculations will normally need to be rounded off: an exam question will tell you how many significant figures or decimal places are required. It is a good idea to show your unrounded answer too.

10 Remember that some exam questions may ask you to leave your answers as exact expressions in terms of π.

Internet Challenge 12

Measuring the Earth

A few hundred years ago, many people thought the Earth was flat. They feared you might fall off the edge if you travelled too far from home!

Most people now accept that the Earth is roughly spherical, with a diameter of roughly 12 800 kilometres.

Use the internet to help research the answers to these questions about the Earth.

1 What observational evidence can you find to support the claim that the Earth is roughly spherical?

2 What organisation claims to have been 'deprogramming the masses since 1547'?

3 Find an accurate value for the Earth's equatorial diameter. Use this figure to calculate the Earth's circumference (around the equator).

4 Find an accurate value for the Earth's polar diameter. Use this figure to calculate the Earth's circumference pole to pole.

5 Find the definition of a Great Circle. Is the equator a Great Circle?

6 Who was the first person to circumnavigate the globe, that is, to travel right around the Earth? How long did the journey take, and when was it completed?

7 Who first circumnavigated the world pole to pole? When?

8 Some adventurous sailors take part in round the world yacht races. How far do they typically travel? Do you think they really do travel around the world, in the strictest sense?

9 What is the origin of our word 'geometry'?

10 The size of the Earth was first measured accurately by Eratosthenes, around 200 BC. Find out as much as you can about Eratosthenes and the methods he used. You might want to collect your findings into a poster for your classroom, or prepare a Powerpoint presentation for your class.

CHAPTER 13

Geometric constructions

In this chapter you will learn how to:

- construct triangles from given information
- carry out standard compass constructions on line segments
- solve geometric problems, including the use of bearings.

You will also be challenged to:

- investigate perspective.

Starter: **Round and round in circles**

Use a sharp pencil, compasses and straight edge to make this drawing.

You may want to colour the diagram after you have made it.

Here are some ideas:

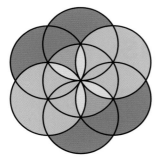
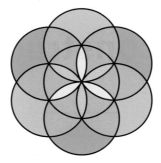

Now try designing your own circle patterns.

13.1 Constructing triangles from given information

Surveyors often use a method called **triangulation** to measure distances. The idea is to measure some combination of angles and distances, and then use them to reconstruct a triangle whose vertex is at the place being surveyed. In fact, many mountaintops in the UK have concrete blocks, or triangulation points, that have been used in this way.

There are several different ways of constructing triangles, depending on the information you are given about them.

1. To construct a triangle, given two sides and the angle in between them (SAS, or side–angle–side)

EXAMPLE

A triangle PQR has side PQ = 9 cm, PR = 5 cm and angle QPR = 55°. Make an accurate construction of this triangle.

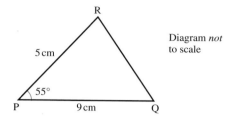

Diagram *not* to scale

SOLUTION

Begin by using your ruler to draw a line segment of length 9 cm, and label the ends P and Q. Then use your protractor to measure an angle of 55° at P.

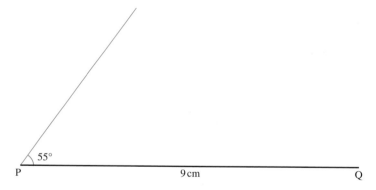

Next, measure a length of 5 cm along the line from P, to locate the point R.

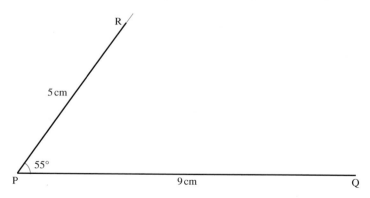

Finally, complete the construction by joining R and Q.

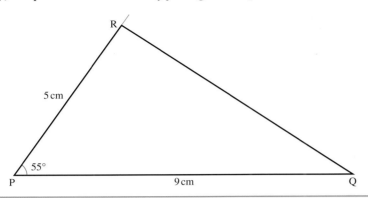

A different procedure is used when you know two angles and an included side.

2. To construct a triangle, given two angles and the side in between them (ASA, or angle–side–angle)

EXAMPLE

A triangle ABC has side AB = 8 cm, angle BAC = 40° and angle ABC = 70°.
Make an accurate construction of this triangle.

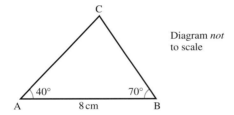

Diagram *not*
to scale

SOLUTION

Begin by constructing a line segment AB, of length 8 cm, and add a line from A
at an angle of 40°, using your protractor to measure this angle.

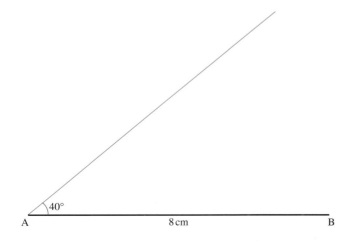

Next, draw a line at an angle of 70° from B.

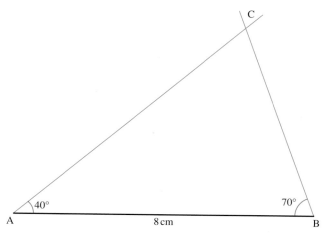

These two lines must intersect at C, so the diagram may be completed:

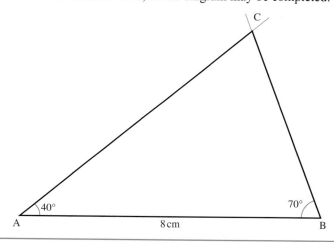

You may be given the values of all three sides but no angles at all. A protractor is now of no use, and you need compasses instead.

3. To construct a triangle, given three sides (SSS, or side–side–side)

EXAMPLE

A triangle LMN has sides LM = 6 cm, LN = 7 cm and MN = 8 cm.
Make an accurate construction of this triangle.

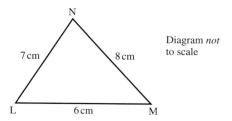

Diagram *not* to scale

SOLUTION

Begin by constructing a line segment LM, of length 6 cm.

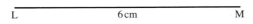

L   6 cm   M

Next, draw an arc of radius 7 cm from L, and another of radius 8 cm from M.

L   6 cm   M

These arcs must intersect at N, so the construction can be completed.

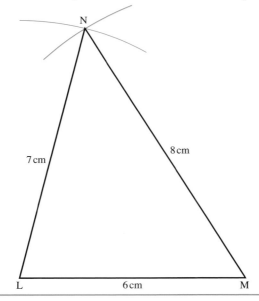

A more confusing scenario is encountered when you are given the values of two sides, and an angle that is *not* included between them. The construction may be *ambiguous*, or even *impossible*!

4. To construct a triangle, given two sides and an angle not between them (SSA, or side–side–angle)

EXAMPLE

In triangle ABC you are given that AB = 7 cm, BC = 5 cm and angle CAB = 40°.

Construct an accurate drawing of this triangle, and show that there are two different solutions based in the given information.

SOLUTION

Begin by drawing a line segment AB of length 7 cm, and construct a line from A at an angle of 40°.

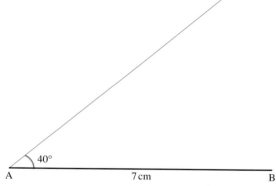

Now open your compasses to a radius of 5 cm, and draw an arc centred on B.

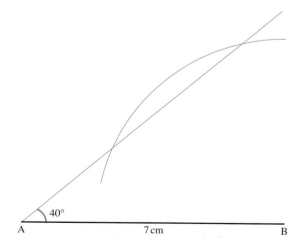

This arc intersects the original line from A in two distinct places, so there are two different ways of completing the construction.

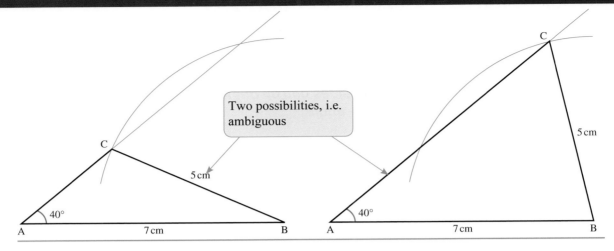

The following exercise gives you some practice at making accurate drawings of triangles. Make sure you leave your construction lines visible, so that your teacher can follow your methods clearly.

EXERCISE 13.1

Make accurate drawings of these triangles, stating which of the various combinations of information you have been given – SAS, ASA, SSS or SSA. If any triangles are ambiguous, draw both possibilities.

1 Draw triangle PQR with PQ = 8 cm, PR = 9 cm, angle QPR = 65°.

2 Draw triangle KLM with KL = 5 cm, angle MKL = 80°, angle KLM = 56°.

3 Draw triangle ABC with AB = 6 cm, AC = 5 cm, angle BAC = 130°.

4 Draw triangle RST with RT = 7.5 cm, RS = 8.5 cm, angle RTS = 90°.

5 Draw triangle PQR with PQ = 8 cm, PR = 7.5 cm, angle PQR = 62°.

6 Draw triangle FGH with HG = 8 cm, HF = 10 cm, FG = 6 cm.

7 The sketch shows a triangle with AB = 85 mm, BC = 55 mm, AC = 70 mm.

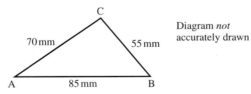

Diagram *not* accurately drawn

Make an accurate diagram of the triangle.

8 The sketch shows a triangle with AB = 75 mm, AC = 70 mm and angle ACB = 90°.

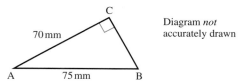

Diagram *not*
accurately drawn

Make an accurate diagram of the triangle.

9 Using compasses, try to make an accurate drawing of triangle PQR with sides PQ = 10 cm, QR = 5 cm, RP = 4 cm. What difficulty do you encounter? Explain why this arises.

10 Triangle JKL is to be constructed with JK = 8 cm, KL = 6.5 cm and angle LJK = 45°.
   a) Try making an accurate construction of this triangle.
   b) What difficulty do you encounter?

13.2 Constructions with line segments

There are three fundamental geometrical constructions that you need to master. Exam questions will expect you to do these with compasses and a straight edge, not with measuring equipment, such as protractors, and ruler measurements would not be permitted. You should leave any construction lines plainly visible.

1. To bisect a given angle

EXAMPLE

Use ruler and compasses to construct the angle bisector of the angle Q shown in the diagram.

SOLUTION

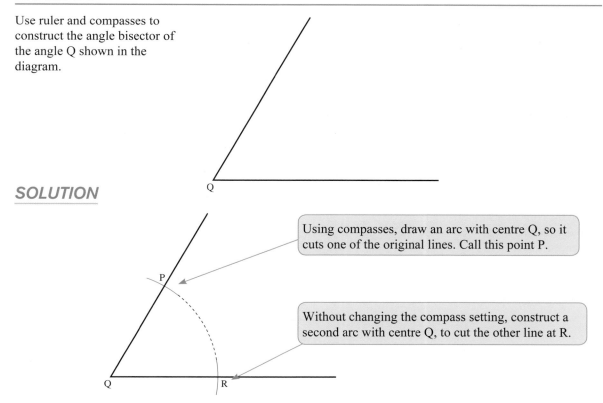

Using compasses, draw an arc with centre Q, so it cuts one of the original lines. Call this point P.

Without changing the compass setting, construct a second arc with centre Q, to cut the other line at R.

237

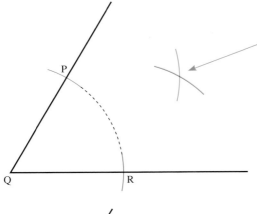

Next, construct two further arcs, centres P and R, to cut here. Once again, do not change the compass setting.

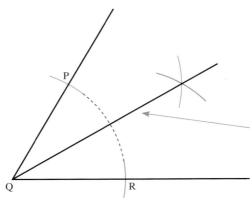

Finally, complete the construction by drawing a straight line from Q to the point of intersection here.

The original angle at Q has now been *divided into two equal parts* – it has been **bisected**.

2. To construct the perpendicular bisector of a given line segment

EXAMPLE

Use straight edge and compasses to construct the perpendicular bisector of the line segment shown in the diagram.

SOLUTION

Open the compasses to more than half the distance from K to L.

From K, construct these two arcs …

… and from L, construct these two.

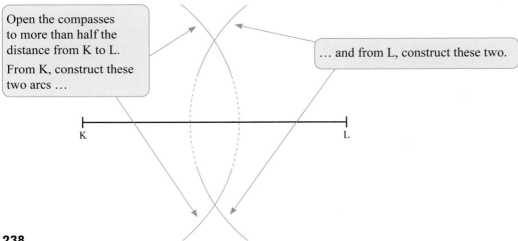

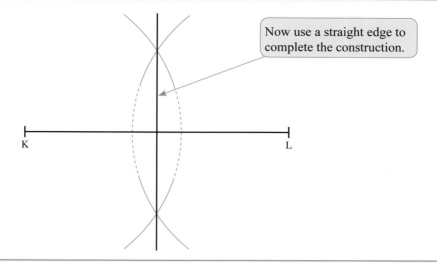

Now use a straight edge to complete the construction.

EXERCISE 13.2

1 Use ruler and compasses to construct the perpendicular bisector of the line segment PQ (PQ is 8 cm long).

2 Use ruler and compasses to construct the bisector of angle ABC.

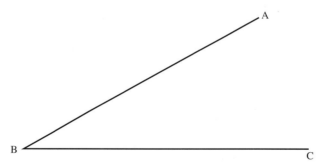

3 Use ruler and compasses to construct the bisector of angle LMN.

4 Use ruler and compasses to construct the perpendicular bisector of the line segment AB (AB is 6 cm long).

5 Using compasses and a straight edge, construct an angle of exactly 30°.
Hint: Draw a line segment about 7 or 8 centimetres long – the exact length is unimportant. Then use compasses and a straight edge to construct an equilateral triangle, using this segment as one side of the triangle. Finally, take one of the angles and bisect it.

6 Using compasses and a straight edge, construct an angle of exactly 45°.
Hint: Make a right angle (e.g. by constructing a perpendicular bisector of a line segment) and then bisect it.

13.3 Bearings

Examination questions may use **bearings** to describe direction.
Here is a reminder of how bearings are used:
- North is taken as the zero angle: 000°.
- Bearings are measured as angles clockwise from North.
- Thus East = 090°, South = 180° and West = 270°.

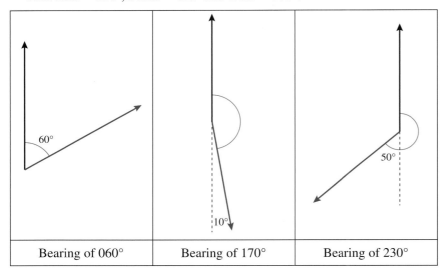

Bearing of 060°	Bearing of 170°	Bearing of 230°

EXAMPLE

a) Measure the bearing of Newtown from Oldtown.

b) Use your answer to calculate the bearing of Oldtown from Newtown.

SOLUTION

a) Step 1 Draw a North line at the place after the word 'from', that is, Oldtown.

Step 2 Draw a line linking the two places.

Step 3 Measure the angle made in a clockwise direction from the North line.

So the bearing of Newtown from Oldtown is <u>050°</u>

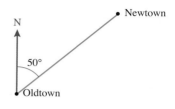

When the angle is less than 100° write a '0' in front so the bearing has three figures.

b) The question asks us to *calculate* so we are not allowed to *measure* the angle.

Step 1 Draw a North line at the place after the word 'from', that is, Newtown.

Step 2 Continue the line linking the two towns beyond Newtown.

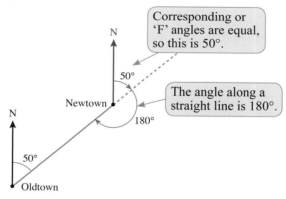

Corresponding or 'F' angles are equal, so this is 50°.

The angle along a straight line is 180°.

The total angle is 50° + 180° = 230°

So the bearing of Oldtown from Newtown is <u>230°</u>

EXERCISE 13.3

1 A ship is 14.5 km away from a lighthouse on a bearing of 230°
   a) Make a scale drawing of the ship and the lighthouse.
      Use a scale of 1 cm to 1 km

   A ferry is 10 km away from the ship on a bearing of 035°
   b) Add the position of the ferry to your scale drawing.
   c) Measure the bearing of the ferry from the lighthouse.

2 Look at this sketch – it is not drawn to scale.
   a) Write down the bearing of B from A.
   b) Work out the bearing of A from B.

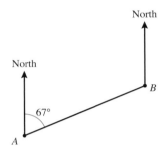

3 Look at the sketch – it is not drawn to scale.
   a) Work out the bearing of R from P.
   b) Work out the bearing of P from R.

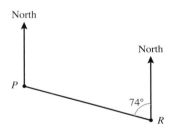

4 Otterbourne is South-west of Queensmead.
   a) What is the bearing of Otterbourne from Queensmead?
   b) What is the bearing of Queensmead from Otterbourne?

5 A helicopter and a plane are flying at the same altitude.
   The bearing of the helicopter from the plane is 147°
   What is the bearing of the plane from the helicopter?

REVIEW EXERCISE 13

1 B is 5 km North of A.
C is 4 km from B.
C is 7 km from A.

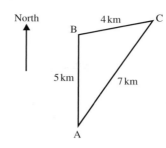

Diagram *not* accurately drawn

a) Make an accurate scale drawing of triangle ABC. Use a scale of 1 cm to 1 km.
b) From your accurate scale drawing, measure the bearing of C from A.
c) Find the bearing of A from C.   [Edexcel]

2 ABCD is a quadrilateral.
AB = 6 cm, AC = 9 cm, BC = 5 cm.
Angle BAD = 66°.
AD = 3.5 cm.
Starting with the line AB, make an accurate drawing of the quadrilateral ABCD.   [Edexcel]

3 Here is a sketch of a triangle.

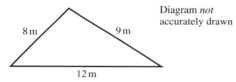

Diagram *not* accurately drawn

The lengths of the sides of the triangle are 8 m, 9 m and 12 m.
Use a scale of 1 cm to 2 m to make an accurate scale drawing of the triangle.   [Edexcel]

4 A map is drawn to a scale of 1 : 25 000
Two schools A and B are 12 centimetres apart on the map.
a) Work out the actual distance from A to B. Give your answer in kilometres.

B is due East of A. C is another school. The bearing of C from A is 064°. The bearing of C from B is 312°.
b) Copy and complete the scale drawing below.
Mark with a cross (×) the position of the school C.

N

A _____ B   [Edexcel]

5 Here is a sketch of a triangle.

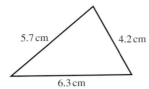

Use ruler and compasses to *construct* this triangle accurately.
You must show all construction lines.

[Edexcel]

[Edexcel]

6 The diagram shows a sketch of triangle ABC.

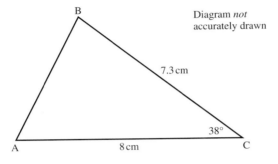

Diagram *not* accurately drawn

a) Make an accurate drawing of triangle ABC.
b) Measure the size of angle A on your diagram.

[Edexcel]

7 The diagram shows the position of each of three buildings in a town.

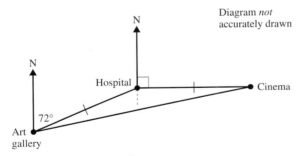

Diagram *not* accurately drawn

The bearing of the Hospital from the Art gallery is 072°.
The Cinema is due East of the Hospital.
The distance from the Hospital to the Art gallery is equal to the distance from the Hospital to the Cinema.
Work out the bearing of the Cinema from the Art gallery.

[Edexcel]

Key points

1 Accurate drawings of triangles may be made with geometrical instruments, provided you are given information about:

- Two sides and an included angle (SAS)
- Two angles and an included side (ASA)
- All three sides (SSS).

2 You can also construct a unique triangle give two sides and a non-included angle, provided the angle is a right angle. If the non-included angle is not a right angle then the information can be ambiguous, which means that two different solutions might be possible (SSA).

3 The examination may ask you to carry out standard geometrical constructions on line segments. In particular you must know how to:

- Bisect a given angle
- Bisect a given line.

4 You might be asked to use these constructions in order to make an angle of 30° (construct an equilateral triangle and then bisect one of its angles) or 45° (construct a right angle and bisect it).

5 Bearings are measured clockwise from North. Bearings are always given as three figures, for example 045° or 196°.

Internet Challenge 18

Investigating polyhedra

Polyhedra are 3-D mathematical shapes made up of a number of 2-D plane faces. Many polyhedra exhibit geometrical symmetries of various kinds. The tetrahedron and the cube are simple examples, but many more exotic polyhedra exist. Constructing models of them can be quite challenging!

Use an internet search engine, such as Google, to look for information about **Platonic solids**. Then answer these questions:

1 How many Platonic solids are known? Are we ever likely to find any more?

2 Why are they called Platonic solids?

3 Design nets for each of the Platonic solids, and trace them on to thin card. Then cut them out, and make some models for your classroom. Remember to include tabs in suitable places.

Modern footballs are assembled from a net of pentagons and hexagons.

4 a Find out how this net is made.

  b Is such a football an example of a Platonic solid?

5 Draw a table for the Platonic solids showing the number of faces, edges and vertices for each one. What rule connects the number of faces, edges and vertices?
Is the rule true for any solid?

The diagram shows a **stellated octahedron**. It is based on an octahedral core, though it also happens to look like two interlocking tetrahedra.

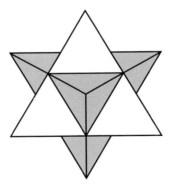

6 Design a net for, and hence build, a model of a stellated octahedron.

7 Use the internet to find examples of other stellated polyhedra.

8 How many different fully symmetric stellations of an icosahedron are known?

9 Find images of Escher's prints *Double Planet* and *Gravity*. On which polyhedra are they based?

10 Use the internet to find out about **fractal polyhedra**. You should be able to find some animated models; add the best ones to your computer's bookmarks.

CHAPTER 14

Transformation and similarity

In this chapter you will learn how to:

- carry out reflections, rotations, translations and enlargements
- use combinations of these transformations
- find missing lengths in 2-D problems using similarity
- find areas and volumes of similar shapes.

You will also be challenged to:

- investigate geometrical definitions.

Starter: **Monkey business**

Nine monkeys have fallen out of their tree. They are not all the same shape and size.

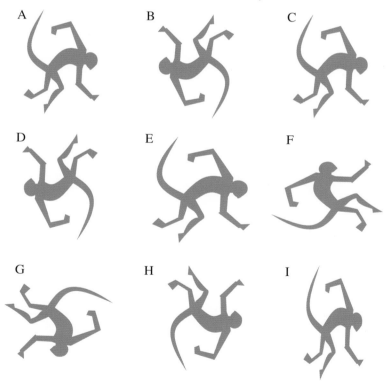

Pick out the monkeys that are not the same shape and size as the rest.

Are there any other differences?

14.1 Reflections

Many objects in mathematics possess **mirror symmetry**, or **reflection symmetry**. 2-D objects will have a mirror line, and this will divide the object into two matching halves, one being a mirror image of the other. The matching halves are **congruent**, i.e. exactly the same shape and size.

You can make a symmetric 2-D shape by reflecting a given shape in a mirror line. This is usually done using a squared coordinate grid. The mirror line may then be described by a simple linear equation. The mirror line might be horizontal (e.g. $y = 3$), vertical (e.g. $x = -2$) or at a 45° angle (e.g. $y = x$).

EXAMPLE

Reflect the given shape in the line $x = 5$

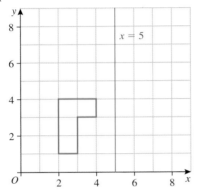

SOLUTION

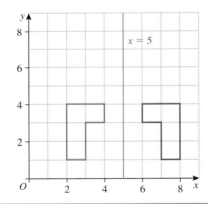

Questions with a diagonal mirror line can be more difficult to visualise. It helps if you rotate your book so that the mirror line is vertical.

EXAMPLE

The diagram shows a triangle P. The triangle has been reflected in a mirror line
to form an image Q.

a) Draw the mirror line on the diagram.

b) Write down the equation of the mirror line.

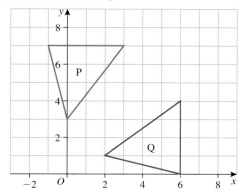

SOLUTION

a)

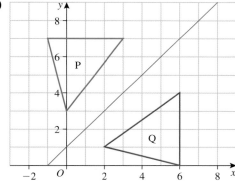

b) The mirror line has equation $y = x + 1$

EXERCISE 14.1

In questions **1** to **4**, draw the reflection of the given shape in the mirror line
indicated. Label the mirror line with its equation in each case.

1

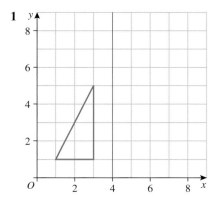

2

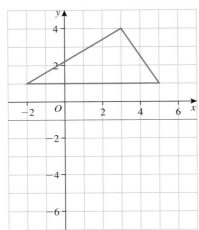

3

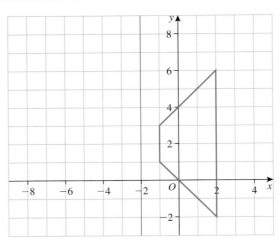

4

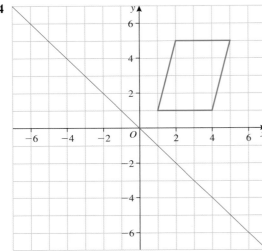

5 The diagram shows a triangle S and its image T after a reflection.

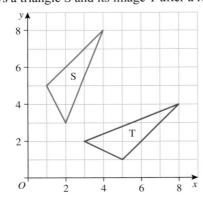

a) Draw the mirror line that has been used for the reflection.

b) Write down the equation of the mirror line.

6 The diagram shows a triangle S and its image T after a reflection.

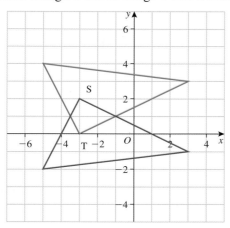

a) Draw the mirror line that has been used for the reflection.
b) Write down the equation of the mirror line.

7 The diagram shows a letter L shape, labelled X.
The shape is to be reflected in a mirror line.
Part of the reflection has been drawn on the diagram.

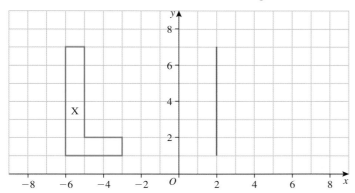

a) Complete the drawing to shown the image. Label it Y.
b) Mark the mirror line, and give its equation.

8 The diagram shows six triangles A, B, C, D, E and F. The six triangles are all congruent to each other.

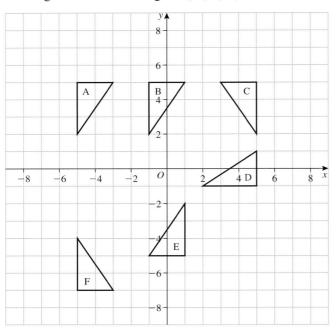

a) Explain the meaning of the word *congruent*.
b) Triangle A can be reflected to triangle F.
State the equation of the mirror line that achieves this.
c) Triangle C is reflected to another triangle using a mirror line $x = 2$.
Which one?
d) Triangle D can be reflected to triangle B using a mirror line.
Give the equation of this line.
e) Triangle D can be reflected to triangle E using a mirror line.
Give the equation of this line.

9 A triangle T is reflected in a mirror line, to form an image, triangle U.
Then triangle U is reflected in the same mirror line, to form an image, triangle V.
What can you deduce about triangle T and triangle V?

14.2 Rotations

A mathematical object may be turned to face in a different direction, while remaining the same shape and size: this is known as **rotation**. An imaginary point acts as a pivot for the rotation: this is the **centre of rotation**. You must remember to specify the size of the turn, or **angle of rotation**, and whether it is **clockwise** or **anticlockwise** as well as specifying the **centre of rotation.**

Positive and negative angles may be used instead of 'clockwise' and 'anticlockwise' to specify the direction of rotation.

• Rotate through **−270°** means rotate **270°** clockwise.

If you find rotations difficult to visualise, ask your teacher for some tracing paper. (This is also permitted in the IGCSE examination.)

EXAMPLE

The diagram shows a rectangle labelled S.

a) Rotate shape S through 90° clockwise, about the origin O. Label the resulting shape T.

b) Now rotate the shape T through 180° about O. Label the resulting shape U.

c) Describe a single rotation that would take S directly to U.

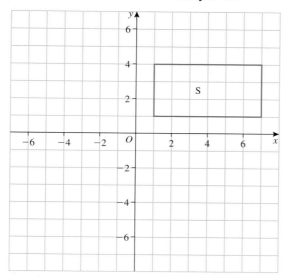

SOLUTION

a)

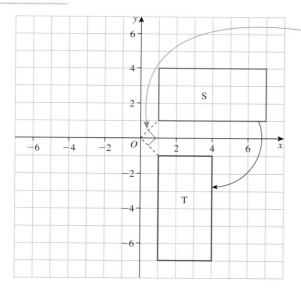

Imagine a ray from the centre (O) to a corner of the shape S.

Rotate this ray to find the new position for the corner.

Repeat as necessary.

b)

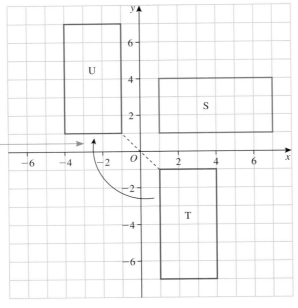

The direction of this second rotation was not specified in the question because 180° clockwise and 180° anticlockwise are exactly the same.

c) U can be obtained directly from S by a <u>90° rotation anticlockwise about *O*</u>.

Rotations are often performed with the point (0, 0), called the origin *O*, as the centre of rotation, but they can be done about other centres.

EXAMPLE

The diagram shows a triangle M drawn on a grid.

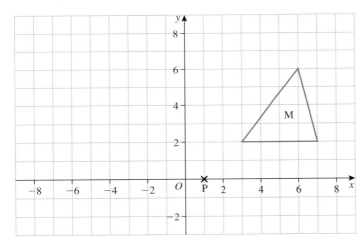

Rotate the triangle M through 90° anticlockwise about the point P (1, 0). Label this new triangle N.

SOLUTION

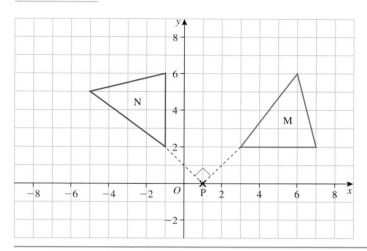

EXERCISE 14.2

Each of these questions requires a coordinate grid in which *x* and *y* can range from −8 to 8.

1 Rotate the trapezium shape 90° clockwise, about *O*.

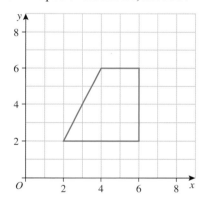

2 Rotate the shape 180°, about *O*.

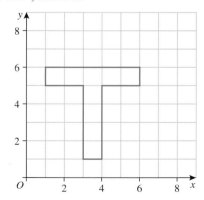

255

3 a) Rotate the triangle T1 90° anticlockwise about *O*. Label the result T2.
  b) Rotate T2 180° about *O*. Label the result T3.
  c) Describe the single rotation that takes T1 directly to T3.

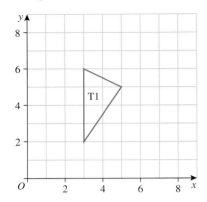

4 a) Rotate shape A 90° anticlockwise about (1, 0). Label the result B.
  b) Rotate shape B 180° about (0, 0). Label the result C.
  c) Describe carefully the single rotation that takes shape C to shape A.

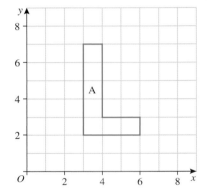

5 a) Rotate shape U 90° anticlockwise about point P (0, 1).
  b) Rotate shape V 90° clockwise about point Q (−1, −1).

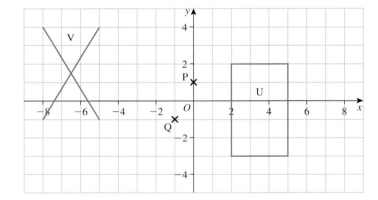

6 a) Rotate the triangle −90° about (1, 1).

b) Now rotate both the new triangle and the original one 180° about (1, 1).

−90° means 90° clockwise.

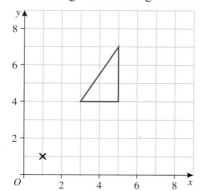

7 The diagram shows an object A and its image B after a rotation.

a) Write down the size and direction of the angle of rotation.

b) Write down the coordinates of the centre of rotation.

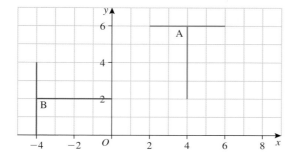

8

If you rotate a shape and then rotate it again, the result is equivalent to a single rotation.

If you reflect a shape and then reflect it again, the result is equivalent to a single reflection.

I'm afraid only one of you is right.

Anita   Bella   Cat

Who is right, and who is wrong?

14.3 Combining transformations

There are three important geometric transformations that preserve congruence, meaning that they do not change the shape or size of an object. These transformations are reflection, rotation and translation.

A **translation** consists simply of sliding an object left/right and/or up/down. You specify a translation by stating how far the object is to be moved in each of the x and y directions, and it can be written as two numbers in a column vector.

For example, $\begin{pmatrix} 5 \\ 2 \end{pmatrix}$ indicates a translation of 5 units to the right and two units up.

$\begin{pmatrix} -3 \\ 0 \end{pmatrix}$ indicates a translation of 3 units left and 0 units up/down

$\begin{pmatrix} -2 \\ -1 \end{pmatrix}$ indicates a translation of 2 units left and 1 units down

In the IGCSE examination you may be required to combine two transformations. Transformations are said to **map** one shape to another, that is, turn one shape into the other.

EXAMPLE

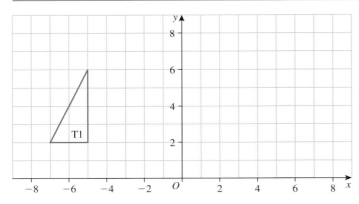

a) Reflect the given triangle T1 in the line $x = -4$, and label the result T2.
b) Reflect T2 in the line $x = 1$, and label the result T3.
c) What single transformation maps T1 directly to T3?

SOLUTION

a)

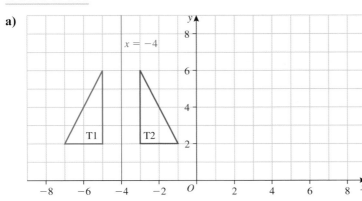

b)

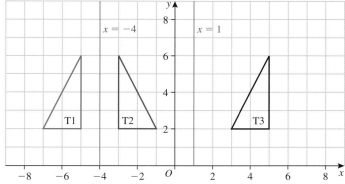

c) T3 is 10 units to the right of T1, so the transformation that maps T1 to T3 is a translation of $\begin{pmatrix} 10 \\ 0 \end{pmatrix}$.

EXERCISE 14.3

1 a) Translate triangle A by the vector $\begin{pmatrix} -4 \\ -2 \end{pmatrix}$
Label the image B.

b) Translate triangle B by the vector $\begin{pmatrix} 0 \\ -5 \end{pmatrix}$
Label the image C.

c) Translate triangle C by the vector $\begin{pmatrix} 4 \\ 0 \end{pmatrix}$
Label the image D.

d) Describe the single translation that would map triangle D onto triangle A.

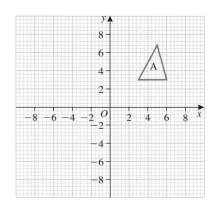

Each of questions **2** to **6** requires a coordinate grid in which x and y can range from -8 to 8.

2

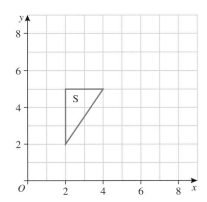

a) Reflect triangle S in the line $x = -1$. Label the new triangle T.
b) Reflect triangle T in the x axis. Label the new triangle U.
c) Describe the *single* transformation that maps S to U.

3 The diagram shows a triangle, T.

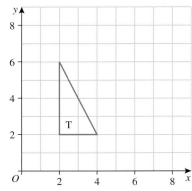

a) Translate triangle T by $\begin{pmatrix} -6 \\ 0 \end{pmatrix}$. Label its image triangle U.

b) Rotate triangle U by 180° about O. Label the result triangle V.

c) Describe the *single* transformation that maps T to V.

4 The diagram shows a triangle, S.

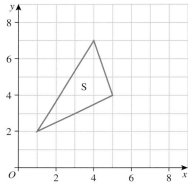

a) Reflect triangle S in the *y* axis. Label this image triangle T.

b) Reflect triangle S in the line *y* = 1. Label this image triangle U.

c) Describe the *single* transformation that maps T directly to U.

5 The diagram shows a set of points that make a letter F shape. The shape is labelled F1.

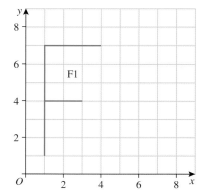

a) Reflect the shape F1 in the *x* axis. Label the result F2.

b) Reflect F2 in the line *y* = *x*. Label the result F3.

c) Describe the single transformation that would take shape F3 to shape F1.

6 The diagram shows a quadrilateral F.

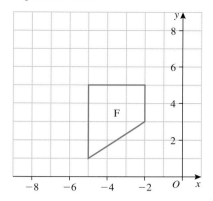

a) Rotate quadrilateral F through 90° anticlockwise about *O*. Label the result G.
b) Rotate quadrilateral G through 90° clockwise about (4, −4). Label the result H.
c) Describe a single transformation that would take F to H.

7 The diagram shows a triangle A.

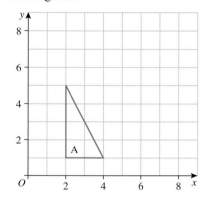

a) Rotate triangle A 90° anticlockwise about (0, 3). Label this image B.
b) Rotate triangle A 180° about the origin *O*. Label this image C.
c) Describe the *single* transformation that transforms triangle B to triangle C.

14.4 Enlargements

You should already be familiar with the idea of enlarging shapes on grids. This should be done in a specific way, using a **centre of enlargement** and a **scale factor**. Rays may be drawn from the centre of enlargement, to show how the transformation is operating.

There are two different scenarios, depending on the value of the scale factor:

- Scale factor *greater than 1*:   simple enlargement – the object gets bigger.
- Scale factor *between 0 and 1*:   the enlargement is a reduction – the object gets smaller.

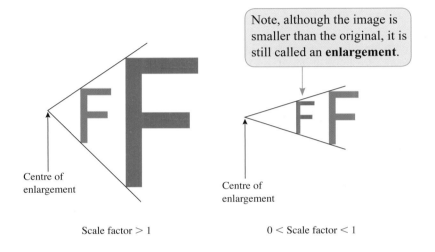

Note, although the image is smaller than the original, it is still called an **enlargement**.

Centre of enlargement

Centre of enlargement

Scale factor > 1

0 < Scale factor < 1

EXAMPLE

The diagram shows a letter F shape and two points, P and Q.

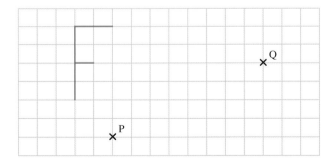

a) Enlarge the letter F by scale factor 2, using P as the centre of enlargement.
b) Enlarge the letter F by scale factor $\frac{1}{2}$, using Q as the centre of enlargement.

SOLUTION

a)

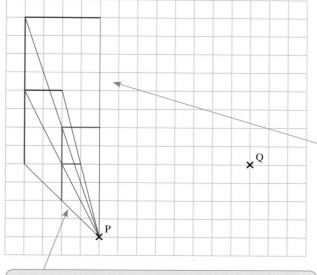

The final F is not the same size as the original, so the two shapes are not congruent, but they are the same shape. The two shapes are said to be mathematically similar.

Draw rays from P to each corner of the original F shape. Then extend these rays so they are twice their original length (factor is ×2).

The rays will locate the corners of the enlarged shape.

b)

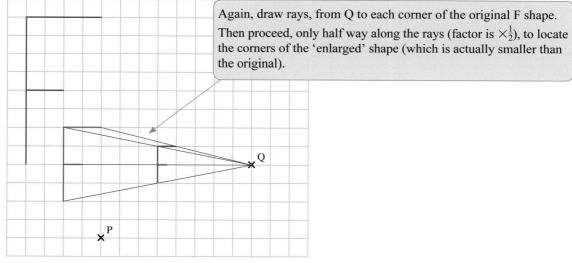

Again, draw rays, from Q to each corner of the original F shape.

Then proceed, only half way along the rays (factor is $\times\frac{1}{2}$), to locate the corners of the 'enlarged' shape (which is actually smaller than the original).

EXERCISE 14.4

1   The diagram shows a shape A.

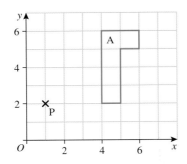

a)   Make a copy of this diagram on squared paper, in which x can run from 0 to 20 and y from 0 to 15.

b)   Enlarge shape A by scale factor 2, centre P. Label the new shape B.

c)   Enlarge shape A by scale factor 3, centre P. Label the new shape C.

d)   Are shapes B and C congruent? Are they similar?

2   The diagram shows a triangle, and a centre of enlargement, P.

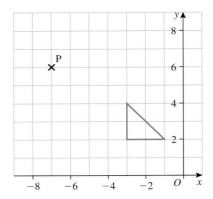

a)   Make a copy of this diagram on squared paper, in which x can run from −5 to 15 and y from −10 to 10.

b)   Enlarge the shape by scale factor $2\frac{1}{2}$, centre P.

3 The diagram shows a shape, A, and two centres P and Q marked with crosses.
   a) Make a copy of this diagram on squared paper, in which x can run from -5 to 10 and y from 0 to 10.
   b) Enlarge shape A, with scale factor 2, centre P. Label the result B.
   c) Enlarge shape B, with scale factor $\frac{1}{2}$, centre Q. Label the result C.
   d) Are shapes A and B congruent? Are they similar?
   e) Are shapes A and C congruent?

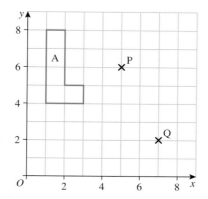

4 The diagram shows an object, A, and its image B after an enlargement.

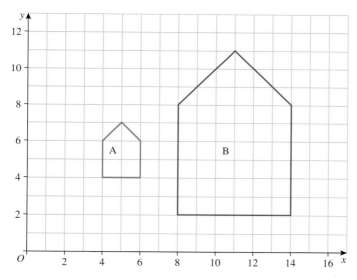

   a) State the scale factor for the enlargement.
   b) Obtain the coordinates of the centre of enlargement.

14.5 Similar shapes and solids

If two objects are similar, then they are exactly the same shape, but one of them is an enlargement of the other. If the scale factor is n, then:

- corresponding lengths are in the ratio $1 : n$
- corresponding areas are in the ratio $1 : n^2$
- corresponding volumes are in the ratio $1 : n^3$.

EXAMPLE

In the diagram, AB and CD are parallel.
AB = 6 cm, CD = 10 cm, AE = 3.6 cm and CE = 7 cm.
a) Explain carefully why triangles AEB and DEC are similar.
b) Calculate the length BE.
c) Work out the length DE.

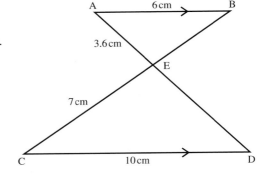

SOLUTION

a) Angles ABE and DCE are equal (alternate angles).
   Angles BAE and CDE are equal (alternate angles).
   Angles AEB and DEC are equal (vertically opposite).
   Thus <u>both triangles contain exactly the same angles,</u> so
   they must be similar.
b) Redrawing the similar triangles so that they are the same way up:

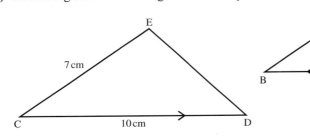

Then, by comparing corresponding sides:
$$\frac{BE}{7} = \frac{6}{10}$$
Thus, cross-multiplying:
$$10 \times BE = 6 \times 7$$
$$10 \times BE = 42$$
$$BE = \frac{42}{10}$$
$$BE = \underline{4.2 \text{ cm}}$$

c) Likewise:
$$\frac{DE}{3.6} = \frac{10}{6}$$
Thus, cross-multiplying:
$$6 \times DE = 10 \times 3.6$$
$$6 \times DE = 36$$
$$DE = \frac{36}{6}$$
$$DE = \underline{6 \text{ cm}}$$

EXAMPLE

The diagram shows two solid cones. They are mathematically similar.

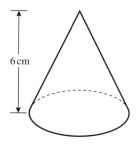
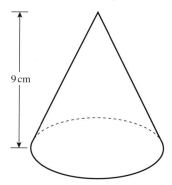

The smaller cone has a curved surface area of 64 cm².
a) Work out the curved surface area of the larger cone.

The two cones are made of the same material. The larger cone has a mass of
1080 grams.
b) Work out the mass of the smaller cone.

SOLUTION

The scale factor is $9 \div 6 = 1.5$
a) Area of larger cone $= 64 \times (1.5^2)$
$$= \underline{144 \text{ cm}^2}$$
b) Mass of smaller cone $= 1080 \div (1.5^3)$
$$= \underline{320 \text{ grams}}$$

EXERCISE 14.5

1 The diagram shows two rectangles. They are mathematically similar.

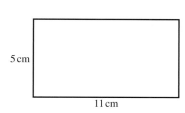

a) Work out the length of the larger rectangle.
b) Work out the ratio of the perimeters of the rectangles, in the form $1 : n$.
c) Find the ratio of the areas of the rectangles.

2 The diagram shows two similar triangles. The smaller triangle has an area of 24 cm^2.

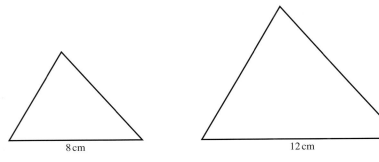

8 cm   12 cm

Work out the area of the larger triangle.

3 The diagram shows two solid cylinders. They are similar. Both cylinders are made of the same material.

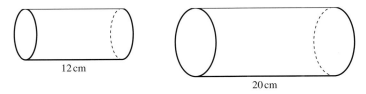

12 cm   20 cm

The larger cylinder has a mass of 40 kg. Work out the mass of the smaller cylinder.

4 The diagram shows five points, P, Q, R, S, T. The line segments PQ and RS are parallel.

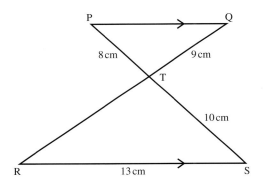

a) Work out the length RT. Hence find the length RQ.
b) Calculate the length PQ.

5 HMS *Cumberland* is a Type 22 frigate.
This ship has a length of 148 metres and a mass of 5300 tonnes.
A marine architect is thinking of designing a larger version of HMS *Cumberland*.
The new ship would be mathematically similar to the original one, but 25% larger in all dimensions.
a) Calculate the length of the new ship design.

The new ship is to be built using the same materials as the original one.
b) Calculate the mass of the new ship.

6 The diagram shows five points, J, K, L, M and N. The line segments JK and MN are parallel.

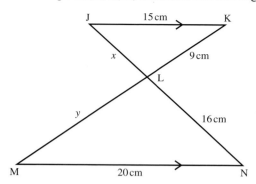

Calculate the lengths *x* and *y*.

7 A garden centre sells two similar statues.
The smaller one is 30 cm tall and weighs 5.5 kg.
The larger one is 40 cm tall. Work out its weight.

8 The diagram shows five points, P, Q, R, S and T. The line segments PQ and ST are parallel.
PR = 8 cm, PS = 4 cm, QR = 10 cm, ST = 18 cm.

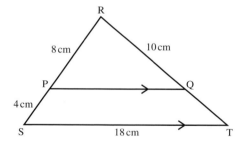

a) Explain fully why triangles RPQ and RST are similar.
b) Work out the length PQ.
c) Work out the length RT.

9 A model aircraft is $\frac{1}{4}$ of full size.
a) The real aircraft is 6.56 metres long. How long is the model?
b) The model has a wing area of 0.925 m². Find the wing area of the real aircraft.

10 Two chocolate bars are mathematically similar. They weigh 250 grams and 500 grams respectively.
The 250 gram bar is 12 cm long. Calculate the length of the 500 gram bar.

REVIEW EXERCISE 14

1 Cylinder **A** and cylinder **B** are mathematically similar.

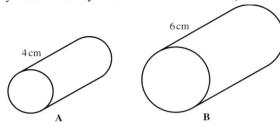

Diagram *not* accurately drawn

The length of cylinder **A** is 4 cm and the length of cylinder **B** is 6 cm.
The volume of cylinder **A** is 80 cm³.
Calculate the volume of cylinder **B**.   [Edexcel]

2

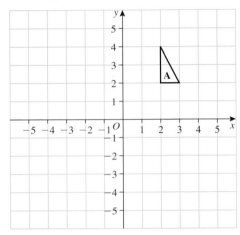

a) On a copy of the grid, rotate triangle **A** 180° about *O*. Label your new triangle **B**.
b) On the grid, enlarge triangle **A** by scale factor $\frac{1}{2}$, centre *O*. Label your new triangle **C**.   [Edexcel]

3 Shape A is enlarged by scale factor 2 to obtain shape B. Shape B is then enlarged by scale factor 3 to obtain shape C. State the single scale factor that would transform shape A to shape C.

4 Two cuboids are mathematically similar. The smaller one has a shortest edge of 5 cm, and its surface area is 400 cm². The larger one has a shortest edge of 8 cm.
Find the surface area of the larger cuboid.

5 Enlarge the shaded triangle by a scale factor $1\frac{1}{2}$, centre P.

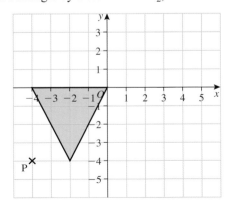

[Edexcel]

6

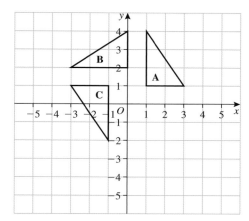

Shape **A** is rotated 90° anticlockwise, centre (0, 1), to shape **B**.
Shape **B** is rotated 90° anticlockwise, centre (0, 1), to shape **C**.
Shape **C** is rotated 90° anticlockwise, centre (0, 1), to shape **D**.
a) Mark the position of shape **D**.
b) Describe the single transformation that maps shape **C** to shape **A**. [Edexcel]

7 Triangle **B** is a reflection of triangle **A**.

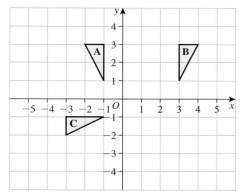

a) **(i)** On a copy of the grid, draw the mirror line for this reflection.
(ii) Write down the equation of the mirror line.
b) Describe fully the single transformation that maps triangle **A** onto triangle **C**. [Edexcel]

8

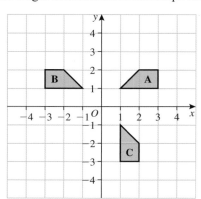

a) Describe fully the single transformation which maps shape **A** onto shape **B**.
b) Describe fully the single transformation which maps shape **A** onto shape **C**. [Edexcel]

271

9

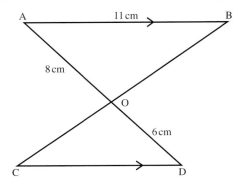

Diagram *not* accurately drawn

AB is parallel to CD.
The lines AD and BC intersect at point O.
AB = 11 cm, AO = 8 cm, OD = 6 cm.
Calculate the length of CD.

[Edexcel]

10

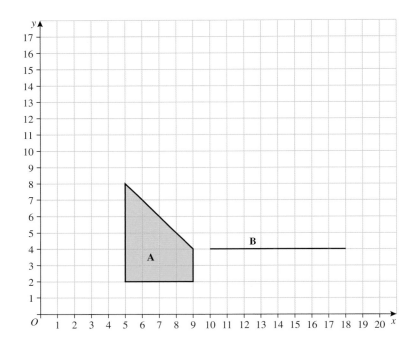

Shape **A** is shown on the grid. Shape **A** is enlarged, centre (0, 0), to obtain shape **B**.
One side of shape **B** has been drawn for you.
a) Write down the scale factor of the enlargement.
b) On a copy of the grid, complete shape **B**.

The shape **A** is enlarged by scale factor $\frac{1}{2}$, centre (5, 16) to give the shape **C**.
c) On the grid, draw shape **C**.

[Edexcel]

11 Martin and Nina have made sandcastles on the beach. Martin's is exactly the same shape as Nina's, but is larger in each dimension. Nina's sandcastle is 24 cm high. It took 12 buckets of sand to make. Martin's sandcastle is 30 cm high.
Work out the number of buckets of sand that Martin needed to make his sandcastle.

12

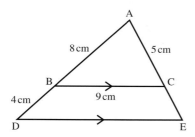

In the triangle ADE, BC is parallel to DE.
AB = 8 cm, AC = 5 cm, BD = 4 cm, BC = 9 cm.
a) Work out the length of DE.
b) Work out the length of CE. [Edexcel]

13 A sheet of drawing paper is mathematically similar to a sheet of A5 paper. A sheet of A5 paper is a rectangle 210 mm long and 148 mm wide. The sheet of drawing paper is 450 mm long.
Calculate the width of the sheet of drawing paper. Give your answer correct to 3 significant figures. [Edexcel]

14

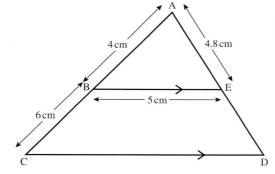

Diagram *not* accurately drawn

BE is parallel to CD. ABC and AED are straight lines.
AB = 4 cm, BC = 6 cm, BE = 5 cm, AE = 4.8 cm.
a) Calculate the length of CD.
b) Calculate the length of ED. [Edexcel]

Key points

1 A reflection is specified by a mirror line.

2 A rotation is specified by a centre of rotation, an angle of rotation, and a direction (clockwise or anticlockwise.)

3 A translation can be expressed in vector form, e.g. $\begin{pmatrix} 2 \\ 3 \end{pmatrix}$ means 2 to the right and 3 up.

4 Two shapes are congruent if they are exactly the same shape and size. Reflections, rotations and translations all preserve congruence.

5 An enlargement is specified by a centre of enlargement and a scale factor. Scale factors larger than 1 actually make the image larger, while scale factors between 0 and 1 cause the image to be reduced so it is smaller in size than the original.

6 Enlargements do not normally preserve congruence. The object and its image will, however, be mathematically similar, i.e. the same shape.

7 When solid objects are enlarged by a scale factor, their perimeters increase by the same ratio. Areas increase according to the square of the scale factor, and volumes by its cube.

Internet Challenge 14

Geometrical definitions

Mathematicians like to attach precise meanings to certain words – these are **definitions**. In the sentences below the letters of the key words have been replaced with ☐ symbols. Find the missing word in each case. (You will know some of these already, but you may need to look up some of the less well-known ones on the internet.)

1 An ☐☐☐☐☐☐☐☐☐☐☐ is a mathematical solid with 20 faces.

2 A ☐☐☐☐☐☐☐☐ is the name for a circular prism.

3 If two objects are the same shape and size they are said to be ☐☐☐☐☐☐☐☐☐.

4 If two shapes are alike in shape but one is larger than the other, they are said to be mathematically ☐☐☐☐☐☐☐.

5 Z-angles are, more properly, called ☐☐☐☐☐☐☐☐☐ angles.

6 ☐☐☐☐☐☐☐☐ lines never touch; they remain at a constant distance apart.

7 A ☐☐☐☐☐ is a solid object in the form of a perforated ring (like a ring doughnut).

8 The interior angles of an ☐☐☐☐☐☐☐ add up to 1080°.

9 A ☐☐☐☐☐☐☐☐☐☐ is exactly half of a sphere.

10 The highest point of a pyramid is known as its ☐☐☐☐.

11 A pyramid with a triangular base is called a ☐☐☐☐☐☐☐☐☐☐☐.

12 ☐☐☐☐☐☐☐ is the correct mathematical name for a 'diamond' with four equal sides.

13 An angle of one-sixtieth of a degree is called a ☐☐☐☐☐☐ of ☐☐☐.

14 An angle of 57.296° is called one ☐☐☐☐☐☐.

15 The diagram below shows a ☐☐☐☐☐☐☐☐☐ cone. This is also a ☐☐☐☐☐☐☐.

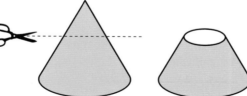

CHAPTER 15

Pythagoras' theorem

In this chapter you will learn how to:

- use Pythagoras' theorem to test whether triangles are right-angled
- use Pythagoras' theorem to find an unknown side in a right-angled triangle
- use Pythagoras' theorem to solve simple three-dimensional problems
- use Pythagoras' theorem to find the distance between two points on a grid.

You will also be challenged to:

- investigate Pythagorean triples.

Starter: Finding squares and square roots on your calculator

When you multiply a number by itself, you are finding its **square**. For example, 3 squared is 9, because $3 \times 3 = 9$. This is usually written $3^2 = 9$. To find the square of 3.1, you would probably prefer to use a calculator:

3.1 $\boxed{x^2}$ = 9.61

The reverse process of squaring is called **square rooting**, or finding the **square root**. This is much harder than squaring, and usually requires the use of a calculator equipped with a square root $\boxed{\sqrt{}}$ key. You may find your calculator screen fills with decimal figures; if so, it is usual to round the answer to 3 significant figures.

Task 1
Look at these numbers. Work out the square of each one. Several of them can be done without a calculator, but you may use a calculator for the harder ones.

  4   7   2.5   1.2   0.8   13   6   16

Task 2
Look at these numbers. Work out the square root of each one. If the answers are not exact then you should round to 3 significant figures.

  13   10   16   22.5   6.25   49   120   121

Task 3
Use your calculator to find the square roots of these numbers, to 3 significant figures where necessary:

  8   9   10   11   12   13   14   15   16   17

Why is it possible to find the square roots of some whole numbers without the need for a calculator?

15.1 Introducing Pythagoras' theorem

Pythagoras' theorem concerns right-angled triangles. Suppose you have a right-angled triangle with sides of lengths a, b and c, with c being the longest side, or **hypotenuse**. Then Pythagoras' theorem states that:

$$c^2 = a^2 + b^2$$

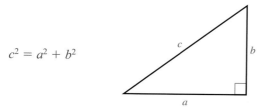

For example, if $a = 4$ cm and $b = 3$ cm then c would be 5 cm, since $5^2 = 25$, $4^2 = 16$, $3^2 = 9$ and $25 = 16 + 9$

$$5^2 = 3^2 + 4^2$$

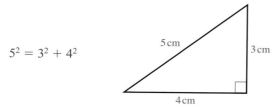

There are many ways to prove Pythagoras' theorem. In fact, Pythagoras' theorem works only in right-angled triangles, so it may be used to check whether a triangle is right angled or not.

In the examples and exercises that follow, capital letters will be used for the vertices of a triangle, such as ABC or PQR. The simplest way of naming an angle is just to use the capital letter of the point at angle – angle A for example. You refer to sides by using two capital letters – the side joining points A and B is written as AB.

Regardless of the letters used for naming corners of the triangle, when substituting into Pythagoras' theorem you may find it convenient to use c for the hypotenuse, and a and b for the other two sides.

EXAMPLE

Use Pythagoras' theorem to check whether each of these triangles is right angled or not.

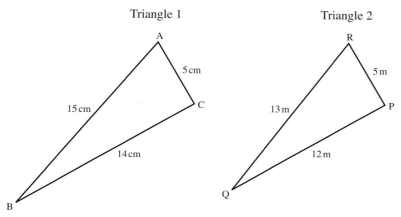

SOLUTION

In triangle 1, the longest side is 15 cm, so try $c = 15$, $a = 14$, $b = 5$

Then $c^2 = 225$ and $a^2 + b^2 = 14^2 + 5^2$
$$= 196 + 25$$
$$= 221$$

Since 225   221, triangle 1 cannot be right angled.

In triangle 2, the longest side is 13 m, so try $c = 13$, $a = 12$, $b = 5$

Then $c^2 = 169$ and $a^2 + b^2 = 12^2 + 5^2$
$$= 144 + 25$$
$$= 169$$

Since c^2 and $a^2 + b^2$ are equal ($= 169$), triangle 2 must be right angled.

EXERCISE 15.1

Look at these triangles, and use Pythagoras' theorem to decide whether they are right angled or not.
Note: The diagrams are not drawn to scale.

1 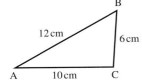 **2** 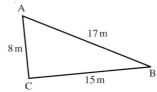 **3**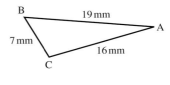

For each of the triangles described below, use Pythagoras' theorem to decide whether it is right angled.
If so, name the angle at which the right angle is located.

4 $AB = 8$ cm, $BC = 6$ cm, $CA = 2.5$ cm

5 $AB = 7.5$ cm, $BC = 4.5$ cm, $CA = 6$ cm

6 $AB = 12$ mm, $BC = 12$ mm, $CA = 5$ mm

7 $PQ = 10.1$ cm, $QR = 7.1$ cm, $RP = 7.2$ cm

8 $PQ = 12$ m, $QR = 16$ m, $RP = 20$ m

9 $PQ = 3.3$ cm, $QR = 5.8$ cm, $RP = 4.5$ cm

10 $AB = 6$ km, $BC = 7$ km, $CA = 8$ km

15.2 Using Pythagoras' theorem to find a hypotenuse

In this section, we shall be working with triangles that are known to be right angled, and will use Pythagoras' theorem to find the hypotenuse. This is the longest side, and is always located directly opposite the right angle.

EXAMPLE

Calculate the length of the side AB, marked x, in the triangle below.

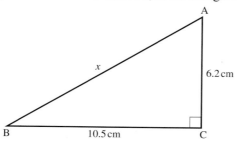

SOLUTION

Pythagoras tells us that:

$x^2 = 6.2^2 + 10.5^2$

$= 38.44 + 110.25$

$= 148.69$

Therefore: $x = \sqrt{148.49}$

$= 12.193\ 850\ 91$

$= \underline{12.2\ \text{cm}}$ (3 s.f.)

> Set out the details of the working clearly…

> … and show your full calculator result…

> … before finally rounding off to 3 significant figures.

EXERCISE 15.2

Find the length of the hypotenuse represented by the letters a to i below. Give your answers to 3 significant figures where appropriate.

1
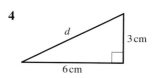

2
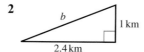

3
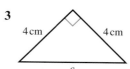

4
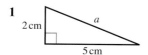

5
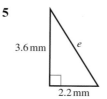

6
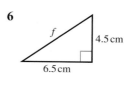

7
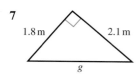

8
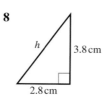

9
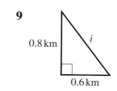

Find the length of the diagonal of each rectangle.

10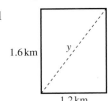

5 cm

6 cm

11

1.6 km

1.2 km

12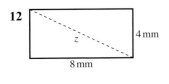

4 mm

8 mm

15.3 Using Pythagoras' theorem to find one of the shorter sides

The method used above may be adapted when the unknown side is not the hypotenuse. In this case, the calculation requires a *subtraction* instead of an *addition*, as shown in the example below.

EXAMPLE

Find the value of y in the right-angled triangle below.

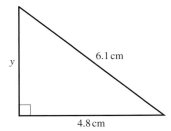

y

6.1 cm

4.8 cm

SOLUTION

By Pythagoras:

$6.1^2 = 4.8^2 + y^2$

Begin by writing Pythagoras' theorem in full. (You could start with the rearranged version.)

Rearranging:

$y^2 = 6.1^2 - 4.8^2$

$= 37.21 - 23.04$

$= 14.17$

Therefore: $y = \sqrt{14.17}$

$= 3.764\ 306\ 045$

$= \underline{3.76\ \text{cm}\ (3\ \text{s.f.})}$

Your numerical answer should be shown:
- as an exact statement $(\sqrt{14.17})$
- then the full calculator result
- and, finally, the rounded value.

Remember to include units (cm) in your answer.

EXERCISE 15.3

Find the length of the side marked by the letters *a* to *i* below.
Give your answers to 3 significant figures. where appropriate.

1

2
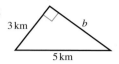

3
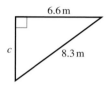

4

5
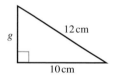

6
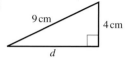

7
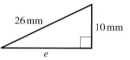

8
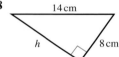

9
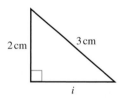

10 A rectangle has length 24 cm and width *x* cm.
Its diagonal is of length 25 cm.
Find the value of *x*.

11 A ship sails due North for 12 km, then turns and sails due East for *y* km.
It ends up 16 km in a direct straight line from its start point.
Find the value of *y*.

The last part of this exercise contains a mixture of questions. Remember to square and add when you are finding a hypotenuse, but square and subtract when finding a shorter side. In both cases, remember to square root at the end.

12 Find x, correct to 3 significant figures.

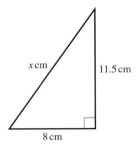

8 cm

x cm

11.5 cm

13 Find y, correct to 3 significant figures.

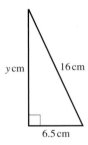

y cm

16 cm

6.5 cm

14 Find z, correct to 3 significant figures.

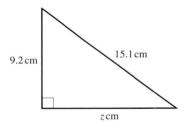

9.2 cm

15.1 cm

z cm

15 Find s, correct to 3 significant figures.

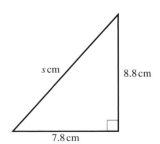

s cm

8.8 cm

7.8 cm

282

15.4 Pythagoras' theorem in three dimensions

EXAMPLE

A room is in the shape of a cuboid measuring 5 m by 7 m by 2.5 m. A string is stretched diagonally across the room, from bottom corner B to the opposite top corner G. Find the length of the string, correct to 3 significant figures.

SOLUTION

To find the length of the string BG we use Pythagoras' theorem twice.

First, in triangle BCD:

$BD^2 = 5^2 + 7^2$

$\quad\ = 25 + 49$

$\quad\ = 74$

Thus $BD = \sqrt{74}$

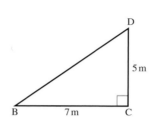

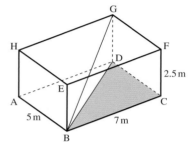

Now in triangle BDG:

$BG^2 = BD^2 + DG^2$

$\quad\ = (\sqrt{74})^2 + 2.5^2$

$\quad\ = 74 + 6.25$

$\quad\ = 80.25$

So $BG = \sqrt{80.25}$

$\quad\quad\ = 8.958\ 2...$

$\quad\quad\ = \underline{8.96\ \text{m}}$ (3 s.f.)

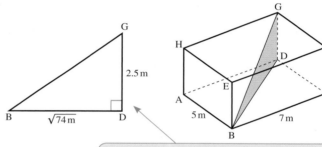

Always make a sketch of the 2-D triangle you have extracted from the 3-D solid object.

EXERCISE 15.4

1   The diagram shows a box in the shape of a cuboid.

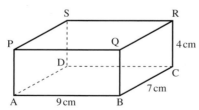

a)   Work out the length AC. Give your answer to 3 significant figures.
b)   Work out the length AR. Give your answer to 3 significant figures.

2   The diagram shows a box in the shape of a cuboid.

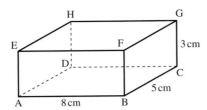

a) Calculate the length BD. Give your answer to 3 significant figures.
b) Calculate the length BH. Give your answer to 3 significant figures.

3   A cuboid measures 4 cm by 10 cm by 12 cm.
a) Make a sketch of the cuboid.
b) Calculate the length of the diagonal, giving your answer correct to 3 significant figures.

4   A postal carton measures 10 cm by 14 cm by 20 cm. Ray wishes to pack a thin brass rod of length 25 cm inside the carton. Use Pythagoras' theorem to explain whether this is possible or not.

5   The diagram shows a wedge. The face ABED is a rectangle, and is at right angles to the face CBEF, which is also a rectangle. AB = 10 cm, BC = 4 cm, BE = 18 cm.

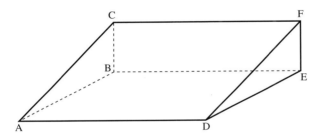

a) Calculate the length AE, correct to 3 significant figures.
b) Calculate the direct distance from A to F, correct to 3 significant figures.

6   A thin straw of length 20.5 cm just fits inside a cylindrical container of length 20 cm.

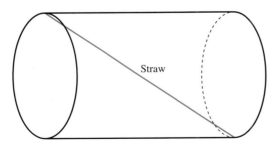

Find the diameter of the cylinder.

7   Find the length of the longest thin rod that will just fit inside a cuboid-shaped box with dimensions 3 cm by 4 cm by 12 cm.

15.5 Pythagoras' theorem on a coordinate grid

Suppose you want to calculate the distance between two points plotted on a coordinate grid. One way of doing this is to use Pythagoras' theorem.

EXAMPLE

Find the distance between the points A $(-2, 1)$ and B $(10, 6)$.

SOLUTION

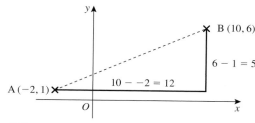

The difference between the x coordinates is $10 - (-2) = 12$.

The difference between the y coordinates is $6 - 1 = 5$.

By Pythagoras' theorem:

$$AB^2 = 12^2 + 5^2$$
$$= 144 + 25$$
$$= 169$$
$$AB = \sqrt{169}$$
$$= \underline{13}$$

It is not necessary to draw a diagram for every problem of this type. You could simply use the formula:

$$\text{Distance} = \sqrt{(\text{difference between } x \text{ coordinates})^2 + (\text{difference between } y \text{ coordinates})^2}$$

Generalising, if (x_1, y_1) and (x_2, y_2) are the two coordinates, the formula is written as:

$$\text{Distance} = \sqrt{(x_2 - x_1)^2 + (y_2 - y_1)^2}$$

EXAMPLE

A triangle has vertices A $(2, 7)$, B $(7, -3)$ and C $(-8, 2)$.

a) Find the lengths of:
   (i) AB   **(ii)** BC   **(iii)** AC
   Give your answers as exact square roots.

b) What kind of triangle is ABC?

SOLUTION

a) (i) $AB = \sqrt{(7-2)^2 + (-3-7)^2} = \sqrt{25 + 100} = \sqrt{125}$

(ii) $BC = \sqrt{(-8-7)^2 + (2--3)^2} = \sqrt{225 + 25} = \sqrt{250}$

(iii) $AC = \sqrt{(-8-2)^2 + (2-7)^2} = \sqrt{100 + 25} = \sqrt{125}$

b) These results show that AB = AC but that BC has a different length.
Therefore the triangle ABC is <u>isosceles</u>.

EXERCISE 15.5

1 Use Pythagoras' theorem to calculate the distance from:
   a) A (4, 1) to B (1, 5)   **b)** P (−5, 5) to Q (3, 20)   **c)** M (−2, 1) to N (6, −3)

2 A triangle ABC has vertices A (−2, −2), B (4, −1) and C (1, 3).
   Dee makes a sketch of the triangle. Dee says that the triangle is isosceles.
   a) Use Pythagoras' theorem to find the lengths of AB, BC and CA.
        Give your answers as exact square roots.
   b) Use your answers to decide whether Dee is right or wrong.

3 A quadrilateral PQRS has vertices P (3, 1), Q (7, 2), R (8, 6) and S (4, 5).
   a) Make a rough sketch of the quadrilateral.
   b) Use Pythagoras' theorem to find the length of the sides PQ, QR, RS and SP.
        Give your answers as exact square roots.
   c) What type of quadrilateral is PQRS?

REVIEW EXERCISE 15

Find the missing lengths, denoted by the letters *a* to *f*.
Round your answers to 3 significant figures where appropriate.

1
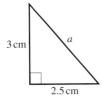
3 cm, *a*, 2.5 cm

2

8.7 cm, *b*, 13.2 cm

3
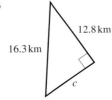
12.8 km, 16.3 km, *c*

4
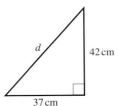
d, 42 cm, 37 cm

5
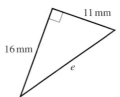
11 mm, 16 mm, *e*

6
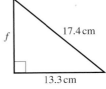
f, 17.4 cm, 13.3 cm

7 The diagram shows two connected right-angled triangles.

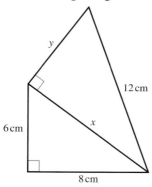

a) Write down the exact value of x, without using a calculator.

b) Use your calculator to find the value of y, giving your answer correct to 2 significant figures.

8 ABCDEFGH is a cuboid, with AD = 2.5 cm, DC = 6.5 cm and CG = 4.5 cm.

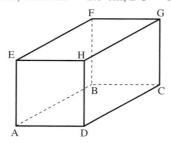

a) Calculate AH, CH and FH, each correct to 3 significant figures.

b) Calculate the distance BH, correct to 3 significant figures.

9 XYZ is a right-angled triangle. XY = 3.2 cm. XZ = 1.7 cm.

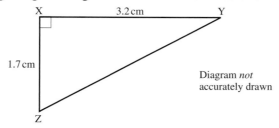

Diagram *not* accurately drawn

Calculate the length of YZ. Give your answer correct to 3 significant figures. [Edexcel]

10 The diagram shows a sketch of a triangle.

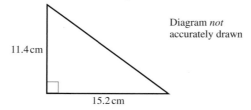

Diagram *not* accurately drawn

a) Work out the area of the triangle. State the units of your answer.

b) Work out the perimeter of the triangle. [Edexcel]

11

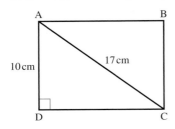

Diagram *not* accurately drawn

ABCD is a rectangle. AC = 17 cm. AD = 10 cm.
Calculate the length of the side CD. Give your answer correct to one decimal place.   [Edexcel]

12

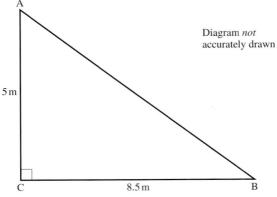

Diagram *not* accurately drawn

ABC is a right-angled triangle. AC = 5 m. CB = 8.5 m.
a) Work out the area of the triangle.
b) Work out the length of AB. Give your answer correct to 2 decimal places.   [Edexcel]

13 Calculate the distance from the point A (2, −1) to the point B (8, 7).

14 A trapezium PQRS is made by joining points P (1, 2), Q (9, 10), R (7, 9) and S (3, 5).
a) Make a sketch of the trapezium.
b) Work out the lengths of PQ and RS.
c) Hence give the ratio PQ : RS in its simplest form.
d) Work out the lengths QR and PS.
e) Explain whether or not the trapezium is isosceles.

15 The diagram represents a cuboid ABCDEFGH. AB = 5 cm. BC = 7 cm. AE = 3 cm.

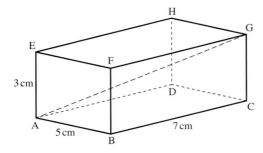

Diagram *not* accurately drawn

Calculate the length of AG. Give your answer correct to 3 significant figures.   [Edexcel]

Key points

1 Suppose a triangle has sides a, b, c, where c is the longest side. If the triangle is right angled, then $c^2 = a^2 + b^2$. This is Pythagoras' theorem.

2 Pythagoras' theorem also works in reverse. Therefore, if $c^2 = a^2 + b^2$ then the triangle must be right angled.

3 To find an unknown hypotenuse, use Pythagoras' theorem in the form $c^2 = a^2 + b^2$

   This can be rearranged to give $c = \sqrt{a^2 + b^2}$

4 To find an unknown shorter side, a, say, use Pythagoras' theorem in the form
   $a^2 = c^2 - b^2$

   This can be rewritten as $a = \sqrt{c^2 - b^2}$

5 Pythagoras problems sometimes involve two stages, especially if they are in three dimensions. Do not round off answers to multi-stage problems until all the calculations have been completed.

6 It is useful to draw the right-angled triangle you are using at each stage of your calculations.

7 In 2-D, the distance between two points (x_1, y_1) and (x_2, y_2) can be found using Pythagoras' theorem:

   $$\text{Distance} = \sqrt{(x_2 - x_1)^2 + (y_2 - y_1)^2}$$

   This can be remembered in words as:

   $$\text{Distance} = \sqrt{(\text{difference between } x \text{ coordinates})^2 + (\text{difference between } y \text{ coordinates})^2}$$

Internet Challenge 15

Investigating Pythagorean triples

Probably the most well-known right-angled triangle has sides in the ratio of 3 : 4 : 5, and is known as the (3, 4, 5) triangle. The numbers (3, 4, 5) form a **Pythagorean triple**, which means that they are whole numbers satisfying $a^2 + b^2 = c^2$. Another Pythagorean triple is (5, 12, 13).

Here are some questions about Pythagorean triples. You may use the internet to help you research some of the answers.

1   Find c such that (8, 15, c) is a Pythagorean triple.

2   How can we easily see that (6, 8, 10) and (10, 24, 26) are Pythagorean triples without doing any detailed calculations?

3   Find all the Pythagorean triples in which each number does not exceed 25.

4   Are there any patterns or formulae for generating them?

5   Are there infinitely many Pythagorean triplets?

6   Are there any **Pythagorean quadruplets**, that is, positive whole numbers a, b, c, d such that $a^2 + b^2 + c^2 = d^2$?

7   Find out as much as you can about Fermat's last theorem. Has it been proved yet?

CHAPTER 16

Introducing trigonometry

In this chapter you will learn how to:

- use sine, cosine and tangent to find unknown lengths in right-angled triangles
- use inverse functions to find unknown angles in right-angled triangles
- solve multi-stage problems using sine, cosine and tangent
- use angle of elevation and angle of depression.

You will also be challenged to:

- investigate famous geometers.

Starter: A triangular spiral

Draw a right-angled triangle with two shortest sides of lengths two units and one unit (use a scale of 2 cm to 1 unit), like this:

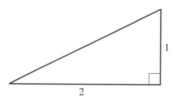

Now construct another right-angled triangle, using the hypotenuse of the first one as one of its short sides, and a length of 1 unit (2 cm) for the other:

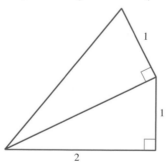

Are these two triangles similar?

Continue the pattern, using a one unit short side each time.
Measure the hypotenuse of the fifth triangle in the pattern. What do you notice?
Try to make as many triangles in the pattern as you can, to make a spiral pattern.

What do you notice about the angles at the centre of the spiral?

16.1 The sine ratio

Consider these two triangles. They both have angles of 30°, 60° and 90°. The two triangles are mathematically similar.

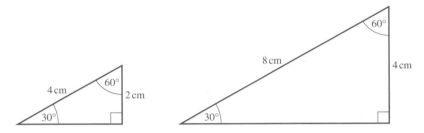

Compare the length of the side opposite the 30° angle, with the length of the hypotenuse. In the smaller triangle this is 2 ÷ 4 = 0.5, and in the larger triangle it is 4 ÷ 8 = 0.5. If you make some other 30° right-angled triangles, you will find the ratio always gives 0.5, regardless of the scale of a particular triangle.

In any right-angled triangle, the answer obtained by dividing the length opposite an angle by the length of the hypotenuse is called the **sine** of that angle. The symbol θ is often used to denote a general angle, so:

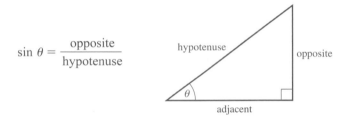

$$\sin \theta = \frac{\text{opposite}}{\text{hypotenuse}}$$

Sines of other angles do not work out to be such convenient quantities. For example, sin 29° = 0.484 809 620 2, correct to 10 decimal places.

Your calculator should contain a sine function button that enables you to obtain the sine of any angle. Make sure that your calculator is set to DEGree mode (rather than RADian or GRADian mode); ask your teacher to check this if you are not sure.

The equation:

$$\sin \theta = \frac{\text{opposite}}{\text{hypotenuse}}$$

can be rewritten as:

$$\text{opposite} = \text{hypotenuse} \times \sin \theta$$

and this allows you to calculate the missing length of a side opposite a given angle.

EXAMPLE

Find the missing length marked x.

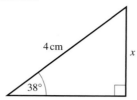

SOLUTION

In this triangle, the side opposite to 38° is x, and the hypotenuse is 4 cm.

$$\text{opposite} = \text{hypotenuse} \times \sin \theta$$
$$x = 4 \times \sin 38°$$
$$= 2.462\ 645\ 901$$
$$= \underline{2.46 \text{ cm}} \text{ (3 s.f.)}$$

> Calculations should be rounded to a sensible number of figures at the end.

You can also find the missing length of a hypotenuse by rearranging the usual equation.

EXAMPLE

Find the missing length marked y.

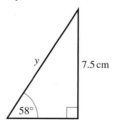

SOLUTION

In this triangle, the side opposite to 58° is 7.5 cm, and the hypotenuse is y.

$$\text{opposite} = \text{hypotenuse} \times \sin \theta$$
$$7.5 = y \times \sin 58°$$
$$\text{so } y = \frac{7.5}{\sin 58°}$$
$$= 8.843\ 838\ 025$$
$$= \underline{8.84 \text{ cm}} \text{ (3 s.f.)}$$

> Show your full calculator value …

> … as well as the rounded value.

Examination questions can ask you to do either type of calculation. As a check, remember that the hypotenuse is always the longest side in a right-angled triangle; in the example above you would check that your value for y is larger than 7.5 cm.

EXERCISE 16.1

Work out the values of the sides represented by letters. Show details of your calculations, and round your final answers correct to 3 significant figures in each case.

1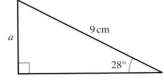

2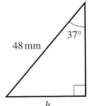

3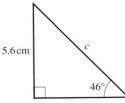

4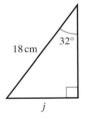

5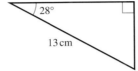

6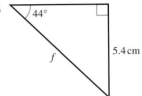

7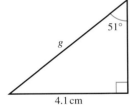

8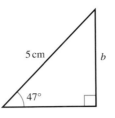

9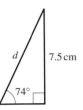

10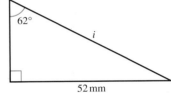

11 In triangle ABC, AB = 12 cm, angle ACB = 90°, angle CAB = 36°. Calculate BC.

12 In triangle PQR, angle PQR = 90°, angle QPR = 29°, PR = 8.8 cm. Calculate QR.

13 In triangle LMN, angle LMN = 90°, angle LNM = 28°, LN = 75 mm. Calculate LM.

14 In triangle ABC, angle CAB = 90°, angle ABC = 44°, AC = 60 mm. Calculate CB.

15 In triangle RST, angle RTS = 90°, angle RST = 17°, RS = 145 mm. Calculate RT.

16.2 The cosine ratio

The **cosine** of an angle is defined in a similar way to the sine function, but using the adjacent side instead of the opposite:

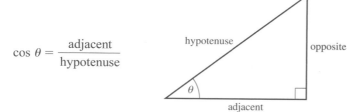

$$\cos \theta = \frac{\text{adjacent}}{\text{hypotenuse}}$$

The equation:

$$\cos \theta = \frac{\text{adjacent}}{\text{hypotenuse}}$$

can be rewritten as:

$$\text{adjacent} = \text{hypotenuse} \times \cos \theta$$

EXAMPLE

Find the missing length marked x.

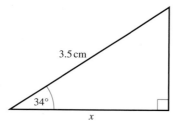

SOLUTION

Labelling the sides as seen from the 34° angle:

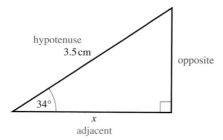

In this triangle, the side adjacent to 34° is x, and the hypotenuse is 3.5 cm.

$$\text{adjacent} = \text{hypotenuse} \times \cos \theta$$
$$x = 3.5 \times \cos 34°$$
$$= 2.901\ 631\ 504$$
$$= \underline{2.90 \text{ cm}} \text{ (3 s.f.)}$$

If you need to find the hypotenuse, division will be necessary.

EXAMPLE

Find the missing length marked as p.

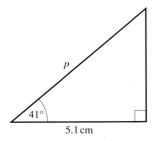

SOLUTION

Labelling the sides as seen from the 41° angle:

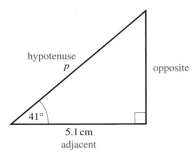

In this triangle, the side adjacent to 41° is 5.1 cm, and the hypotenuse is p cm.

$$\text{adjacent} = \text{hypotenuse} \times \cos \theta$$
$$5.1 = p \times \cos 41°$$
$$p = \frac{5.1}{\cos 41°}$$
$$= \underline{6.76 \text{ cm}} \text{ (3 s.f.)}$$

Some questions might require the use of both sine and cosine. Take care to identify the sides correctly.

EXAMPLE

Find the missing lengths marked as x and y.

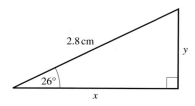

SOLUTION

Labelling the sides as seen from the 26° angle:

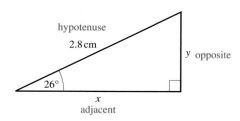

In this triangle, the side *adjacent* to 26° is x, and the hypotenuse is 2.8 cm.

$$\text{adjacent} = \text{hypotenuse} \times \cos \theta$$
$$x = 2.8 \times \cos 26°$$
$$= 2.516\ 623\ 33$$
$$= \underline{2.52 \text{ cm}} \text{ (3 s.f.)}$$

The side *opposite* to 26° is y, so:

$$\text{opposite} = \text{hypotenuse} \times \sin \theta$$
$$x = 2.8 \times \sin 26°$$
$$= 1.227\ 439\ 211$$
$$= \underline{1.23 \text{ cm}} \text{ (3 s.f.)}$$

> You could check these answers by putting them into Pythagoras' theorem:
> $$1.232^2 + 2.522^2 = 7.8633$$
> then $\sqrt{7.8633} = 2.804$ cm (4 s.f.)
> (This check does not give exactly 2.8 because of the rounding that has been used.)

EXERCISE 16.2

Work out the values of the sides represented by letters, giving your final answers correct to 3 significant figures.

1

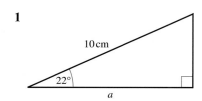

2

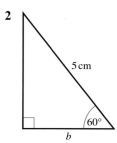

3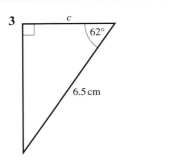

c

62°

6.5 cm

4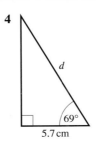

d

69°

5.7 cm

5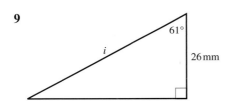

11 cm

23°

e

6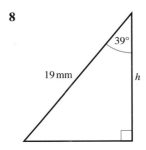

f

37°

6.3 cm

7

58°

g

1.4 cm

8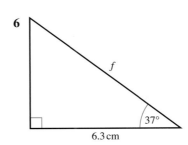

39°

19 mm

h

9

61°

i

26 mm

10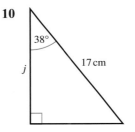

38°

17 cm

j

11 In triangle ABC, AB = 9 cm, angle ACB = 90°, angle CBA = 27°. Calculate BC.

12 In triangle PQR, angle PQR = 90°, angle QPR = 41°, PR = 6.5 cm. Calculate PQ.

13 In triangle ABC, angle ABC = 90°, angle BCA = 66°, BC = 44 mm. Calculate AC.

14 In triangle JKL, angle KJL = 90°, angle JKL = 46°, LK = 87 mm. Calculate JK.

15 In triangle EFG, angle EGF = 90°, angle FEG = 33°, EG = 48 cm. Calculate EF.

16.3 The tangent ratio

Sine and cosine are examples of trigonometrical ratios, that is, ratios that arise from measuring triangles. There is one further important trigonometrical ratio, namely the **tangent** of an angle, defined as follows:

$$\tan \theta = \frac{\text{opposite}}{\text{adjacent}}$$

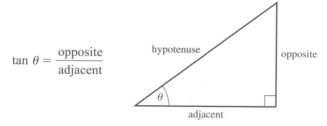

The equation:

$$\tan \theta = \frac{\text{opposite}}{\text{adjacent}}$$

can be rewritten as:

$$\text{opposite} = \text{adjacent} \times \tan \theta$$

EXAMPLE

Find the missing length marked as x.

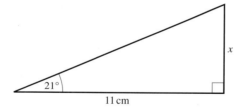

SOLUTION

Labelling the sides as seen from the 21° angle:

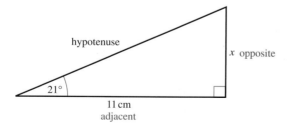

In this triangle, the side opposite to 21° is x, and the adjacent is 11 cm.

$$\text{opposite} = \text{adjacent} \times \tan \theta$$
$$x = 11 \times \tan 21°$$
$$= 4.222\ 504\ 385$$
$$= \underline{4.22} \text{ cm (3 s.f.)}$$

EXERCISE 16.3

Work out the values of the sides represented by letters, giving your final
answers correct to 3 significant figures.

1

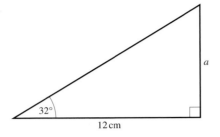

2

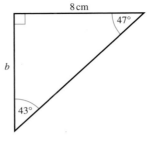

3

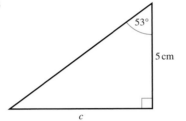

4

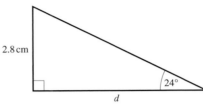

5

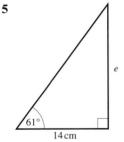

6

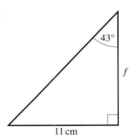

7

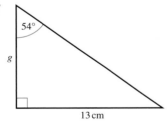

8

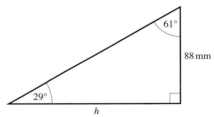

9

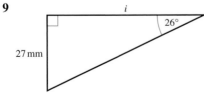

10

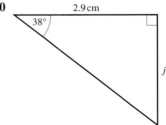

11 In triangle ABC, AB = 21 cm, angle ABC = 90°, angle CAB = 43°. Calculate BC.

12 In triangle PQR, angle PQR = 90°, angle QPR = 17°, QR = 22.5 cm. Calculate PQ.

16.4 Choosing the right trigonometrical function

Exam questions will expect you to recognise which of sine, cosine or tangent is appropriate in a particular setting. To help remember which is which, you might want to use the mathematical 'word' SOHCAHTOA. This is best written across three triangles:

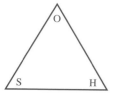 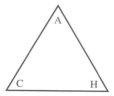 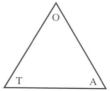

There are two stages to solving a problem. First, decide which of the three ratios is appropriate: for example, if the question refers to the adjacent and the hypotenuse but makes no reference at all to the opposite, then you cannot use either sine or tangent, but you can use cosine. Then cover up the quantity you are looking for, and the triangle tells you whether to multiply or divide the two remaining values. (You may have used a similar technique in a science lesson with distance, speed and time.)

The next two examples will use this method. Do not write a trig ratio such as 'sin' on its own – it must always contain an angle, for example, sin 42° or sin x.

EXAMPLE

Find the missing lengths x and y.

a)

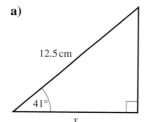

12.5 cm

41°

x

b)

3.7 cm

43°

y

SOLUTION

a)

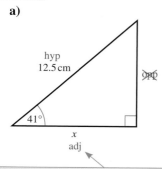

hyp
12.5 cm

opp

41°

x
adj

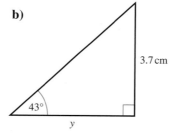

adj = cos × hyp

The adjacent is to be found, and the hypotenuse is known, but the opposite plays no part. Therefore select the cosine function, since it does not use opposite.

301

$$adjacent = \cos \theta \times hypotenuse$$
$$x = \cos 41° \times 12.5$$
$$= 9.433\ 869\ 753$$
$$= \underline{9.43\ cm}\ (3\ s.f.)$$

b)

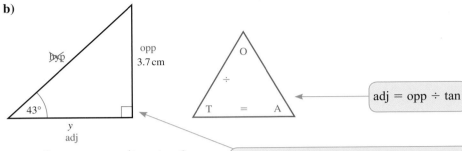

$$adjacent = opposite \div \tan \theta$$
$$y = 3.7 \div \tan 43°$$
$$= 3.967\ 764\ 227$$
$$= \underline{3.97\ cm}\ (3\ s.f.)$$

adj = opp ÷ tan

This time the hypotenuse plays no part, so the tangent function is selected: it uses opposite and adjacent.

EXERCISE 16.4

Find the unknown lengths, represented by letters, correct to 3 significant figures.

1

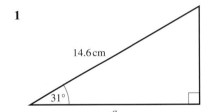

2

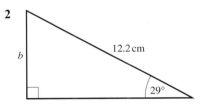

3

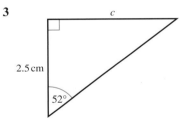

4

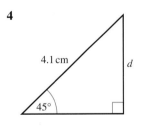

5

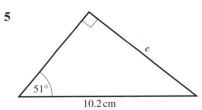

6

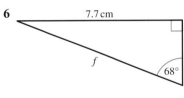

7

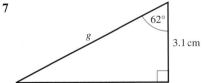

8

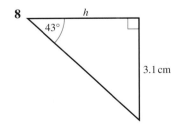

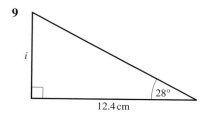

9

10

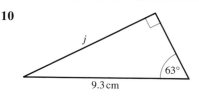

16.5 Finding an unknown angle

So far you have used sine, cosine and tangent to find an unknown side in a right-angled triangle. In such problems you will always be told the value of angle θ – in effect you are using a known angle to help you find an unknown side.

The process can be done in reverse, which means you can use known sides to help you find an unknown angle. When you come to use the sine, cosine or tangent button on your calculator, you must tell the calculator that you are performing the calculation in reverse. This is usually done by pressing the shift or second function key, written as $\sin^{-1}$, $\cos^{-1}$ or $\tan^{-1}$. (You would say *inverse sine*, etc.) Always use these inverse functions when you are expecting the answer to be an angle.

EXAMPLE

A ladder is 4 metres long. It leans against a vertical wall, and reaches 3.5 metres up the wall. Find the angle that the ladder makes with the horizontal ground.

SOLUTION

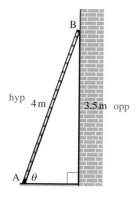

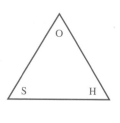

$$\sin \theta = \text{opposite} \div \text{hypotenuse}$$
$$= 3.5 \div 4$$
$$= 0.875$$
$$\theta = \sin^{-1} 0.875$$
$$= 61.044\ 975\ 63$$
$$= \underline{61.0°} \text{ (nearest } 0.1°)$$

EXERCISE 16.5

Work out the values of the angles represented by letters. Show details of your calculations, and round your final answers correct to 1 decimal place.

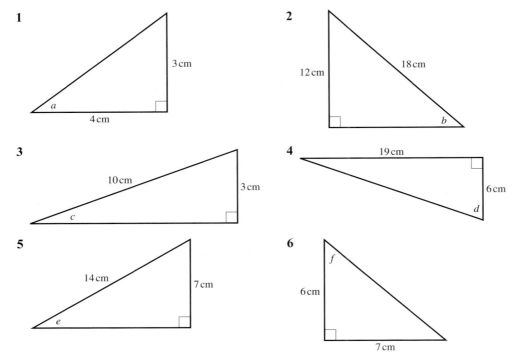

1   3 cm   4 cm   *a*

2   12 cm   18 cm   *b*

3   10 cm   3 cm   *c*

4   19 cm   6 cm   *d*

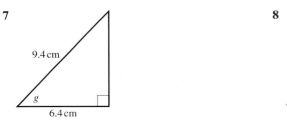

5   14 cm   7 cm   *e*

6   *f*   6 cm   7 cm

7   9.4 cm   *g*   6.4 cm

8   *h*   12.2 cm   8.2 cm

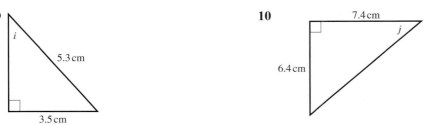

9   *i*   5.3 cm   3.5 cm

10   7.4 cm   *j*   6.4 cm

11 In triangle PQR, PQ = 14 cm, QR = 10 cm, angle PQR = 90°. Calculate angle PRQ.

12 In triangle LMN, angle LMN = 90°, MN = 17 cm, LN = 25 cm. Calculate angle LNM.

16.6 Multi-stage problems

Some questions will ask you to find a missing quantity, and then go on to use this to find another, possibly using Pythagoras' theorem as well as trigonometry. *Do not round off too early* – whenever you carry out a new calculation you should use the full calculator value of any previous calculations; otherwise inaccuracies can creep into your work.

EXAMPLE

The diagram shows a flagpole CB. It is supported by a wire, AB, 25 metres long. A is 21 metres from the base C of the flagpole. The flagpole is supported by a second wire, BD, which makes an angle of 65° with the horizontal ground.

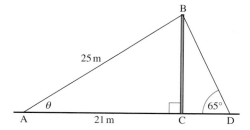

a) Calculate the height BC of the flagpole, correct to 3 significant figures.

b) Calculate the length BD.

c) Work out the value of the angle marked θ on the diagram.

SOLUTION

a) By Pythagoras' theorem, in triangle ABC:

$$BC^2 = 25^2 - 21^2$$
$$= 184$$
$$BC = \sqrt{184}$$
$$= \underline{13.6 \text{ m}} \text{ (3 s.f.)}$$

b) Now in triangle BCD:

$$\text{hypotenuse} = \frac{\text{opposite}}{\sin \theta}$$

$$BD = \frac{13.564\ 659\ 97}{\sin 65°}$$

$$= 14.966\ 946\ 28$$

$$= \underline{15.0 \text{ m}} \text{ (3 s.f.)}$$

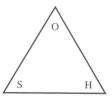

c) For the angle θ, AC = 21 m (adj) and AB = 25 m (hyp), so use cosine:

$$\cos \theta = \frac{\text{adjacent}}{\text{hypotenuse}}$$
$$= \frac{21}{25}$$
$$= 0.84$$
$$\theta = \cos^{-1} 0.84$$
$$= 32.859\,880\,38$$
$$= \underline{32.9°} \text{ (nearest 0.1°)}$$

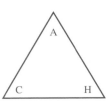

Trigonometry questions can include references to bearings.

EXAMPLE

A ship leaves its harbour and sails due south for 10 km.
It then sails due East for 20 km, then stops.

a) How far is the ship from its harbour?

b) The ship wishes to return directly to its harbour.
On what bearing must it sail?

SOLUTION

a)

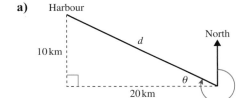

By Pythagoras' theorem:

$$d^2 = 10^2 + 20^2$$
$$= 100 + 400$$
$$= 500$$
$$d = \sqrt{500}$$
$$= \underline{22.4 \text{ km}} \text{ (3 s.f.)}$$

b) The angle θ is found using the tangent function:

$$\tan \theta = \frac{10}{20}$$
$$= 0.5$$
$$\theta = \tan^{-1} 0.5$$
$$= 26.6°$$

The bearing is $270° + 26.6° = \underline{296.6°}$

EXERCISE 16.6

1 A ship sails 30 km due East, then 45 km due North.
  a) Illustrate this information on a sketch.
  b) How far, to the nearest 0.1 km, is the ship from its starting point?
  c) What bearing, to the nearest degree, should it steer to return directly to its start point?

2 In the diagram, PRS is a straight line.
  PQ = 18 cm, PR = 13 cm, QS = 16 cm.
  QR is perpendicular to PS.

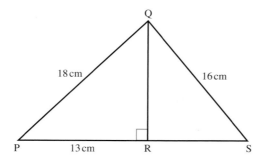

  a) Calculate the length QR. Give your answer to 3 significant figures.
  b) Calculate the area of triangle PQR. Give your answer to 3 significant figures.
  c) Calculate angle RQS. Give your answer to the nearest 0.1°.

3 The diagram shows a cross-section of a tent.
  PQ = QR = 1.9 m. PR = 1.6 m.
  M is the midpoint of the line segment PR.

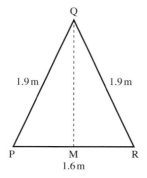

  a) Explain why angle PMQ must be a right angle.
  b) Calculate the angle QPM. Give your answer to the nearest 0.1°.
  c) Calculate the height of the tent. Give your answer to the nearest centimetre.

4 ABCD is a kite.
  AB = AD = 10 cm. BC = DC = 7 cm.
  Angle ABC = angle ADC = 90°.
  a) Illustrate this information on a sketch.
  b) Calculate the length AC, correct to the nearest millimetre.
  c) Find angle BAD, correct to the nearest degree.

5 In the diagram, EHG is a straight line.
Angle EHF = 90°, angle FGH = 34°.
EF = 11 cm, HG = 14 cm.

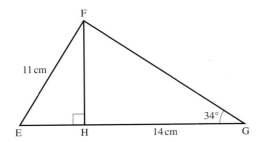

a) Work out the length of FH, correct to 3 significant figures.

b) Work out the value of angle FEH, correct to the nearest 0.1°.

6 A ship leaves port and sails on a bearing of 075° for 20 km.
It then sails on a bearing of 345° for 8 km, before stopping because of engine failure.

a) Draw a sketch to show this information.

b) How far is the ship from port when it stops? Give your answer to 3 significant figures.

c) A helicopter leaves port with spare parts to repair the ship's engines. Calculate the bearing that the helicopter should fly on in order to reach the ship by the shortest route. Give your answer to the nearest degree.

16.7 Angles of elevation and depression

When looking at a distant object, we often measure by how many degrees the line of sight to it lies above or below the horizontal. If it is *above the horizontal*, we call this the **angle of elevation**; if is *below the horizontal* then it is known as the **angle of depression**.

EXAMPLE

A man is standing on level ground, 50 metres away from an office block. The office block is 105 metres tall.

a) Work out the distance from the man to the top of the office building.

b) Work out the angle of elevation of the top of the building as seen by the man.

SOLUTION

a)

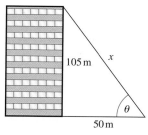

Using Pythagoras' theorem,
$x^2 = 105^2 + 50^2$
$= 13\,525$
$x = \sqrt{13\,525}$
$= 116.297$
$= \underline{116 \text{ metres}}$ (3 s.f.)

> The angle of elevation, θ, is the angle between the horizontal and the line of sight to the top of the building.

b) Let the angle of elevation be θ.

$\tan \theta = \dfrac{105}{50}$
$= 2.1$
$\theta = \tan^{-1} 2.1$
$= 64.536\,654\,94$
$= \underline{64.5°}$ (3 s.f.)

EXAMPLE

A coastguard is standing on top of a cliff at a height of 200 metres above sea level. He looks out to sea, and observes a small boat at an angle of depression of 21 degrees.

a) Calculate the distance from the coastguard to the boat. Give your answer to 3 significant figures.

b) Calculate the distance of the boat from the foot of the cliff. Give your answer to 3 significant figures.

SOLUTION

a) Coastguard

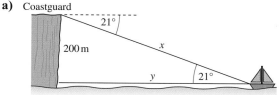

> The angle of *depression* at the coastguard is equal to the angle of *elevation* at the boat – these are alternate angles.

Using trigonometry (sine) we have

$x = \dfrac{200}{\sin 21}$
$= 558.0856…$
$= \underline{558 \text{ metres}}$ (3 s.f.)

b) Using trigonometry (tangent) we have

$$y = \frac{200}{\tan 21}$$
$$= 521.0178\ldots$$
$$= \underline{\underline{521 \text{ metres}}} \text{ (3 s.f.)}$$

EXERCISE 16.7

1 The diagram shows a vertical mast viewed from a point P on level ground. The angle of elevation of the top T of the mast from P is 27 degrees. The distance TP is 120 metres.

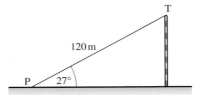

Calculate the height of the mast. Give your answer correct to 3 significant figures.

2 A climber is sitting on the summit S of a mountain. He looks down and sees his camp C in the valley below. The direct distance from the summit to the camp is 2200 metres. The summit is at an altitude 420 metres higher than the camp.

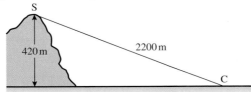

a) Calculate the angle of elevation of the summit as seen from the camp. Give your answer to the nearest 0.1°.

b) Write down the angle of depression of the camp as seen from the summit.

3 From a point P on level ground, the angle of elevation of a vertical mast is 19°. The distance from P to the foot of the mast is 150 metres.

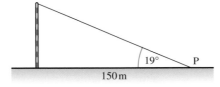

Calculate the height of the mast. Give your answer to 3 significant figures.

4 A radar station R detects an aircraft A flying at an altitude of 1500 metres. The horizontal distance from R to a point directly below the aircraft is 2000 metres.

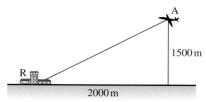

a) Calculate the direct distance RA between the radar station and the aircraft.

b) Calculate the angle of elevation of the aircraft from the radar station. Give your answer correct to 0.1°.

5 A vertical mast AB is supported by two straight cables AX and AY. The points X, Y and B are on level ground. AX = 34 m, BX = 16 m and AY = 42 m.

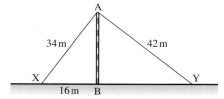

a) Calculate the height of the mast, AB.

b) Calculate the angle of elevation of the top of the mast as seen from Y. Give your answer correct to the nearest 0.1°.

REVIEW EXERCISE 16

1 Find the unknown lengths, marked with letters. Give your answers correct to 3 significant figures.

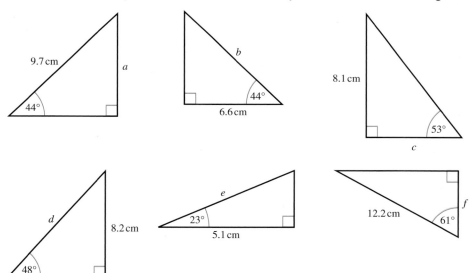

2 Find the unknown angles, marked with letters. Give your answers to the nearest 0.1°.

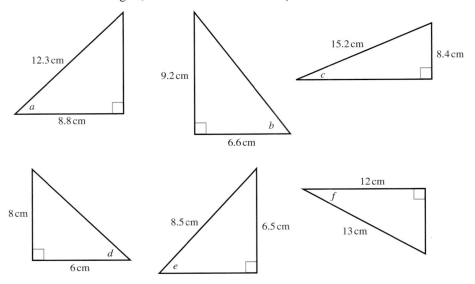

3 From a point P at ground level, the angle of elevation of the top of a lighthouse is 42°. The distance from P to the foot of the lighthouse is 39 metres.

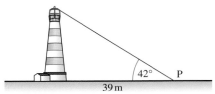

Calculate the height of the lighthouse. Give your answer to 3 significant figures.

4 From the top of a cliff, a man looks out to sea and sees a small boat at a direct distance of 350 metres. The boat is 310 metres from the foot of the cliff.

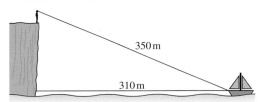

Calculate the angle of depression of the boat as seen by the man. Give your answers to the nearest 0.1°.

5 Mr Jones puts his ladder against the wall of his house.

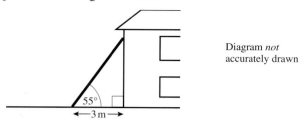

Diagram *not* accurately drawn

The angle the ladder makes with the ground is 55°. The foot of the ladder is 3 metres from the base of the wall of the house. Work out how far up the wall the ladder reaches. Give your answer, in metres, correct to 3 significant figures.

[Edexcel]

6 ABCD is a quadrilateral.

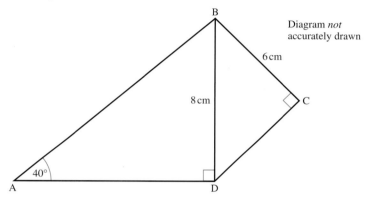

Diagram *not* accurately drawn

Angle BDA = 90°, angle BCD = 90°, angle BAD = 40°. BC = 6 cm, BD = 8 cm.
a) Calculate the length of DC. Give your answer correct to 3 significant figures.
b) Calculate the size of angle DBC. Give your answer correct to 3 significant figures.
c) Calculate the length of AB. Give your answer correct to 3 significant figures.

[Edexcel]

7 The diagram shows the positions of three schools P, Q and R.
School P is 8 kilometres due West of School Q.
School R is 3 kilometres due North of School Q.

a) Calculate the size of the angle marked $x°$. Give your answer correct to one decimal place.

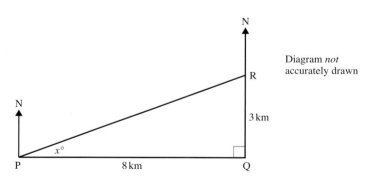

Diagram *not* accurately drawn

Simon's house is 8 kilometres due East of School Q.
b) Calculate the bearing of Simon's house from School R.

[Edexcel]

8 The diagram shows a trapezium.

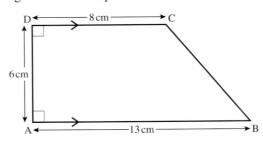

Diagram *not* accurately drawn

AB is parallel to DC. Angle A = 90°. AB = 13 cm, AD = 6 cm and CD = 8 cm.
Calculate the size of the angle B. Give your answer correct to one decimal place.   **[Edexcel]**

9 ABD and DBC are two right-angled triangles.
AB = 9 m. Angle ABD = 35°. Angle DBC = 50°.
Calculate the length of DC.
Give your answer correct to 3 significant figures.

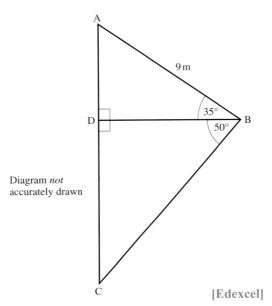

Diagram *not* accurately drawn

[Edexcel]

10 AB and BC are two sides of a rectangle.

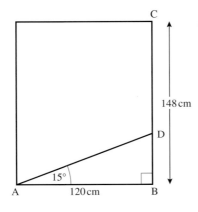

Diagram *not* accurately drawn

AB = 120 cm and BC = 148 cm. D is a point on BC. Angle BAD = 15°.
Work out the length of CD. Give your answer correct to 3 significant figures.   **[Edexcel]**

11 The diagram represents a vertical flagpole, AB. The flagpole is supported by two ropes, BC and BD, fixed to the horizontal ground at C and at D.

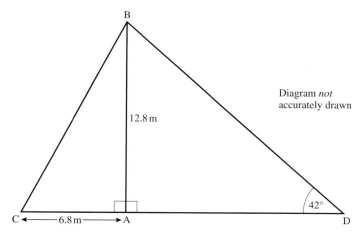

Diagram *not* accurately drawn

AB = 12.8 m, AC = 6.8 m, angle BDA = 42°.
a) Calculate the size of angle BCA. Give your answer correct to 3 significant figures.
b) Calculate the length of the rope BD. Give your answer correct to 3 significant figures.   [Edexcel]

12 DE = 6 m. EG = 10 m. FG = 8 m.
Angle DEG = 90°. Angle EFG = 90°.
a) Calculate the length of DG.
   Give your answer correct to 3 significant figures.
b) Calculate the size of the angle marked $x°$.
   Give your answer correct to one decimal place.

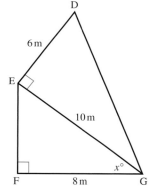

Diagram *not* accurately drawn

[Edexcel]

Key points

1 Trigonometry can be used to find unknown lengths or angles in right-angled triangles. Calculators have several different angle modes available: make sure yours is set to DEG mode before doing any trigonometrical calculations.

2 The three basic trigonometrical ratios are sine (sin), cosine (cos) and tangent (tan). Their definitions are readily remembered using SOHCAHTOA, which can be written in triangular form:

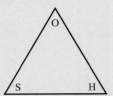 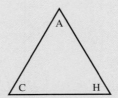 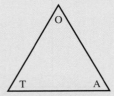

3 When these ratios are used to find missing lengths, the calculator will usually generate large numbers of decimal figures. Examination questions will typically ask you to round off to 3 significant figures.

4 If you are given some sides of a triangle, and asked to find a missing angle, then sin, cos or tan are used in reverse. Remember to press the 'Inv' or 'Second Function' key on your calculator, to tell it that the function is being inverted. Examination questions will typically ask for angles to be rounded to one decimal place of a degree or, sometimes, 3 significant figures.

5 In the examination, you may have to use Pythagoras' theorem in combination with trigonometry. Use your full calculator value throughout any multi-stage calculation, and save all the rounding until the end.

6 The examination might also ask you to use trigonometry to solve problems with bearings. Remember that these are measured using three figures, so that North is 000°, East 090°, South 180° and West 270°.

7 The angle that a line of sight makes *above* the horizontal is called the *angle of elevation*. If the line of sight to an object lies *below* the horizontal then this is the *angle of depression*.

Internet Challenge 16

Famous geometers

Here are some clues about people who are famous for their work in geometry. Use the internet to find out their names. Try to find some interesting facts about each one.

1 This person wrote a set of 13 geometry books, called 'The Elements', in the third century BC.

2 A famous geometry theorem, named after him, was well known to the ancient Egyptians and Chinese. He was born on the Greek island of Samos.

3 'Clouds are not spheres, mountains are not cones …' said this 20th century mathematician, a pioneering thinker behind the development of fractal geometry.

4 The 'Great geometer', he lived from about 262 BC to 190 BC.

5 This German mathematician and astronomer used geometrical methods to understand the movement of the planets in their orbits. His three laws of planetary motion were published in 1609 and 1619.

6 'If I have seen further, it is by standing on the shoulders of giants' is a quote attributed to this English mathematical genius, who worked in geometry, algebra and calculus.

7 This outstanding Swiss geometer lived from 1707 to 1783.

8 He developed a geometry in which angles in a triangle do not need to add up to 180°. He suffered from poor health throughout his life, dying of tuberculosis in Italy in 1866.

9 The tetrahedron, cube, octahedron, dodecahedron and icosahedron are collectively named after this philosopher/mathematician and pupil of Socrates.

10 This German mathematician had a 'bottle' with no inside named after him.

CHAPTER 17

Circle theorems

In this chapter you will learn how to:

- use correct vocabulary associated with circles
- use tangent properties to solve problems
- prove and use various theorems about angle properties inside a circle
- prove and use the alternate segment (intersecting tangent and chord) theorem
- prove and use the intersecting chords theorem.

You will also be challenged to:

- investigate the nine-point circle theorem.

Starter: **Circle vocabulary**

Here are some words you will often encounter when working with circles:

    Centre   Radius   Chord   Diameter   Circumference

    Tangent   Arc   Sector   Segment

Match the correct words to the nine diagrams below:

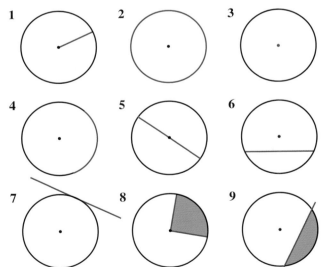

10 a) The French call a rainbow an '*arc-en-ciel*' (*ciel* = sky).
    Do you think this is a good name?

  b) The English call a piece of an orange a 'segment'.
    Do you think this is a good name?

17.1 Tangents, chords and circles

In this section you will learn some theorems about circles, and then use them to solve problems. The theorems are concerned with tangents and chords.
A **chord** is a line segment joining two points on the circumference of a circle.
A **tangent** is a straight line that touches a circle only once.

A line segment drawn from the centre of a circle to the midpoint of a chord will intersect the chord at right angles.

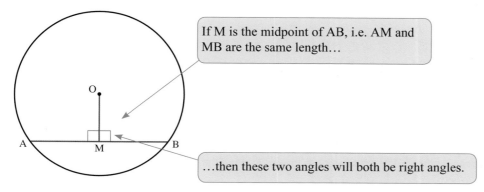

If M is the midpoint of AB, i.e. AM and MB are the same length…

…then these two angles will both be right angles.

A tangent and radius meet at right angles.

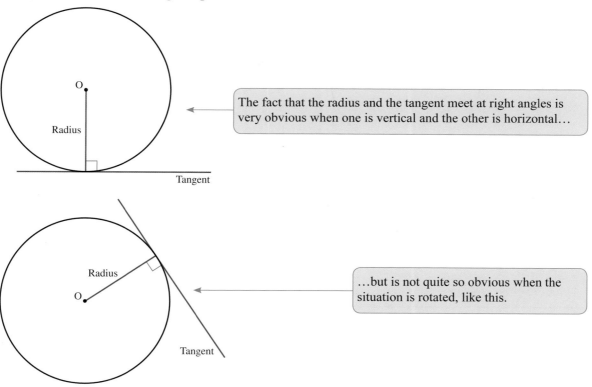

The fact that the radius and the tangent meet at right angles is very obvious when one is vertical and the other is horizontal…

…but is not quite so obvious when the situation is rotated, like this.

The two external tangents to a circle are equal in length.

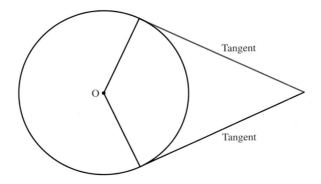

These theorems may be used to help you determine the values of missing angles in circles. When you use them, remember to tell the examiner which theorem(s) you have used.

EXAMPLE

The diagram shows a circle, centre O. PT is a tangent to the circle. Find the value of x.

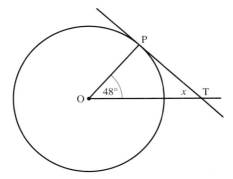

SOLUTION

Angle OPT = 90° (angle between the radius and tangent is 90°).

The angles in triangle TOP add up to 180°, so:

$$x + 48° + 90° = 180°$$
$$x + 138° = 180°$$
$$x = 180° - 138°$$
$$\underline{\underline{x = 42°}}$$

EXAMPLE

The diagram shows a circle, centre Q.
AB is a chord across the circle.
M is the midpoint of AB.
Find the value of y.

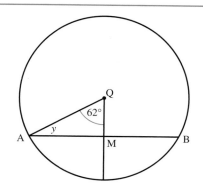

SOLUTION

Angle AMQ = 90° (radius bisecting chord).

The angles in triangle AMQ add up to 180°.

So: $y + 62° + 90° = 180°$
$y + 152° = 180°$
$y = 180° - 152°$
$y = 28°$

Since a radius bisects a chord at right angles, there is often an opportunity to use Pythagoras' theorem.

EXAMPLE

The diagram shows a radius OT that bisects the chord AB at M. MB = 12 cm.
The radius of the circle is 13 cm.
Work out the length MT.

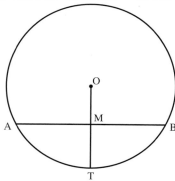

SOLUTION

First join OB:

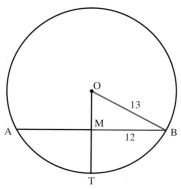

Now apply Pythagoras' theorem to triangle OBM:

$$OM^2 = 13^2 - 12^2$$
$$= 169 - 144$$
$$= 25$$
$$OM = \sqrt{125}$$
$$= 5 \text{ cm}$$

The distance OT is a radius, that is, 13 cm.
Thus MT = 13 − 5
= 8 cm

EXERCISE 17.1

1 PT is a tangent to the circle, centre O. Angle PTO = 29°.

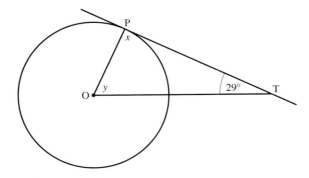

   a) State, with a reason, the value of the angle marked *x*.
   b) Work out the value of the angle marked *y*.

2 PT is a tangent to the circle, centre O. PT = 24 cm. OP = 7 cm.

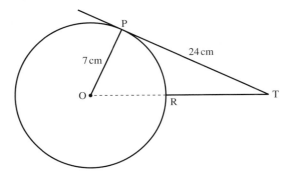

   The line OT intersects the circle at R, as shown. Work out the length of RT.

3 TP and TR are tangents to the circle, centre O. Angle PTR is 44°.

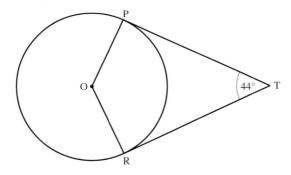

   a) Work out the size of angle POR. Give reasons.
   b) What type of quadrilateral is OPTR? Explain your reasoning.

4 TP and TR are tangents to the circle, centre O. Angle POR is 130°.

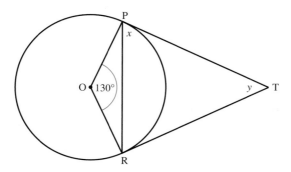

a) What type of triangle is triangle OPR?
b) Work out the value of x.
c) Work out the value of y.

5 The diagram shows a circle, centre O. The radius of the circle is 5 cm.
M is the midpoint of EF. OM = 3 cm.

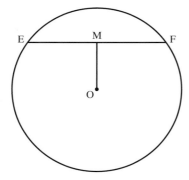

Calculate the length of EF.

6 The diagram shows a circle, centre O. AB = 34 cm. M is the midpoint of AB.
OM = 8 cm.

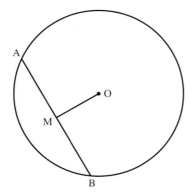

Work out the radius of the circle.

7 From a point T, two tangents TP and TQ are drawn to a circle, centre O.
   a) Make a sketch to show this information.

   The length TQ is measured, and found to be exactly the same as the length PO.
   b) What type of quadrilateral is TPOQ?

8 The diagram shows a circle, centre O.
   AB and CD are chords.
   The radius OT passes through the midpoints M and N of the chords.
   OM = 8 cm, NT = 3 cm, AB = 30 cm.

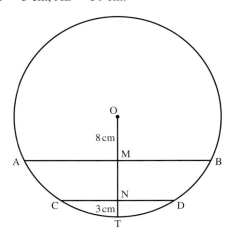

   a) Explain why angle AMO = 90°.
   b) Use Pythagoras' theorem to calculate the distance AO.
      Show your working.
   c) Calculate the distance MN.
   d) Calculate the length of the chord CD.

17.2 Angle properties inside a circle

There are several important theorems about angles inside a circle.
You will need to learn these, and use them to solve numerical problems.
You may also be asked to prove why they are true.

Consider two points, A and B say, on the circumference of a circle.
The angle subtended by the arc AB at the centre is angle AOB.

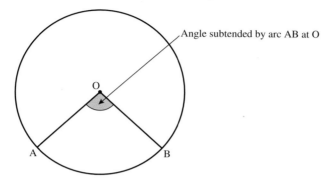

Angle subtended by arc AB at O

The angle subtended by the arc AB at a point X on the circumference is angle AXB.

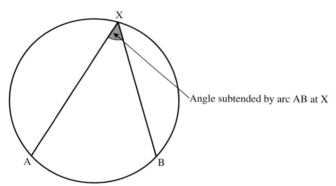

Angle subtended by arc AB at X

There is a theorem in circle geometry which states that angle AOB is exactly twice angle AXB.

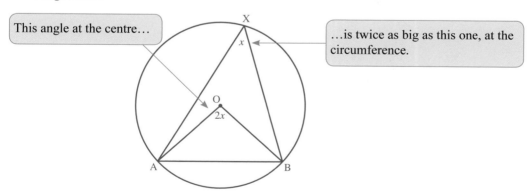

This angle at the centre…

…is twice as big as this one, at the circumference.

This result is quite easy to prove, and is the basic theorem upon which several other circle theorems are built.

THEOREM

The angle subtended by an arc at the centre of a circle is twice the angle subtended by the same arc at the circumference of the circle.

PROOF

Make a diagram to show the arc AB, the centre O, and the point X on the circumference of the circle:

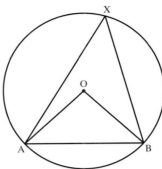

From X, draw a radius to O, and produce it, which means extend it slightly:

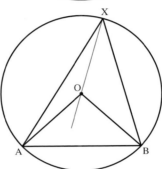

Triangle AOX is isosceles, since both OA and OX are radii of the same circle. Therefore angles OAX and OXA are equal. These are marked on the diagram with a letter *a*:

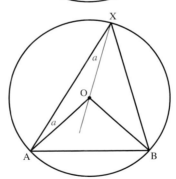

Likewise the triangle BOX is isosceles, since both OB and OX are radii of the same circle. Therefore angles OBX and OXB are equal. These are marked on the diagram with a letter *b*:

The angle at the circumference is angle AXB = *a* + *b*.

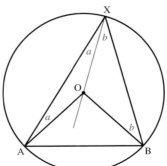

To obtain an expression for the angle at the centre, look at this magnified copy of the diagram:

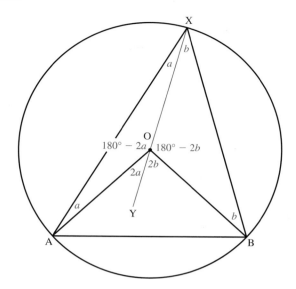

Angle AOX = $180 - a - a = 180 - 2a$

So: angle AOY = $180 - (180 - 2a) = 2a$

Likewise:

angle BOX = $180 - b - b = 180 - 2b$

So: angle BOY = $180 - (180 - 2b) = 2b$

Thus the angle at the centre is:

Angle AOB = $2a + 2b$

            = $2(a + b)$

But angle AXB = $a + b$, from above.

Therefore angle AOB = $2 \times$ angle AXB.

Thus, the angle subtended by an arc at the centre of a circle is twice the angle subtended by the same arc at the circumference of the circle.

Two further theorems can be deduced immediately from this first one.
You can quote the previous theorem to justify these proofs.

THEOREM

Angles subtended by an arc in the same segment of a circle are equal.

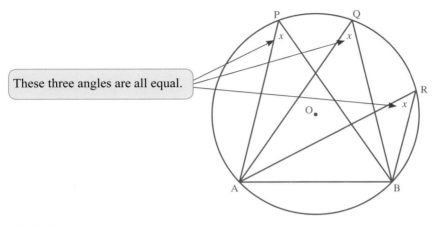

These three angles are all equal.

PROOF

Join AO and OB so that they form an angle at the centre:

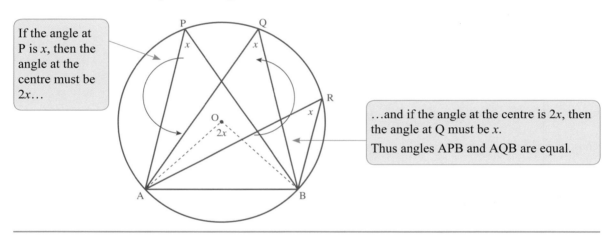

If the angle at P is x, then the angle at the centre must be $2x$...

...and if the angle at the centre is $2x$, then the angle at Q must be x.

Thus angles APB and AQB are equal.

THEOREM

The angle subtended in a semicircle is a right angle.

PROOF

Since AB is a diameter, AOB is a straight line.

Thus angle AOB = 180°.

Using the result that the angle at the circumference is half that at the centre:

Angle APB = 180° ÷ 2 = <u>90°</u>

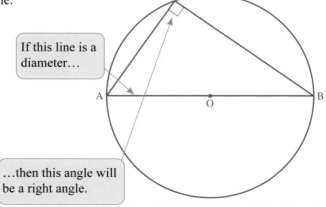

If this line is a diameter...

...then this angle will be a right angle.

EXAMPLE

Find the values of the angles marked x and y. Explain your reasoning in both cases.

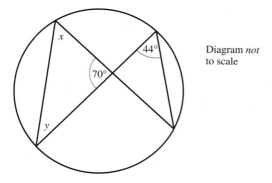

Diagram *not* to scale

SOLUTION

Angle $x = 44°$ (angles in the same segment are equal).

$x + 70° + y = 180°$ (angles in a triangle add up to $180°$) and $x = 44°$, so:

$$44° + 70° + y = 180°$$
$$114° + y = 180°$$
$$y = 180° - 114°$$
$$y = 66°$$

EXAMPLE

Find the values of the angles marked x and y. Explain your reasoning in both cases.

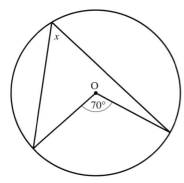
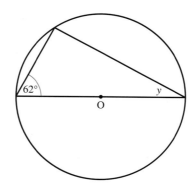

SOLUTION

$x = 70 \div 2 = 35°$ (angle at centre $= 2 \times$ angle at circumference)

$y = 180 - 90 - 62 = 28°$ (angle in a semicircle is a right angle and angles in a triangle add up to $180°$)

EXERCISE 17.2

Find the missing angles in these diagrams, which are not drawn to scale. Explain your reasoning in each case.

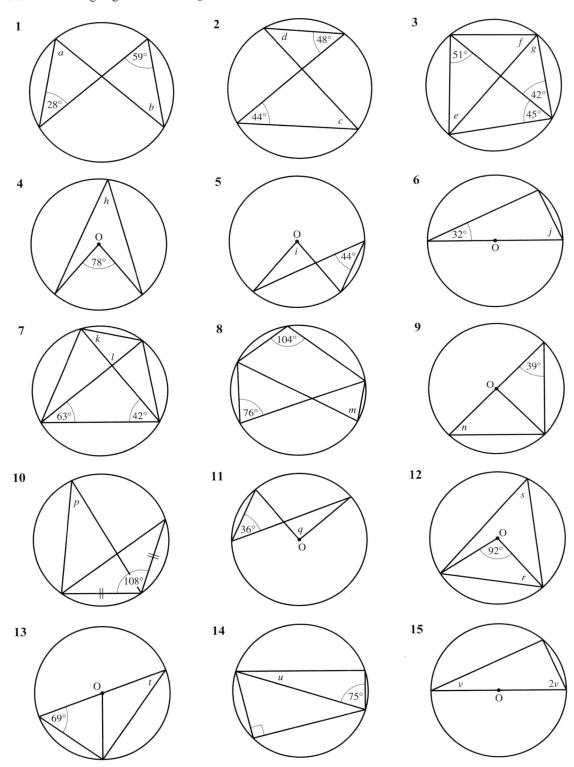

17.3 Further circle theorems

THEOREM

The angles subtended in opposite segments add up to 180°.

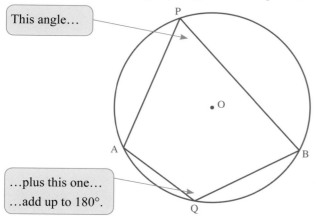

This angle…

…plus this one…
…add up to 180°.

PROOF

Denote the angles APB and AQB as p and q respectively.

Then the angles at the centre are twice these, that is, $2p$ and $2q$.

Angles at point O add up to 360°, so:

$$2p + 2q = 360°$$

Thus $2(p + q) = 360°$

So $p + q = 180°$

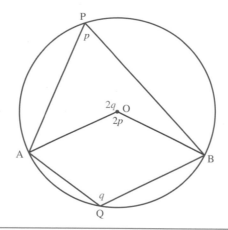

The points A, P, B and Q form a quadrilateral whose vertices lie around a circle; it is known as a **cyclic quadrilateral**. Thus the theorem may also be stated as:

Opposite angles of a cyclic quadrilateral add up to 180°

EXAMPLE

Find the angles x and y.

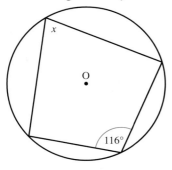 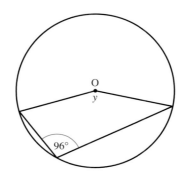

Diagram *not* to scale

SOLUTION

For angle x, we have $x + 116° = 180°$ (angles in opposite segments)

Thus $x = 180° - 116°$
$\underline{= 64°}$

For angle y, two construction lines are needed:

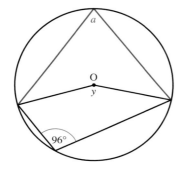

Use angles in opposite segments, $a = 180 - 96 = 84°$.

Using the angle at centre is twice the angle at circumference:

$y = 84 \times 2 = \underline{168°}$

THEOREM

The angle between a tangent and chord is equal to the angle subtended in the opposite segment. (This is often called the **alternate segment** theorem.)

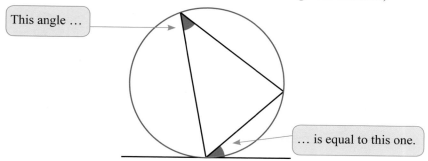

This angle …

… is equal to this one.

PROOF

First, consider the special case of a tangent meeting a diameter:

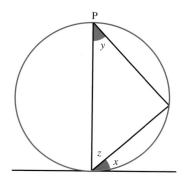

Since the angle in a semicircle is 90°, the other two angles in the triangle add up to 90°.

Hence $y + z = 90°$.

Since a radius and tangent meet at 90°:

$$x + z = 90°$$

Hence $x + z = y + z$.
From which it follows that $x = y$.

Now move P around the circle to Q, say, so that it is no longer on the end of a diameter. The angle at Q is equal to the angle at P, as they are angles in the same segment.
Thus the theorem is proved.

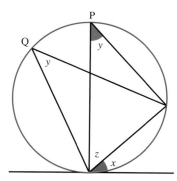

EXERCISE 17.3

Find the missing angles in these diagrams, which are not drawn to scale.
Explain your reasoning in each case.

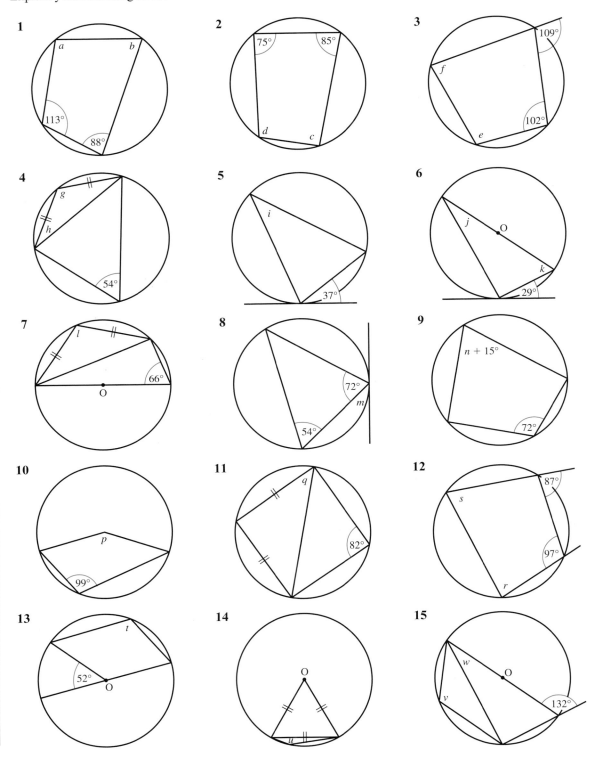

17.4 Intersecting chords

THEOREM

Consider a pair of chords AB and CD intersecting at a point X inside a circle, as shown in this diagram:

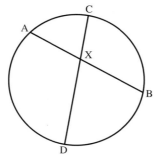

Then the lengths AX, BX, CX and DX are related by the following result:

$$AX \times BX = CX \times DX$$

PROOF

Join AC and DB to complete two triangles AXC and DXB as below:

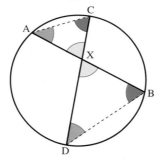

Then angles CAX and BDX are equal (angles in the same segment).

Angles ACX and DBX are equal (angles in the same segment).

Angles AXC and DXB are equal (vertically opposite).

Thus the triangles ACX and DBX are mathematically similar, so triangle DBX is an enlargement of triangle ACX.

The enlargement factor is $\dfrac{BX}{CX}$, but it is also $\dfrac{DX}{AX}$, and these must be equal, so

$$\frac{BX}{CX} = \frac{DX}{AX}$$

Cross-multiplying,

$$AX \times BX = CX \times DX$$

and the result is proved.

EXAMPLE

Find the missing length x in the diagram below.

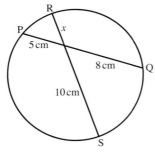

SOLUTION

Using the result for intersecting chords we have

$$10 \times x = 5 \times 8$$
$$= 40$$
$$x = \frac{40}{10}$$
$$= 4$$

So the missing length is $\underline{x = 4 \text{ cm}}$.

The same principle is valid even when the two chords cross over outside the circle, as in the next example.

EXAMPLE

PQ and SR are chords of a circle, and, when produced, intersect at X.

QX = 10 cm. SR = 6 cm. RX = 9 cm.

a) Write down the length SX.
b) Work out the length PX.
c) Work out the length PQ.

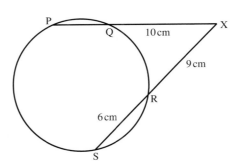

SOLUTION

a) SX = 6 + 9 = $\underline{15 \text{ cm}}$

b) PX × QX = RX × SX

PX × 10 = 9 × 15

$$= 135$$

$$PX = \frac{135}{10}$$

$$= \underline{13.5 \text{ cm}}$$

c) PQ = 13.5 − 10 = $\underline{3.5 \text{ cm}}$

EXERCISE 17.4

1 Find the missing length, x cm, in this diagram.

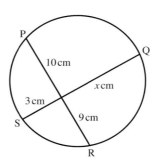

2 Find the missing length, y cm, in this diagram.

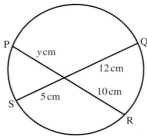

3 Chords AB and DC, when produced, meet at point X.
AB = 3 cm. BX = 12 cm. CX = 10 cm.

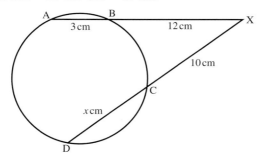

   a) Work out the length DX.
   b) Hence work out the length DC, marked as x cm.

4 In the diagram below, PQ is a diameter of the circle.
PX = 2 cm. RX = 6 cm. SX = 4 cm. QX = x cm.

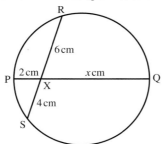

   a) Work out the value of x.
   b) Hence work out the radius of the circle.

5 Chords PQ and RS, when produced, meet at point W.
PQ = 7 cm. QW = 8 cm. SW = 10 cm.

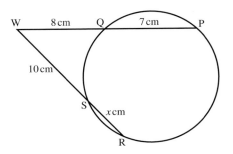

  a) Work out the length WR.

  b) Hence work out the length SR, marked as *x* cm.

REVIEW EXERCISE 17

1 A circle of diameter 10 cm has a chord drawn inside it. The chord is 7 cm long.

  a) Make a sketch to show this information.

  b) Calculate the distance from the midpoint of the chord to the centre of the circle.
     Give your answer correct to 3 significant figures.

2 The diagram shows a circle, centre O.
PT and RT are tangents to the circle. Angle POR = 144°.

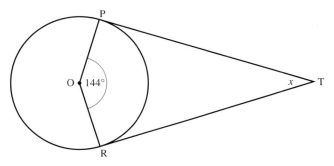

  a) Work out the size of angle PTR, marked *x*.

  b) Is it possible to draw a circle that passes through the four points P, O, R and T?
     Give reasons for your answer.

3 The diagram shows a circle, centre O.
PT and RT are tangents to the circle. Angle PTR = 32°.

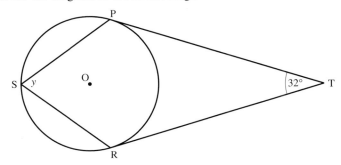

  a) Work out the size of angle PSR, marked *y*.
     Hint: Draw in OP and OR.

  b) Is it possible to draw a circle that passes through the four points P, S, R and T?
     Give reasons for your answer.

4 In the diagram, A, B and C are points on the circle, centre O.
Angle BCE = 63°. FE is a tangent to the circle at point C.

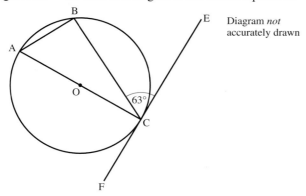

Diagram *not*
accurately drawn

a) Calculate the size of angle ACB. Give reasons for your answer.
b) Calculate the size of angle BAC. Give reasons for your answer.

[Edexcel]

5 P, Q, R and S are points on the circumference of a circle, centre O.
PR is a diameter of the circle. Angle PSQ = 56°.

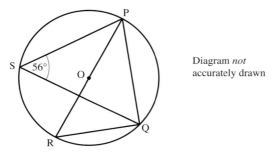

Diagram *not*
accurately drawn

a) Find the size of angle PQR. Give a reason for your answer.
b) Find the size of angle PRQ. Give a reason for your answer.
c) Find the size of angle POQ. Give a reason for your answer.

[Edexcel]

6 A, B, C and D are four points on the circumference of a circle.
ABE and DCE are straight lines. Angle BAC = 25°. Angle EBC = 60°.
a) Find the size of angle ADC.
b) Find the size of angle ADB.

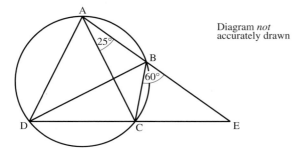

Diagram *not*
accurately drawn

Angle CAD = 65°. Ben says that BD is a diameter of the circle.
c) Is Ben correct? You must explain your answer.

[Edexcel]

7 The diagram shows a circle, centre O. AC is a diameter. Angle BAC = 35°.
D is the point on AC such that angle BDA is a right angle.

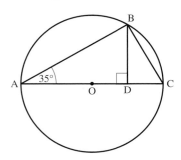

Diagram *not*
accurately drawn

a) Work out the size of angle BCA. Give reasons for your answer.

b) Calculate the size of angle DBC.

c) Calculate the size of angle BOA. [Edexcel]

8 A, B, C and D are four points on the circumference of a circle.
TA is the tangent to the circle at A. Angle DAT = 30°. Angle ADC = 132°.

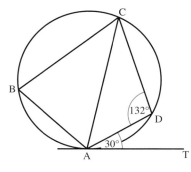

Diagram *not*
accurately drawn

a) Calculate the size of angle ABC. Explain your method.

b) Calculate the size of angle CBD. Explain your method.

c) Explain why AC cannot be a diameter of the circle. [Edexcel]

9 Points A, B and C lie on the circumference of a circle with centre O.
DA is the tangent to the circle at A. BCD is a straight line. OC and AB intersect at E.

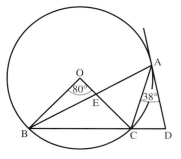

Diagram *not*
accurately drawn

Angle BOC = 80°. Angle CAD = 38°.

a) Calculate the size of angle BAC.

b) Calculate the size of angle OBA.

c) Give a reason why it is not possible to draw a circle with diameter ED through the point A. [Edexcel]

10 A, B, C and D are points on the circumference of a circle centre O.
A tangent is drawn from E to touch the circle at C.
Angle AEC = 36°. EAO is a straight line.

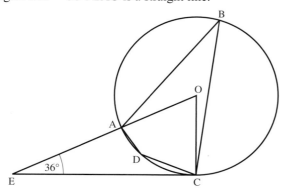

Diagram *not*
accurately drawn

a) Calculate the size of angle ABC. Give reasons for your answer.
b) Calculate the size of angle ADC. Give reasons for your answer.

[Edexcel]

11 P, Q and R are points on a circle. O is the centre of the circle.
RT is the tangent to the circle at R. Angle QRT = 56°.

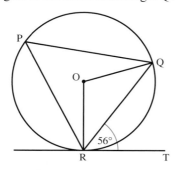

Diagram *not*
accurately drawn

a) Find **(i)** the size of angle RPQ and **(ii)** the size of angle ROQ.

A, B, C and D are points on a circle. AC is a diameter of the circle.
Angle CAD = 25° and angle BCD = 132°.

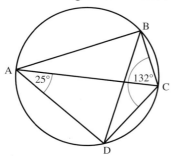

Diagram *not*
accurately drawn

b) Calculate **(i)** the size of angle BAC and **(ii)** the size of angle ABD.

[Edexcel]

12 In the diagram, PQ and RS are chords of the circle.
PX = 4 cm. QX = 6 cm. RX = 3 cm. SX = x cm.

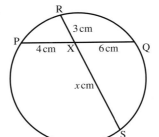

Work out the value of x.

13 A, B and C are points on a circle, centre O.
BCD is a straight line.

Work out the value of x.

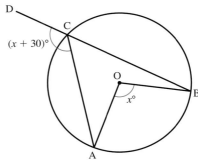

Diagram *not* accurately drawn

14 A, B and Y are points on a circle, centre O.

AB is a chord of the circle.

X is the midpoint of AB.

XOY is a straight line.

XY = 15 cm.

AB = 12 cm.

Calculate the diameter of the circle.

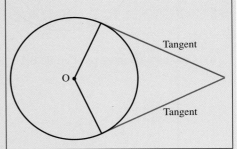

Key points

Basic circle properties

A line segment drawn from the centre of a circle to the midpoint of a chord will intersect the chord at right angles.	A tangent and radius meet at right angles.	The two external tangents to a circle are equal in length.
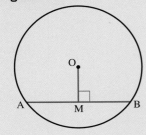	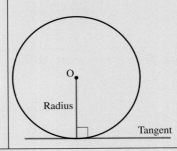	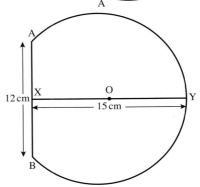

Circle theorems

The angle subtended by an arc at the centre of a circle is twice the angle subtended by the same arc at the circumference of the circle.	Angles subtended by an arc in the same segment of a circle are equal.	The angle subtended in a semicircle is a right angle.

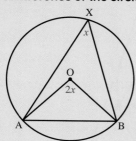
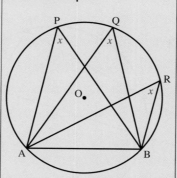
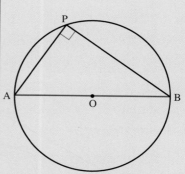

The angles subtended in opposite segments add up to 180°.	The angle between a tangent and chord is equal to the angle subtended in the opposite segment.

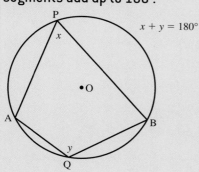

$x + y = 180°$

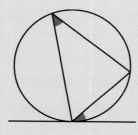

Intersecting chords (internal)	Intersecting chords (external)

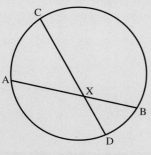

$AX \times BX = CX \times DX$

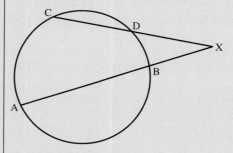

$AX \times BX = CX \times DX$

The nine-point circle theorem

This diagram shows the nine-point circle. Here are the instructions to make it.

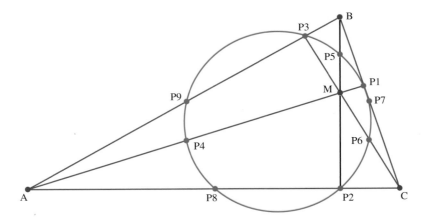

Start with any general triangle, whose vertices are A, B and C.

Construct points P1, P2, P3. Can you see what rule is used to locate them?

Construct the point M. Can you see how P1, P2, P3 are used to do this?

Construct points P4, P5, P6. Can you see what rule is used to locate them?

Construct points P7, P8, P9. Can you see what rule is used to locate them?

Then it should be possible to draw a circle that passes through all nine of the points:
P1, P2, P3, P4, P5, P6, P7, P8 and P9.

Look at the diagram, and see if you can figure out how the various points are constructed.
Use the internet to check that your deductions are correct.

Try to make a nine-point circle of your own, using compass constructions.
You might also try to do this using computer graphics software.

Which mathematician is thought to have first made a nine-point circle?

Can you find a proof that these nine points all lie on the same circle?

CHAPTER 18

Sets

In this chapter you will learn how to:

- use set notation to record and describe simple sets
- find the union and intersection of two given sets
- display sets using Venn diagrams
- work with complementary sets.

You will also be challenged to:

- investigate Russell's paradox.

Starter: **Does it all add up?**

In Class 1 there are 13 boys and 17 girls.
Andy says, 'There must be 30 children altogether in Class 1.'

In Class 2 there are 15 children who can write with their right hand and
15 children who can write with their left hand.
Britney says, 'There must be 30 children altogether in Class 2.'

In Class 3 there are 12 children who say their favourite sport is cricket and
18 children who say their favourite sport is tennis.
Carlo says, 'There must be 30 children altogether in Class 3.'

In Class 4 there are 10 children who can play the piano and 20 children who can
play the violin. No one plays both these instruments.
Donna says, 'There must be 30 children altogether in Class 4.'

Decide whether each person's statement is right or wrong, explaining your
answer carefully.

18.1 Introducing set notation

In mathematics a **set** is a collection of objects. A set may be described by a
rule, or by listing the members of the set in brackets. The members of a set are
usually called its **elements**. Here are some examples of sets:

A = {even numbers from 2 to 10 inclusive} = {2, 4, 6, 8, 10}

B = {prime numbers from 2 to 11 inclusive} = {2, 3, 5, 7, 11}

C = {all quadrilaterals}

D = {Sunday, Monday, Tuesday, Wednesday, Thursday, Friday, Saturday}

EXAMPLE

a) The set A = {1, 3, 5, 7, 9}. Suggest a rule that describes set A.

b) The set B = {months of the year beginning with the letter M}.
List the elements of B.

SOLUTION

a) A = {odd numbers from 1 to 9 inclusive}

b) B = {March, May}

The symbol $\in$ stands for 'is an element of' and is used to indicate membership of a set.

Thus kite $\in$ {all quadrilaterals}.

The crossed-out symbol $\notin$ stands for 'is not an element of'.

Thus 3 $\notin$ {even numbers from 2 to 10 inclusive}.

EXAMPLE

Insert the symbol $\in$ or $\notin$ to complete each statement below.

a) 17 ☐ {all prime numbers}

b) 60 ☐ {all square numbers}

c) pentagon ☐ {all quadrilaterals}

SOLUTION

a) 17 $\in$ {all prime numbers}

b) 60 $\notin$ {all square numbers}

c) pentagon $\notin$ {all quadrilaterals}

> Since 17 **is a member** of the set of all primes, use the $\in$ symbol.

> Since pentagon **is not a member** of the set of all quadrilaterals, use the $\notin$ symbol.

If an element x is a member of both set A and set B, it is said to belong to the set 'A intersection B' written A $\cap$ B. The symbol $\cap$ denotes **intersection**.

EXAMPLE

A = {3, 6, 9, 12} and B = {1, 3, 5, 7, 9, 11}. List the members of A $\cap$ B.

SOLUTION

The numbers 3 and 9 appear in both sets, so

A $\cap$ B = {3, 9}

If an element x is a member of either set A or set B (or both), it is said to belong to the set 'A union B', written $A \cup B$. The symbol $\cup$ denotes **union**.

EXAMPLE

$A = \{3, 6, 9, 12\}$ and $B = \{1, 3, 5, 7, 9, 11\}$. List the members of $A \cup B$.

SOLUTION

The numbers 3, 6, 9 and 12 appear in A, and, in addition, 1, 5, 7 and 11 appear in B, so

$$A \cup B = \{1, 3, 5, 6, 7, 9, 11, 12\}$$

The **number of elements** in a set A is denoted by the symbol $n(A)$. Thus if $A = \{3, 6, 9, 12\}$ then $n(A) = 4$, since the set contains four elements. If a set has no elements at all then it is **empty**. The **empty set** is also called the **null set**, and may be written as $\{ \}$ to show that there are no elements in it; it may also be written as $\varnothing$.

Examination questions often refer to two or more sets drawn from a **background set** or **universal set** $\mathscr{E}$.

EXAMPLE

$\mathscr{E} = \{$whole numbers from 1 to 20 inclusive$\}$
$A = \{$even numbers$\}$, $B = \{$square numbers$\}$
a) Write down the values of $n(\mathscr{E})$, $n(A)$ and $n(B)$.
b) Find the value of $n(A \cap B)$.

SOLUTION

a) Since there are 20 numbers in the universal set, $n(\mathscr{E}) = 20$.
$\qquad A = \{2, 4, 6, 8, 10, 12, 14, 16, 18, 20\}$ so $n(A) = 10$.
$\qquad B = \{1, 4, 9, 16\}$ so $n(B) = 4$.
b) $A \cap B = \{4, 16\}$ so $n(A \cap B) = 2$.

When a set contains lists of numbers it may be more efficient to describe it using an algebraic rule. Such rules often use inequality signs.

EXAMPLE

$\mathscr{E} = \{$integers$\}$
$A = \{x: 10 < x < 15\}$
$B = \{x: 5 \leqslant x \leqslant 7\}$
List the members of: **a)** A **b)** B.

> The colon : stands for 'such that'.
> Set A contains integers x such that $10 < x < 15$.

SOLUTION

a) The elements of A are integers above 10 and below 15, so
$\qquad A = \{11, 12, 13, 14\}$.
b) The elements of B are integers from 5 to 7 inclusive, so $B = \{5, 6, 7\}$.

EXERCISE 18.1

1 A = {odd numbers from 21 to 31 inclusive}
   a) List the members of set A.
   b) Write down the value of $n(A)$.

2 B = {5, 10, 15, 20, 25}. Suggest a rule for set B.

3 $\mathcal{E}$ = {integers}
   A = {x: $2 \leqslant x \leqslant 7$}
   B = {x: $5 < x < 12$}
   C = {x: $20 < x \leqslant 25$}

   List the members of:
   a) A   **b)** B   **c)** C

4 Insert the symbol $\in$ or $\notin$ to complete each statement below.
   a) 45 ☐ {multiples of 7}
   b) 24 ☐ {factors of 144}
   c) 2 ☐ {all prime numbers}

5 $\mathcal{E}$ = {items belonging to John}, A = {hardback books}, B = {fiction books}
   Describe carefully the set A $\cap$ B.

6 P = {all even numbers}, Q = {all odd numbers}
   a) Describe the set P $\cup$ Q.
   b) Describe the set P $\cap$ Q.

7 $\mathcal{E}$ = {whole numbers between 1 and 50 inclusive}
   A = {multiples of 3}, B = {multiples of 5}
   a) Find the values of:
      (i) $n(A)$ **(ii)** $n(B)$
   b) Describe carefully the set A $\cap$ B.

8 $\mathcal{E}$ = {polygons}
   A = {quadrilaterals}, B = {triangles}, C = {shapes with all sides equal in length}
   a) Explain why A $\cap$ B = $\varnothing$.
   b) Describe carefully the members of B $\cap$ C.
   c) Describe carefully the members of A $\cap$ C.

9 $\mathcal{E}$ = {positive integers less than 30}
   A = {multiples of 6}, B = {4, 8, 12, 16}, C = {multiples of 5}
   a) Find the values of:
      (i) $n(A)$ **(ii)** $n(B)$ **(iii)** $n(A \cap C)$
   b) What is the set B $\cap$ C?

10 $\mathcal{E}$ = {integers}
   A = {x: $10 < x < 20$}
   B = {x: $5 \leqslant x \leqslant 12$}
   a) List the members of A $\cap$ B.
   b) Write an algebraic rule for the members of A $\cup$ B.

18.2 Venn diagrams

The relationships between different sets can be illustrated by drawing **Venn diagrams**. In a Venn diagram each set is represented by a circle or oval, contained within a rectangle for the universal set. Depending on the elements in the sets it might be appropriate to have the sets overlapping, or even for one set to be completely contained inside another. These examples show some of the possibilities.

EXAMPLE

$\mathcal{E}$ = {positive integers less than 20}, A = {3, 6, 9, 12}, B = {5, 10, 15}
Illustrate these sets with a Venn diagram.

SOLUTION

Since there is no element common to both A and B, there is no overlap between the two sets.

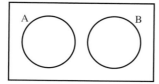

EXAMPLE

$\mathcal{E}$ = {all positive integers}, A = {2, 4, 6, 8, 10}, B = {5, 10, 15}
Illustrate these sets with a Venn diagram.

SOLUTION

Since 10 is in both A and B, there is now an overlap between the two sets.

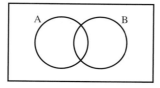

A set P is a **subset** of another set Q if every member of P is also a member of Q. Subsets should be illustrated on Venn diagrams by drawing them completely inside the set of which they are a subset. The notation P ⊂ Q is used to write 'P is a subset of Q'.

Venn diagrams need not be restricted just to two sets; in the examination you may need to work with three.

EXAMPLE

$\mathcal{E}$ = {all positive integers}, A = {1, 3, 5, 7, 9, 11, 13}, B = {1, 3, 5},
C = {all prime numbers}
Illustrate these sets with a Venn diagram.

SOLUTION

First, look at sets A and B. Every element of B is also in A, so B is a subset of A.

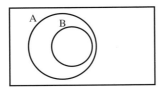

Now add set C. There are primes within B, and also within A (but not B), and also outside A. The set C must overlap all of these regions, thus:

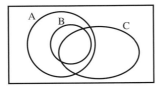

Sometimes numbers are marked on Venn diagrams, to show how many elements there are in each part of the diagram. These are useful for solving practical problems about overlapping sets.

EXAMPLE

$\mathcal{E}$ = {pupils in Year 7}, P = {piano players}, V = {violin players},
C = {cello players}
The Venn diagram shows the numbers of pupils playing various combinations of instruments.

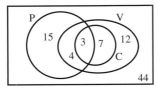

a) How many Year 7 pupils play the piano?
b) How many Year 7 pupils are there in total?
c) How many of the violin players also play the piano?
d) Fred says, 'Everyone in Year 7 who plays the cello also plays the violin.'
   Explain briefly whether Fred is right or wrong.

SOLUTION

a) Number of piano players $= 15 + 4 + 3$

$$= \underline{22}$$

b) Total number of Year 7 pupils $= 15 + 4 + 3 + 7 + 12 + 44$

$$= \underline{85}$$

c) Number of violin players who also play piano $= 4 + 3$

$$= \underline{7}$$

d) The diagram shows that C is a subset of V, so <u>Fred is right</u>.

If some of the information appears to be missing, it may be helpful to use algebra. The next example shows you how this might be done.

EXAMPLE

---

30 people took part in a transport survey. 16 of them said they owned cars, and 12 said they owned bicycles. 6 said they owned neither a car nor a bicycle.
a) Illustrate this information on a Venn diagram.
b) Work out the number of people who said they owned both a car **and** a bicycle.

SOLUTION

---

a) Let $\mathscr{E} = \{$people in survey$\}$, $C = \{$car owners$\}$, $B = \{$bicycle owners$\}$.
Let the number who said they owned both be x. Then the Venn diagram is:

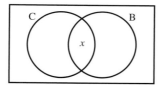

Since the total for C is 16, the figure for the part of C outside B is $16 - x$.
Similarly, the figure for the part of B outside C is $12 - x$.
The Venn diagram may now be completed:

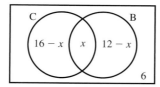

b) Since there were 30 people in the survey altogether, then

$$(16 - x) + x + (12 - x) + 6 = 30$$

$$34 - x = 30$$

$$x = 4$$

Thus <u>four people</u> said they owned both a car and a bicycle.

EXERCISE 18.2

1 The diagram shows two sets A and B.
On a copy of the diagram, draw a third set C
so that $C \subset A$ and $B \cap C = \varnothing$.

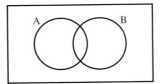

2 The diagram shows two sets P and Q.
On a copy of the diagram, draw a third set R
so that $R \subset (P \cap Q)$.

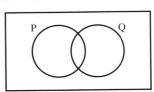

3 In class 4K there are 16 children who play hockey,
and 10 who play tennis. 4 children play both hockey and tennis.
There are 25 children in class 4K altogether.
  a) Copy and complete this Venn diagram to show the number of
    children in each region.
  b) How many children in class 4K play neither hockey nor tennis?

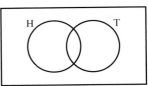

4 The diagram shows two sets A and B.
On a copy of the diagram, draw a third set C
so that $A \cap C = \varnothing$ and $B \cap C = \varnothing$.

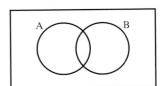

5 The diagram shows two sets P and Q.
On a copy of the diagram, draw a third set R
so that $P \cap R = \varnothing$ but $Q \cap R \neq \varnothing$.

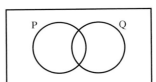

6 The 40 members of staff at a local company
were asked whether they liked Mexican, Chinese
or Indian food. The partly completed Venn
diagram shows information about their responses.
You are given the following additional information:

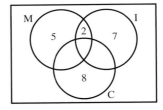

      4 people said they liked all three types of food.
      17 people said they liked Mexican food.
      20 people said they liked Chinese food.

  a) Mark this additional information of the diagram.
  b) How many people said they liked Indian food?
  c) How many people said they did not like all three types of food?

7 $\mathscr{E} = \{$integers from 1 to 20 inclusive$\}$
  $P = \{$prime numbers$\}$
  $Q = \{$even numbers$\}$
  $R = \{$square numbers$\}$
  a) Illustrate these sets with a Venn diagram.
  b) Describe the set $P \cap Q$.
  c) Describe the set $P \cap R$.

8 The 26 members of a scout group were asked whether they liked Ice Skating and whether they liked Bowling. 10 said they liked Ice Skating, 12 said they liked Bowling and 9 said they did not like either.
 a) Use this information to work out how many of the scouts liked both Ice Skating and Bowling.
 b) Illustrate this information on a Venn diagram.
 c) How many of the scouts liked exactly **one** out of Ice Skating and Bowling?

9 The diagram shows two sets P and Q.
 a) On a copy of the diagram, draw a set R so that $P \cup Q \subset R$.
 b) On a second copy of the diagram, draw a set T so that $T \subset Q$ and $T \cap P = \varnothing$.

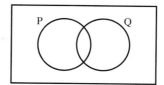

10 50 members of the public were asked which forms of transport they use regularly. (Each member of the public may state more than one form of transport if appropriate.) The partly completed Venn diagram shows information about their responses.

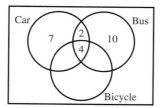

You are given the following additional information:

   18 people said they used bicycles.

   20 people said they used buses.

   15 people said they used cars.

 a) Copy and complete the diagram.
 b) Fred says, '13 people don't use any form of transport at all.'
   Explain carefully whether Fred's statement is correct or not.

18.3 Further Venn diagrams

The **complement** of a set A comprises all those elements which are not members of A, written A'.

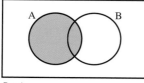

Set A

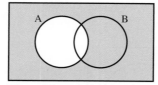

The complement A' of set A

Complements are sometimes encountered with brackets. Care must be taken to work from the insides of the brackets first.

EXAMPLE

The diagram shows sets A and B.

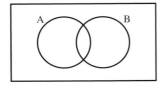

On two copies of this diagram, shade
a) A′ ∩ B′   **b)** (A ∩ B)′.

SOLUTION

a) A′ ∩ B′ is the intersection of the complement of A with the complement of B, that is, the region external to both A and B:

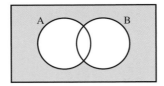

b) For (A ∩ B)′ it is necessary to find A ∩ B first, and then take the complement of that region, to obtain this result:

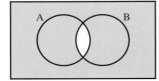

Harder problems might involve working with three sets. It is best to work in stages.

EXAMPLE

The diagram shows sets A, B and C. Shade the region corresponding to A′ ∩ (B ∩ C).

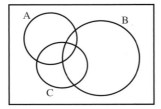

SOLUTION

First, shade A' on one copy of the diagram, and B ∪ C on another:

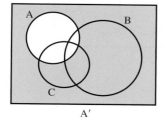

$$A'$$

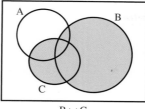

$$B \cup C$$

To find A' ∩ (B ∪ C) simply take the intersection between these two diagrams:

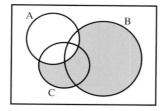

EXAMPLE

$\mathscr{E}$ = {positive integers less than 20}, A = {3, 6, 9, 12}, B = {5, 10, 15},
C = {6, 12, 18}
Illustrate these sets with a Venn diagram.

SOLUTION

Since there is no element common to both A and B, there is no overlap between
the two sets.

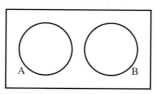

Now we need to add set C. C does have some elements in common with A,
but not with B:

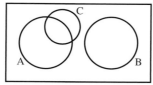

EXERCISE 18.3

1 Make two copies of the Venn diagram opposite.
Label them **a)** and **b)**.
   a) On diagram **a)**, shade the region $P \cup Q$.
   b) On diagram **b)**, shade the region $P' \cup Q$.

2 $\mathscr{E} = \{$integers from 1 to 15 inclusive$\}$
   A = $\{$even numbers$\}$
   B = $\{$multiples of 3$\}$

   a) List the members of A.
   b) Describe accurately the members of A'.
   c) List the members of $A \cap B'$.

3 Make three copies of the Venn diagram opposite.
Label them **a)**, **b)** and **c)**.
   a) On diagram **a)**, shade the region $A \cap B$.
   b) On diagram **b)**, shade the region $A' \cap B$.
   c) On diagram **c)**, shade the region $(A \cap B)'$.

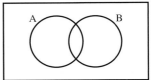

4 $\mathscr{E} = \{$letters of the alphabet$\}$
   P = $\{$vowels$\}$
   Q = $\{$A, B, C, D, E$\}$

   a) Describe accurately the members of P'.
   b) The set R is defined as $R = P' \cap Q$. State the value of $n(R)$.

5 Make a copy of the Venn diagram opposite,
and shade the region $A \cap B'$.

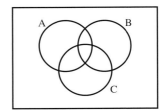

6 Make a copy of the Venn diagram opposite,
and shade the region $P' \cup Q'$.

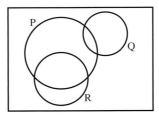

7 Make a copy of the Venn diagram opposite,
and shade the region $A' \cap B$.

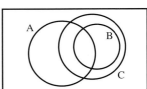

8 Make a copy of the Venn diagram below, and shade the region (P ∪ Q)′ ∩ R.

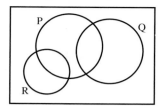

REVIEW EXERCISE 18

1 ℰ = {1, 2, 3, 4, 5, 6, 7, 8, 9, 10}
   A = {1, 3, 5, 6}
   B = {4, 5, 6, 7, 8}

  a) List the members of the set A′.
  b) List the members of the set A ∩ B.

2 Look at the five statements in the box.

> P ∩ Q = P   P ∪ Q = Ø   P ∩ Q = Ø
>
>   P ⊂ Q   Q ⊂ P

Choose a statement from the box that describes the relationship between the sets P and Q in each case.

a)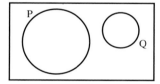

b)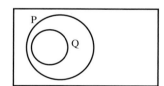

3 The diagram shows some information about the students in Year 10 at a small school.
   Some of them belong to the Chess Club, some to the Drama Club and some to the Astronomy Club.

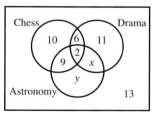

  a) How many Year 10 students belong to the Chess Club?
  b) There are 24 Year 10 students in the Drama Club. Find the value of x.
  c) There are 22 Year 10 students in the Astronomy Club. Find the value of y.
  d) Work out the total number of students in Year 10.

4 $\mathscr{E} = \{x: x \text{ is an integer and } 1 \leqslant x \leqslant 30\}$
A = {multiples of 3}
B = {multiples of 4}

a) List the members of the set A.
b) Explain whether it is true that $7 \in A \cup B$.
c) List the members of the set $A \cap B$.

5 $\mathscr{E} = \{\text{whole numbers}\}$
A = {multiples of 2}
B = {multiples of 3}

a) Make a copy of the Venn diagram below.
Draw a small cross to show which region would contain the number 15.

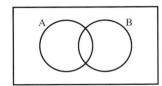

b) C = {multiples of 12}. Draw a circle on your diagram to show the set C.

6 A = {1, 2, 3, 4, 5, 6}
B = {2, 4, 6, 8}

a) List the members of $A \cup B$.
b) List the members of $A \cap B$.
c) The element x is such that $x \in B$ but $x \notin A$. Find x.

7 Make a copy of this Venn diagram showing the set A.

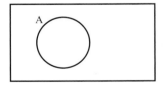

The set B is such that $A \cap B = \varnothing$.
The set C is such that $A \cup B$ is a subset of C.
Complete the Venn diagram to show the sets B and C.

8 $\mathscr{E} = \{\text{counties in England}\}$
A = {counties beginning with the letter C}
B = {counties with a coastline}

a) Describe carefully the set B′.
b) The set $A \cap B$ is not empty. Explain this statement as simply as you can.

9 Make a copy of the Venn diagram below.

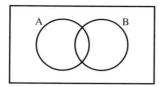

On your diagram add a set C so that A ∩ C = ∅ and B ∪ C = B.

10 ℰ = {whole numbers from 1 to 100 inclusive}
A = {multiples of 5}
B = {multiples of 7}

a) $n(A) = 20$ and $n(B) = 14$. Explain carefully what these statements mean.
b) C = A ∪ B. Explain why $n(C)$ 34.
c) List the members of A ∩ B.

11 A = {1, 2, 3, 4}
B = {1, 3, 5}

a) List the members of:
(i) A ∩ B
(ii) A ∪ B
b) Explain clearly the meaning of $3 \in A$.

[Edexcel]

12 Set P is shown on the Venn diagram.
Two sets Q and R are such that:

R ⊂ P

Q ∩ R = ∅

P ∪ Q = P

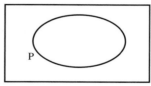

Complete the Venn diagram to show sets Q and R.

[Edexcel]

13 The universal set, ℰ = {whole numbers}
A = {multiples of 5}
B = {multiples of 3}

Sets A and B are represented by the circles in the diagram.

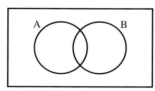

a) (i) On a copy of the diagram shade the region that represents the set A ∩ B′.
(ii) Write down three members of the set A ∩ B′.
C = {multiples of 10}
b) (i) On a copy of the diagram draw a circle to represent the set C.
(ii) Write down three members of the set A ∩ B ∩ C′.

[Edexcel]

Key points

1   A set is a collection of objects, called elements. The elements of a set can be described either by a rule or by listing them all.

2   Sets are usually defined with respect to a background or universal set, denoted by $\mathscr{E}$.

3   The relationship between sets may be shown using a Venn diagram.

4   The intersection of two sets A and B comprises all the elements which are members of both A and B, and is written A ∩ B.

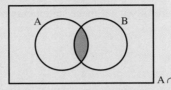

5   The union of two sets P and Q comprises all the elements which are members of either P or Q (or both), and is written P ∪ Q.

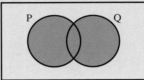

6   The complement of a set A comprises all those elements which are not members of A, and is written A′.

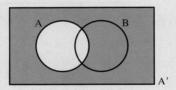

7   Summary of important symbols:

∈   is a member of

∉   is not a member of

n(A)   the number of elements in set A

∪   union

∩   intersection

⊂   is a subset of

∅   empty set

$\mathscr{E}$ universal set or background set

A′   complement of set A

Internet Challenge 18

Who shaves the barber?

In a small village all of the men are clean shaven. There is one village barber, and he shaves all the men (but only those men) who do not shave themselves.

So – who shaves the barber?

It would seem logical to suppose that the men in the village comprise two sets: set A comprises the men who shave themselves, while set B consists of the men who are shaved by the barber. Try drawing a Venn diagram to show these two sets. Should they overlap, or not?

You should find that it is not possible to answer the question of who shaves the barber, because the way in which the problem is stated contains an inbuilt contradiction. This is Russell's paradox, and serves to warn mathematicians that great care is needed when defining sets.

Use the internet to read more about Russell's paradox, and then find the answers to these questions.

1 What was Russell's full name?

2 When was he born?

3 Which Cambridge college did Russell enter, in 1890?

4 Russell wrote a famous book – the *Principia Mathematica* – with another author. Who was Russell's co-author?

5 The title *Principia Mathematica* was also used by another famous mathematician for his great work, published in 1687. Who was this?

6 What important result did Russell prove on page 362 of his *Principia Mathematica*?

7 What prestigious prize did Russell win in 1950?

8 In 1958 Russell became the first President of which fledgling organisation?

9 How many times did Russell marry?

10 When did he die?

CHAPTER 19

Working with data

In this chapter you will learn how to:

- use mean, median, mode and range
- find the interquartile range
- construct and interpret cumulative frequency diagrams and histograms.

You will also be challenged to:

- investigate correlation coefficients.

Starter: **Lies, damned lies and statistics**

Some people think that statistics can be used to deliberately mislead the reader.

Each of these statistical diagrams is misleading. See if you can spot how.

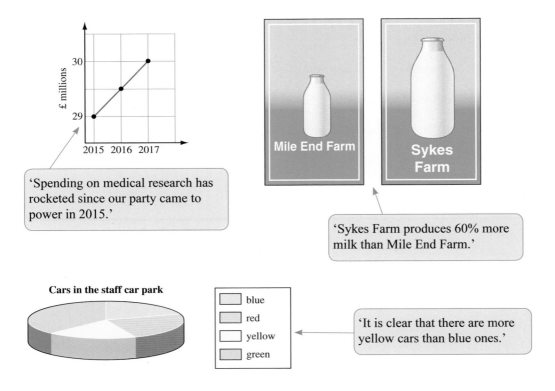

'Spending on medical research has rocketed since our party came to power in 2015.'

'Sykes Farm produces 60% more milk than Mile End Farm.'

'It is clear that there are more yellow cars than blue ones.'

19.1 Calculations with frequency tables

Some statistical data is **discrete**, that is, the values can be recorded exactly. If you asked members of your class how many brothers and sisters they have, their answers would be numbers like 0, 1, 2, 3 and so on. Other data may be **continuous**, that is, the data may take any value within a certain range. If you asked members of your class how tall they are, the answers could only be recorded to the nearest cm or nearest mm; you would never be able to write down someone's *exact* height. Discrete data usually arises when you gather data by counting, whereas continuous data tends to result when you are measuring something with a ruler, scales or a clock.

If a data set contains only a small number of discrete alternatives, with many repeats, then a stem and leaf diagram is inappropriate, and it is better to use a **frequency table** instead; this is simply a table that tells you how many times each value occurs in the data set. Frequency tables are also used to handle **grouped data** (both discrete and continuous), in which the values are banded together into **groups**, or **classes**.

You can use frequency tables to calculate three different types of **average**:
- the **mean** – the total of all the values, divided by how many there are;
- the **median** – the middle number in an ordered data set;
- the **mode** – the value that occurs most often.

You can also measure the spread of a data set by working out:
- the **range** – the difference between the highest and lowest values.
- the **interquartile range** – the range of the middle 50% of the data.

To work out the interquartile range, you need to find the **upper** and **lower quartiles** first; these are the values located one-quarter and three-quarters of the way through the data set, once it has been ordered.

The precise way in which these calculations are done will depend on whether you are working with discrete or continuous data, and also whether it has been grouped or not; in practice there are three slightly different cases you may encounter.

Case 1: Simple discrete frequency table

EXAMPLE

Boxes of matches are supposed to contain an average of 50 matches. Sophie decides to check this figure. She takes a sample of 20 boxes, and counts the matches in them.

a) Calculate the value of the mean.
b) Write down the mode.
c) Work out the range of the data.
d) Find the value of the median.
e) Find the lower and upper quartiles. Hence find the interquartile range.

Number of matches, x	Frequency, f
48	1
49	5
50	7
51	0
52	5
53	1
54	1

SOLUTION

a) For the value of the mean:

Number of matches, x		Frequency, f		x × f
48	×	1	=	48
49	×	5	=	245
50	×	7	=	350
51		0		
52		5		
53		1		
54		1		

Multiply x and f together:
$48 \times 1 = 48$
$49 \times 5 = 245$, etc.

Number of matches, x	Frequency, f	x × f
48	1	48
49	5	245
50	7	350
51	0	0
52	5	260
53	1	53
54	1	54
	20	1010

Then sum the numbers in each of the last two columns.

$$\text{Mean} = \frac{1010}{20}$$

$$= \underline{50.5 \text{ matches}}$$

Divide 1010 by 20 to get the mean.

b) For the mode:

Number of matches, x	Frequency, f
48	1
49	5
50	**7**
51	0
52	5
53	1
54	1

The highest frequency is 7, so the corresponding figure for the mode is 50.

The mode is <u>50 matches</u>

c) Regarding the range of the data, the lowest value is 48 and the highest is 54. Therefore the range is $54 - 48 = \underline{6\ matches}$

d) For the value of the median:

Number of matches, x	Frequency, f
48	1
49	5
50	7
51	0
52	5
53	1
54	1

$1 + 5 = 6$, which is not half way through the list of items, so the median is not 48 or 49.

$1 + 5 + 7 = 13$, which is more than half way through the list of items, so the median must be 50.

Alternatively, for n items, work out: $\dfrac{n+1}{2} = \dfrac{20+1}{2} = 10\frac{1}{2}$, so the median lies between the 10th and 11th values.

The median is $\underline{50\ matches}$

e) Using a similar counting method, one-quarter of the way through the list of items gives 49, and three-quarters gives 52, so:

Lower quartile $= \underline{49\ matches}$. Upper quartile $= \underline{52\ matches}$

Interquartile range $= 52 - 49 = \underline{3\ matches}$

The methods of the previous example may be adapted to deal with grouped discrete data. You will find, however, that your calculations might only provide estimates of the average or spread, if the raw values have been lost in the grouping.

Case 2: Grouped discrete frequency table

EXAMPLE

The marks for Year 10's end of term mathematics test are shown in the table.

Mark, m	Frequency, f
0 to 4	0
5 to 9	1
10 to 15	5
16 to 19	23
20 to 24	24
25 to 29	18
30 to 34	15
35 to 39	14
40 or more	0

a) Calculate an estimate of the value of the mean.
b) Write down the modal class.
c) Work out an estimate for the range of the data.
d) Suggest a value of the median.
e) Estimate the lower and upper quartiles, and the interquartile range.

SOLUTION

a)

Mark, m	Frequency, f		Midpoint, x		$x \times f$
0 to 4	0	×	2	=	0
5 to 9	1	×	7	=	7
10 to 14	5	×	12	=	60
15 to 19	23				
20 to 24	24				
25 to 29	18				
30 to 34	15				
35 to 39	14				
40 or more	0				

The 5 values between 10 and 14 are treated as if they are all equal to the midpoint, 12.

Mark, m	Frequency, f	Midpoint, x	$x \times f$
0 to 4	0	2	0
5 to 9	1	7	7
10 to 14	5	12	60
15 to 19	23	17	391
20 to 24	24	22	528
25 to 29	18	27	486
30 to 34	15	32	480
35 to 39	14	37	518
40 or more	0	–	0
	100		2470

As before, add up the totals in the f and the $x \times f$ columns.

$$\text{Estimated mean} = \frac{2470}{100}$$
$$= \underline{24.7 \text{ marks}}$$

The mean is only an estimate because midpoints have been used in the calculations. The raw values are not available. *Do not round off* to the nearest whole number!

b) The modal class is found by looking for the highest frequency:

Mark, m	Frequency, f
0 to 4	0
5 to 9	1
10 to 15	5
16 to 19	23
20 to 24	**24**
25 to 29	18
30 to 34	15
35 to 39	14
40 or more	0

24 is the highest frequency, so the modal class is 20 to 24.

The data is grouped, so you can't work out a single value as the mode, which is why the modal class is used instead.

The modal class is <u>20 to 24 marks</u>

c) Estimated range = $37 - 7 = $ <u>30 marks</u>

Use the midpoints of the highest and lowest class to estimate the range.

d) By counting through the table, the frequencies show that the 50.5th number occurs towards the end of the 20 to 24 group.

Thus the estimated median is <u>24 marks</u>

$$\frac{n+1}{2} = \frac{100+1}{2} = 50.5$$

so median lies between the 50th and 51st values.

e) For the quartiles, we need to look through the table to identify the locations of the 25.5th and 75.5th numbers. These are about 16 and 30, respectively.

Thus the estimated interquartile range is $30 - 16 = $ <u>14 marks</u>

Essentially the same method is used when dealing with grouped continuous data, though the notation used for writing the class intervals is a little different.

Case 3: Grouped continuous frequency table

EXAMPLE

The resistances (in ohms) of a sample of 50 electronic components are measured. The table shows the results:

Resistance, R, in ohms	Frequency, f
$80 \leqslant R < 90$	10
$90 \leqslant R < 100$	23
$100 \leqslant R < 110$	11
$110 \leqslant R < 130$	6

a) Calculate an estimate of the mean resistance.
b) State the class interval that contains the median resistance.
c) Explain whether it is possible for the range to be 48.

SOLUTION

a)

Resistance, R	Frequency, f	Midpoint, x	$x \times f$
$80 \leqslant R < 90$	10	85	850
$90 \leqslant R < 100$	23	95	2185
$100 \leqslant R < 110$	11	105	1155
$110 \leqslant R < 130$	6	120	720
	50		4910

> Once again, midpoint values must be used for each class.

$$\text{Estimated mean} = \frac{4910}{50}$$
$$= 98.2 \text{ ohms}$$

> $\dfrac{50+1}{2} = 25.5$
> so median lies between the 26th and 25.5th values.

b) The median will be in the 25.5th position, and, from the table:

  median lies in $90 \leqslant R < 100$ ohms

c) It *is* possible for the range to be 48 ohms, for example, $128 - 80$

EXERCISE 19.1

1   Tanya surveys cars driving past her school into town in the morning. She counts the number of occupants of each car. The frequency table shows her results.

Number of occupants	Frequency
1	7
2	13
3	11
4	9
5 or more	0

  a) How many cars did Tanya survey?
  b) Work out the mean number of occupants per car.

2   Maurizio records the number of people at work in his department each day. The frequency table shows his results.

Number of people at work	Frequency	Midpoint
25 to 29	3	
30 to 34	7	
35 to 39	11	
40 to 44	4	

a) Calculate an estimate of the mean number of people at work each day.

b) State the modal class.

c) Benoit says, 'The range is 20.' Explain why Benoit must be *wrong*.

3 Meera has measured the lengths of a sample of cucumbers from her stall. The frequency table shows her results.

Length of cucumber, L, in cm	Frequency
$20 \leqslant L < 22$	2
$22 \leqslant L < 24$	5
$24 \leqslant L < 26$	8
$26 \leqslant L < 28$	4
$28 \leqslant L < 30$	1

Work out an estimate for the mean length of a cucumber in this sample. Show all your working clearly.

4 A registrar records the ages of men who were married in her office during one week. The frequency table shows her results.

Age, A, in years	Frequency
$20 \leqslant A < 30$	4
$30 \leqslant A < 40$	7
$40 \leqslant A < 50$	6
$50 \leqslant A < 60$	3
$60 \leqslant A < 70$	1

Work out an estimate of the mean age of the men who married that week.
Give your answer of years correct to 3 significant figures.

5 Sean sets his friends a puzzle, which they solve under timed conditions. The frequency table shows the times taken.

Time, T, in minutes	Frequency
$5 \leqslant T < 10$	1
$10 \leqslant T < 15$	4
$15 \leqslant T < 20$	12
$20 \leqslant T < 25$	1
$25 \leqslant T < 30$	2

a) Calculate an estimate of the mean time taken to solve the puzzle.

b) Explain briefly why your answer can only be an estimate.

19.2 Solving problems involving the mean

Suppose 5 people have a mean age of 26. They meet a friend who is aged 32.

The mean age of all 6 people together is *not* $\dfrac{26 + 32}{2} = 29$. This figure is too

high. It does not take into account that there are 5 people with a mean age of 26 and only 1 person aged 32. The correct mean must be lower than 29.

The best way of solving problems like this is to work with overall totals. The following example shows you how.

EXAMPLE

Suppose 5 people have a mean age of 26. They meet a friend who is aged 32.
Work out the mean age of all 6 people.

SOLUTION

The total age of the 5 people is $26 \times 5 = 130$.
The sixth person is aged 32, so the total age of all 6 people is $130 + 32 = 162$.

Therefore the mean age for all 6 people is $\dfrac{162}{6} = \underline{27}$.

This method is also useful for finding the combined mean of two unequal-sized groups.

EXAMPLE

An office employs 12 men and 18 women. The mean age of the 12 men is 33.5 years. The mean age of the 18 women is 29 years. Work out the mean age of all 30 men and women.

SOLUTION

The total age of the men is $12 \times 33.5 = 402$.
The total age of the women is $18 \times 29 = 522$.
The total age of all 30 people is $402 + 522 = 924$.

Therefore the mean age of all 30 people is $\dfrac{924}{30} = \underline{30.8}$.

EXERCISE 19.2

1 5 babies have a mean weight of 8 kg. Another baby has a weight of 14 kg. Work out the mean weight of all 6 babies.

2 6 cats have a mean weight of 7 kg. 9 dogs have a mean weight of 12 kg. Work out the mean weight of all 15 animals.

3 There are 11 players in a cricket team. 5 of the players have a mean score of 27 runs. The other 6 players have a mean score of 32.5 runs. Find the mean score for all 11 players.

4 There are 330 students at a Sixth form college. The students take an IQ test designed to measure their intelligence. The table shows the results.
Calculate the mean IQ for all 330 students.

	Number of students	Mean IQ
Lower Sixth students	150	112.5
Upper Sixth students	180	118

5 Billy is measuring the lengths of ten steel rods. He measures them in two batches of 5. The mean length of the first 5 rods is 12.2 cm. The mean length of the second 5 rods is 12.8 cm.

Billy says, 'The mean length of all ten rods is $\dfrac{12.2 + 12.8}{2} = 12.5$.'

Explain carefully whether Billy is right or wrong.

6 The mean age of the 15 children in Class 2W is 11 years. When their teacher is included the mean increases to 13 years. Work out the age of the teacher.

7 Jim has taken 5 Mathematics examination papers, and has a mean score of 78 marks. How many marks does he need to score in the sixth paper, in order for his overall mean to increase to 80?

8 Nina plays 15 holes on a golf course. The mean number of strokes it takes her to complete the 15 holes is 3.6. She then plays another 3 holes. The mean number of strokes for all 18 holes is 4. Calculate Nina's mean score for the last 3 holes.

19.3 Histograms

The way in which statistical data is spread out is known as its **distribution**. Some data sets are distributed uniformly, but others have distinctive peaks and troughs.

The *shape* of a distribution may be studied by looking at a histogram. A histogram is used to display grouped continuous data. Each group is represented by a bar but the class intervals and bars may have unequal widths.

The height of each bar is called the frequency density where:

$$\text{Frequency density} = \frac{\text{frequency}}{\text{class width}}$$

A histogram should have:
- a continuous numerical scale on **BOTH** axes
- a vertical axis labelled 'frequency density'
- no gaps between the bars.

Rearranging the definition gives   frequency = frequency density × class width

For each bar,   area of bar = frequency density × class width

So the area of each bar gives the frequency.

EXAMPLE

The table shows information about the ages of all the residents of
Limetree Avenue.

Age (A years)	Frequency
$0 < A \leqslant 5$	5
$5 < A \leqslant 15$	12
$15 < A \leqslant 20$	6
$20 < A \leqslant 30$	3
$30 < A \leqslant 50$	15
$50 < A \leqslant 70$	8
$70 < A \leqslant 100$	4

a) Calculate the frequency density for each age group.

b) Draw a histogram for the data.

c) Use your histogram to calculate an estimate for the percentage of residents
over 40 years old. Give your answer correct to the nearest 1%.

SOLUTION

a) $\text{Frequency density} = \dfrac{\text{frequency}}{\text{class width}}$

> Notice the frequency densities for these groups are the same: 1.2 people per year in each age group.

Age (A years)	Class width (years)	Frequency	Frequency density
$0 < A \leqslant 5$	5	6	$6 \div 5 = 1.2$
$5 < A \leqslant 15$	10	12	$12 \div 10 = 1.2$
$15 < A \leqslant 20$	5	9	$9 \div 5 = 1.8$
$20 < A \leqslant 30$	10	3	$3 \div 10 = 0.3$
$30 < A \leqslant 50$	20	15	$15 \div 20 = 0.75$
$50 < A \leqslant 70$	20	8	$8 \div 20 = 0.4$
$70 < A \leqslant 100$	30	6	$6 \div 30 = 0.2$

b)

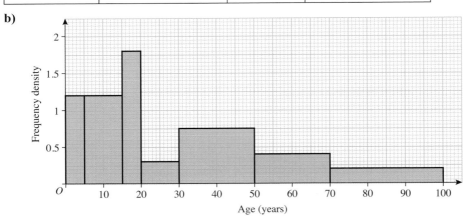

c) Since the area of each bar gives the frequency, you need to work out the blue shaded area on the diagram.

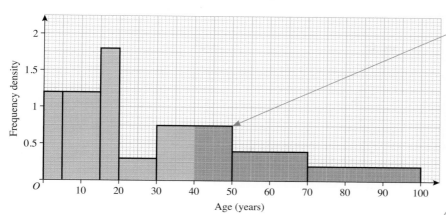

Assume that half the residents in this group are aged between 40 and 50; so you just need half of this bar.

Because you made this assumption, the answer to part c) can only be an estimate.

Number of residents over 40 is $10 \times 0.75 + 20 \times 0.4 + 30 \times 0.2 = 21.5$

Total number of residents is 59

Add the frequencies in the table or work out the area of the whole histogram.

So percentage of residents over 40 is $\dfrac{21.5}{59} \times 100 = 36\%$ to the nearest 1%

EXAMPLE

The resistances in ohms of a sample of 50 electronic components are measured. The table shows the results.

a) Calculate the frequency density for each group.
b) Draw a histogram for the data.

Resistance, R	Frequency, f
$80 \leqslant R < 90$	10
$90 \leqslant R < 100$	23
$100 \leqslant R < 110$	11
$110 \leqslant R < 130$	6

SOLUTION

a) The first three groups are of width 10, and the fourth is of width 20:

Resistance, R	Frequency, f	Group width	Frequency density
$80 \leqslant R < 90$	10	10	**1**
$90 \leqslant R < 100$	23	10	**2.3**
$100 \leqslant R < 110$	11	10	**1.1**
$110 \leqslant R < 130$	6	20	**0.3**

b)

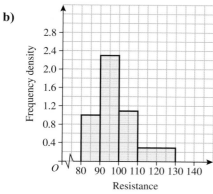

373

You can check the accuracy of the histogram by working out the area of each rectangle. For example, the first rectangle is $10 \times 1 = 10$, which matches the frequency given in the original table.

EXAMPLE

The histogram shows information about the time, in hours, students spent on homework one weekend.

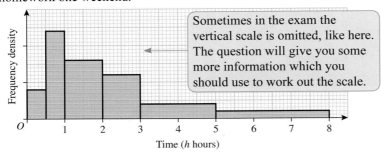

Sometimes in the exam the vertical scale is omitted, like here. The question will give you some more information which you should use to work out the scale.

4 students spent between 3 and 5 hours on their homework.

a) How many students spent less than 2 hours on their homework?
b) Calculate an estimate for the number of students who spent more than 6 hours on their homework.
c) In which interval does the median lie?

SOLUTION

a) Class width $= 5 - 3 = 2$

$$\text{Frequency density} = \frac{\text{frequency}}{\text{class width}} = \frac{4}{2} = 2$$

Use the information that '*4 students spent between 3 and 5 hours on their homework*' to work out the scale on the vertical axis.

The height of the 5th bar is 2 squares so each square on the vertical axis corresponds to a frequency density of 1.

The height of the 1st bar is 4 squares and its class width is $0.5 - 0 = 0.5$

The height of the 2nd bar is 12 squares and its class width is $1 - 0.5 = 0.5$

The height of the 3rd bar is 8 squares and its class width is $2 - 1 = 1$

Now use frequency = frequency density × class width to find the frequencies of the first 3 bars.

Total frequency is $(4 \times 0.5) + (12 \times 0.5) + (8 \times 1) = 2 + 6 + 8 = 16$

16 students spent less than 3 hours on their homework.

6–8 hours is $\frac{2}{3}$ of this class width.

The height of this bar is 1.

b) The final bar represents 5 – 8 hours and has a class width of $8 - 5 = 3$

The frequency of this group is $1 \times 3 = 3$.

$\frac{2}{3}$ of $3 = 2$ students

So you would expect $\frac{2}{3}$ of this group to spend more than 6 hours on their homework.

c) To find the median, you first need to work out the total frequency – it is easier to draw up a table.

Time spent (*h* hours)	Frequency
$0 < h \leq 0.5$	2
$0.5 < h \leq 0.5$	6
$1 < h \leq 2$	8
$2 < h \leq 3$	$1 \times 6 = 6$
$3 < h \leq 5$	4
$5 < h \leq 8$	3
Total frequency	29

From part a).

Given in the question.

From part b).

So the median is the $\dfrac{29+1}{2}$ = 15th value

Find a running total of the frequencies:
$2 + 6 + 8 = 16$

The 15th value lies in the $\underline{1 < h \leq 2}$ interval.

EXERCISE 19.3

1 The unfinished histogram and table give information about the heights in centimetres of the Year 11 students at Mathstown High School.

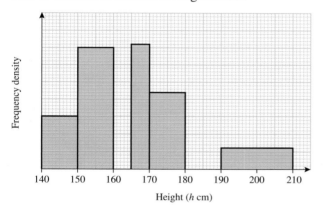

Height (*h* cm)	Frequency
$140 \leq h < 150$	15
$150 \leq h < 160$	
$160 \leq h < 165$	20
$165 \leq h < 170$	
$170 \leq h < 180$	
$180 \leq h < 190$	12
$190 \leq h < 210$	

a) Use the histogram to complete a copy of the table.

b) Use the table to complete a copy of the histogram.

[Edexcel]

2   One Monday, Victoria measured the time in seconds that individual birds spent on her bird table.
She used this information to complete the frequency table shown.

Time (t seconds)	Frequency
$0 < t \leqslant 10$	8
$10 < t \leqslant 20$	16
$20 < t \leqslant 25$	15
$25 < t \leqslant 30$	12
$30 < t \leqslant 50$	6

a) Use the table to complete a copy of the histogram.

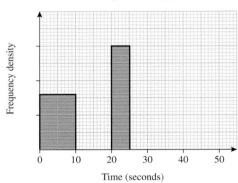

On Tuesday she conducted a similar survey and drew the following histogram from her results.

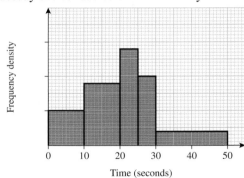

b) Use the histogram to complete a copy of the table.

Time (t seconds)	Frequency
$0 < t \leqslant 10$	10
$10 < t \leqslant 20$	
$20 < t \leqslant 25$	
$25 < t \leqslant 30$	
$30 < t \leqslant 50$	

[Edexcel]

3 The unfinished table and histogram show information about the time, in hours, for which cars were parked in a short-stay airport car park.

Time (t hours)	Frequency
$0 < t \leqslant 1$	20
$1 < t \leqslant 2$	28
$2 < t \leqslant 4$	34
$4 < t \leqslant 8$	52
$8 < t \leqslant 16$	

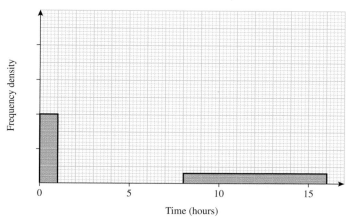

Time (hours)

a) On graph paper, use the information in the table to complete a copy of the histogram.
b) Use the information in the histogram to complete a copy of the table.

[Edexcel]

4 The histogram shows information about the weights of some potatoes.

The vertical scale has been omitted.

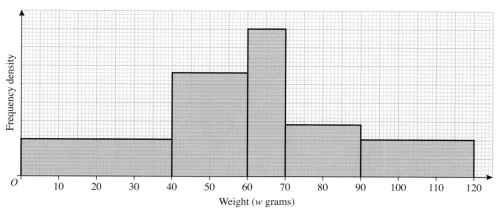

Weight (w grams)

15 potatoes weigh more than 90 grams.
a) Find the total number of potatoes.
b) Work out an estimate for the number of potatoes which weigh less than 50 grams.
c) In which interval does the median lie?

19.4 Cumulative frequency

You can obtain important information about a frequency distribution by keeping a running total of the frequencies. This is known as **cumulative frequency**. A graph of these running totals, known as a **cumulative frequency graph**, is a convenient way of determining the median and the quartiles of the distribution. From these, the interquartile range can be calculated.

EXAMPLE

The ages of the teachers on the staff at a school are shown in the table.

a) Draw up a cumulative frequency table for these data.
b) Construct a cumulative frequency graph, and use it to find values for the median, the lower quartile and the upper quartile.
c) Give the interquartile range.
d) Estimate the number of teachers who are older than 45.

Age (A) in years	Frequency
$20 \leqslant A < 30$	14
$30 \leqslant A < 40$	31
$40 \leqslant A < 50$	13
$50 \leqslant A < 60$	12
$60 \leqslant A < 70$	2

SOLUTION

a) The cumulative frequencies are shown in the table:

Age (A) in years	Cumulative frequency
$0 \leqslant A < 20$	0
$0 \leqslant A < 30$	14
$0 \leqslant A < 40$	45
$0 \leqslant A < 50$	58
$0 \leqslant A < 60$	70
$0 \leqslant A < 70$	72

0 staff are aged under 20
14 staff are aged 20–30
31 staff are aged 30–40
So the cumulative total is
$0 + 14 + 31 = 45$

Just keep adding each additional frequency on to the running total as you go down this column.

b)

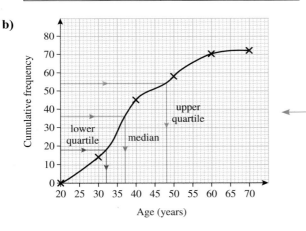

Three-quarters of $72 = 54$
Half of $72 = 36$
One-quarter of $72 = 18$
Use these values to read off the quartiles and the median.

From the graph: Upper quartile $= 48$ Median $= 37$ Lower quartile $= 32$

c) Interquartile range $= 48 - 32 = \underline{16}$

d)

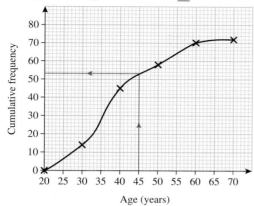

From the graph, the number of teachers aged 45 or under $= 53$
Therefore the number who are over 45 is $72 - 53 = \underline{19}$

EXERCISE 19.4

1 Fred carried out a survey of the time, in seconds, between one car and the next car on a road. His results are shown in the cumulative frequency graph below.

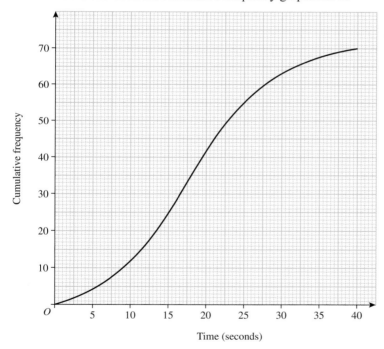

a) How many cars were in the survey?
b) Use the graph to estimate the median time.
c) Use the graph to estimate the percentage of times that were greater than 25 seconds. [Edexcel]

2   2400 people took an examination paper. The maximum mark for this paper was 80. The cumulative frequency graph below gives information about the marks. The pass mark was 44 marks.

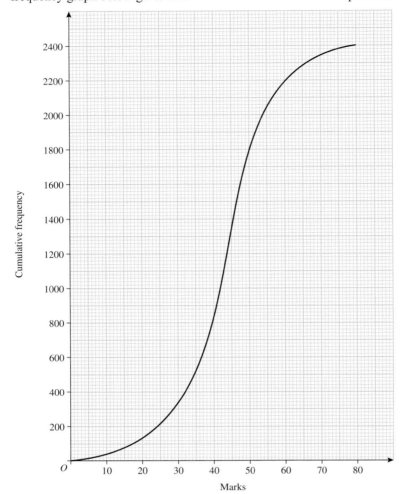

a) Use the cumulative frequency graph to estimate the number of people who did *not* pass this paper.

The same 2400 people took a second examination paper. The maximum mark was 80. This table gives information about the marks for the second paper.

b) On a copy of the grid, draw a cumulative frequency graph to show this information.

The same number of people did not pass this paper.

c) Use your cumulative frequency graph to estimate the pass mark for the second paper.   **[Edexcel]**

Mark	Cumulative frequency
0–10	20
0–20	80
0–30	200
0–40	500
0–50	900
0–60	1800
0–70	2200
0–80	2400

3 At a supermarket, members of staff
recorded the lengths of time that
80 customers had to wait in the queues
at the check-outs.
The waiting times are given in the
frequency table on the right.

Waiting time (t seconds)	Frequency
$0 \leqslant t < 50$	4
$50 \leqslant t < 100$	7
$100 \leqslant t < 150$	10
$150 \leqslant t < 200$	16
$200 \leqslant t < 250$	30
$250 \leqslant t < 300$	13

a) Complete a copy of the cumulative
frequency table.

Waiting time (t seconds)	Cumulative frequency
$0 \leqslant t < 50$	
$0 \leqslant t < 100$	
$0 \leqslant t < 150$	
$0 \leqslant t < 200$	
$0 \leqslant t < 250$	
$0 \leqslant t < 300$	

b) On a copy of the grid, draw a
cumulative frequency graph for
these data.
Use your graph to work out an
estimate for:
(i) the median waiting time,
(ii) the number of these
customers who had to wait
for more than 3 minutes.

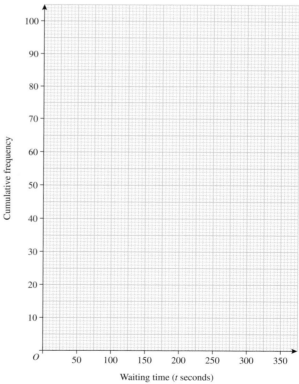

Waiting time (t seconds)

[Edexcel]

19.5 Median and quartiles for a discrete data set

If you are given a small list of numerical values, you can find the median and quartiles by counting. Here are the steps to follow:

- Arrange the values in order of size, smallest to largest.
- Count the number of items, and call this n.
- Add 1, to get $n + 1$ and then:

    the **lower quartile** is in the $\frac{1}{4}(n + 1)$th position

    the **median** is in the $\frac{1}{2}(n + 1)$th position

    the **upper quartile** is in the $\frac{3}{4}(n + 1)$th position.

If you are asked for the **interquartile range**, remember this is the upper quartile minus the lower quartile.

The interquartile range is a measure of spread. A low interquartile range shows that the data are close together and a high interquartile range shows that the data are spread out.

EXAMPLE

The time, in seconds, taken by a group of children to run around a short obstacle course are as follows:

Amy	Ben	Chay	Dyf	Eoin	Flora	Gaby	Hanita	Izi	Jai	Kit
26	19	22	27	23	21	18	27	21	24	21

a) Find the median time.
b) Find the interquartile range of the times.

The interquartile range for a second group of children is 3 seconds.

c) Which group of children will finish closer together?
Give a reason for your answer.

SOLUTION

Arranging the items in order we obtain 18, 19, 21, 21, 21, 22, 23, 24, 26, 27, 27 and there are 11 items in the list.
$n = 11$, so $n + 1 = 12$

$\frac{1}{4}$ of 12 $= 3$; $\frac{1}{2}$ of 12 $= 6$; $\frac{3}{4}$ of 12 $= 9$.

Thus the quartiles are in the 3rd and 9th positions, and the median is in the 6th position:

18, 19, 21, 21, 21, 22, 23, 24, 26, 27, 27

     ↑   ↑   ↑

    LQ   m   UQ

a) The median is 22 seconds.
b) The interquartile range is $26 - 21 = 5$ seconds.
c) The second group of children have a lower interquartile range, so the times for this group are more consistent and they will finish the race closer together.

EXERCISE 19.5

1 The lengths of seven films are 109, 115, 121, 124, 125, 133 and 151 minutes.
   a) Write down the median of these times.
   b) Find the upper and lower quartiles, and the interquartile range.

2 The voltages of 11 batteries are measured. Here are the results:

   1.23, 1.16, 0.98, 1.22, 1.45, 1.22, 1.17, 1.33, 1.28, 1.32, 1.29

   a) Arrange the values in order of size, and hence find the median.
   b) Work out the upper and lower quartiles.

3 A group of boys took a mathematics test. Here are their scores:

   63, 81, 42, 24, 38, 45, 60, 56, 62, 75, 92, 71, 85, 68, 88

   a) Work out the median score.
   b) Work out the interquartile range of the scores.

   A group of girls took the same test as the boys. The median score for the girls was 62. The interquartile range for the girls was 17.
   c) Compare the boys' scores and the girls' scores. Make two comments about their similarities or differences.

4 The maximum daily temperatures in °C were recorded at a weather station for the month of January. Here are the results:

   6, 3, 10, 11, 10, 6, 5, 2, −1, 0, −3, 1, 2, 2, 5, 8, 7, 4, 5, 5, 1, 0, 1, 3, 2, 1, 1, 2, 4, 1, 2

   a) Find the median temperature.
   b) Find the upper and lower quartiles.

5 Here are the numbers of runs scored by the batsmen of a cricket team during an innings:

   56, 7, 24, 11, 62, 47, 3, 0, 12, 13, 1

   a) Work out the median.
   b) Find the interquartile range.

REVIEW EXERCISE 19

1 75 boys took part in a darts competition.
Each boy threw darts until he hit the centre of the dartboard.
The number of darts thrown by the boys are grouped in this frequency table.
   a) Work out the class interval which contains the median.
   b) Work out an estimate for the mean number of darts thrown by each boy.

Number of darts thrown	Frequency
1 to 5	10
6 to 10	17
11 to 15	12
16 to 20	4
21 to 25	12
26 to 30	20

[Edexcel]

2 Ben asked 50 people how much they paid for a new computer.
The results are shown in this frequency table.

Price (£P)	Number of computers
$0 < P \leqslant 500$	7
$500 < P \leqslant 1000$	20
$1000 < P \leqslant 1500$	11
$1500 < P \leqslant 2000$	9
$2000 < P \leqslant 2500$	3

383

Calculate an estimate for the mean price paid for a new computer. [Edexcel]

3   Sybil weighed some pieces of cheese.
    The table gives information about her results.

Weight (w) grams	Frequency
$90 < w \leqslant 94$	1
$94 < w \leqslant 98$	2
$98 < w \leqslant 102$	6
$102 < w \leqslant 106$	1

Work out an estimate of the mean weight. [Edexcel]

4   The grouped frequency table shows information about the number of hours worked
    by each of 200 headteachers in one week.

Number of hours worked (t)	Frequency
$0 < t \leqslant 30$	0
$30 < t \leqslant 40$	4
$40 < t \leqslant 50$	18
$50 < t \leqslant 60$	68
$60 < t \leqslant 70$	79
$70 < t \leqslant 80$	31

a) Work out an estimate of the mean number of hours worked by the headteachers that week.

b) Complete a copy of the cumulative frequency table:

Number of hours worked (t)	Cumulative frequency
$0 < t \leqslant 30$	0
$0 < t \leqslant 40$	
$0 < t \leqslant 50$	
$0 < t \leqslant 60$	
$0 < t \leqslant 70$	
$0 < t \leqslant 80$	

c) On a copy of the grid on the right, draw a cumulative frequency diagram for your table.

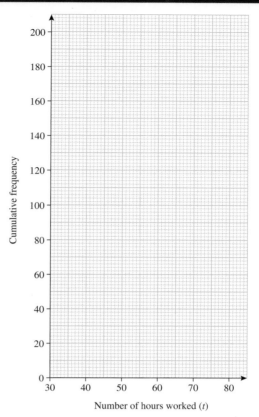

d) Use your graph to find an estimate for the interquartile range of the number of hours worked by the headteachers that week. Show your method clearly. **[Edexcel]**

5 Here are the marks scored in a maths test by the students in two classes:

| Class A | 2 | 13 | 15 | 16 | 4 | 6 | 19 | 10 | 11 | 4 | 5 | 15 | 4 | 16 | 6 |
| Class B | 12 | 11 | 2 | 5 | 19 | 14 | 6 | 6 | 10 | 14 | 9 |

a) Work out the interquartile range of marks for each class.
b) Use your answers to give one comparison between the marks of Class A and Class B. **[Edexcel]**

6 There are 15 students in Class A. In a test the students gained these marks:

2   1   2   5   5   6   9   2   5   6   7   5   6   5   6

a) Find the interquartile range of these marks.
b) The students in Class B took the same test. Their marks had a median of 7 and an interquartile range of 2.
Make two comparisons between the marks of the two classes. **[Edexcel]**

7 27 boys and 34 girls took the same test.
The mean mark of the boys was 76. The mean mark of the girls was 82.
Calculate the mean mark of all these students. Give your answer correct to 1 decimal place. **[Edexcel]**

8 A shop employs eight men and two women. The mean weekly wage of the ten employees is £396.
The mean weakly wage of the 8 men is £400.
Calculate the mean weekly wage of the 2 women. **[Edexcel]**

9 **a)** A youth club has 60 members. 40 of the members are boys. 20 of the members are girls.
   The mean number of videos watched last week by all 60 members was 2.8.
   The mean number of videos watched last week by the 40 boys was 3.3.
   Calculate the mean number of videos watched last week by the 20 girls.

   b) Ibrahim has two lists of numbers. The mean of the numbers in the first list is p. The mean of the
   numbers in the second list is q. Ibrahim combines the two lists into one new list of numbers.

   Ibrahim says, 'The mean of the new list of numbers is equal to $\dfrac{p+q}{2}$.'

   One of two conditions must be satisfied for Ibrahim to be correct.
   Write down each of these conditions.   [Edexcel]

10 The table shows the number of students in three groups attending Maths City High School last Monday.
   No student belonged to more than one group.

Group	Number of students
A	135
B	225
C	200

   Mrs Allen carried out a survey about the students' travelling times from home to school last Monday.
   Mrs Allen worked out that:
   • The mean time for Group A students was 24 minutes.
   • The mean time for Group B students was 32 minutes.
   • The mean time for Group C students was the same as the mean time for all 560 students.

   Work out the mean time for all 560 students.   [Edexcel]

11 The histogram gives information about the
   times, in minutes, 135 students spent on
   the internet last night.

   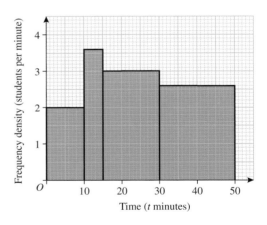

   Use the histogram to complete a copy
   of the table.

Time (t minutes)	Frequency
$0 < t \leqslant 10$	
$10 < t \leqslant 15$	
$15 < t \leqslant 30$	
$30 < t \leqslant 50$	
Total	135

   [Edexcel]

12 Mrs Smith asked the Year 11 students at her school how long they had spent revising Maths the evening before their Maths exam. The unfinished histogram and frequency table give information about their responses.

Revision time (t minutes)	Frequency
$20 \leqslant t < 25$	20
$25 \leqslant t < 40$	
$40 \leqslant t < 60$	
$60 \leqslant t < 85$	
$85 \leqslant t < 95$	32

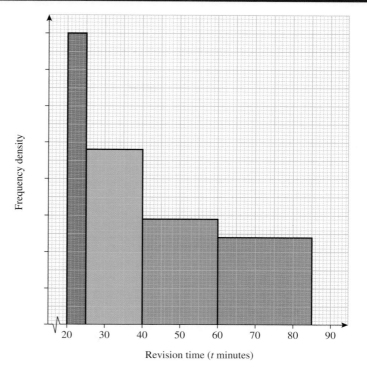

No student revised for less than 20 minutes. No student revised for 95 minutes or more.

a) Use the histogram to complete a copy of the table.

b) Use the table to complete a copy of the histogram.

[Edexcel]

13 The incomplete table and histogram give some information about the ages of people who live in a village.

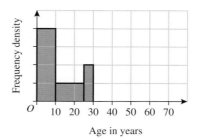

a) Use this information to complete a copy of the frequency table below.

Age (x) in years	Frequency
$0 < x \leqslant 10$	160
$10 < x \leqslant 25$	
$25 < x \leqslant 30$	
$30 < x \leqslant 40$	100
$40 < x \leqslant 70$	120

b) Complete a copy of the histogram.

[Edexcel]

14 The histogram shows information about the time, in minutes, some students spent travelling to school one morning.
The vertical scale on the histogram has been omitted.

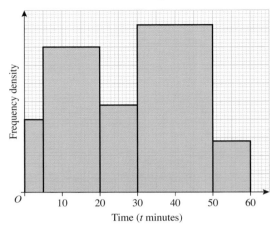

Time (*t* minutes)

Five students had a journey time of less than 5 minutes.
a) Find the total number of students.
b) Calculate an estimate for the number of students who had a journey time of between 10 minutes and 40 minutes.
c) In which interval does the median lie?

Key points

1 A good way to organise data is by using a frequency table. This works for discrete data, grouped discrete data and grouped continuous data, so it is a very flexible technique. You can calculate the mean (or an estimate of the mean) from a frequency table, but the methods differ slightly according to which of the three types of table it is.

2 A histogram uses rectangles to represent each group (or class) interval. The rectangles should touch, with no gaps between them. If you multiply the height of each rectangle (its frequency density) by the group width, the answer should be the same as the frequency for that group:

Frequency = frequency density × group width

Exam questions often ask you to use a given histogram to fill in a frequency table, or vice versa (or both), so make sure you know how to do this, using the formula above.

3 Cumulative frequency tables are made by keeping a running total of the frequencies in an ordinary table. You make a cumulative frequency graph (or diagram) by plotting the cumulative frequencies against the top ends of the class intervals, and not their midpoint values. A cumulative frequency graph allows you to read off values for the median and the upper and lower quartiles. You can then use these to work out the interquartile range.

4 You already know how to find the median of a discrete data set given in a list by counting the number of items, n, and then adding one to obtain $n + 1$. The median is then found by counting to the $\frac{1}{2}(n + 1)$th position, when the items are in order of size. In a similar way, for a discrete data set the lower quartile (LQ) is in the $\frac{1}{4}(n + 1)$th position and the upper quartile (UQ) is in the $\frac{3}{4}(n + 1)$th position.

Internet Challenge 19

Statistical quotes

In the Starter for this chapter, you saw ways in which data may be presented in a misleading way.

Some people think that statistics can be used deliberately to mislead.

Here are some famous quotations on this theme.

Use the internet to find out who said or wrote each one.

1 'There are three kinds of lies: lies, damned lies and statistics.'

2 'First get your facts; then you can distort them at your leisure.'

3 'The pure and simple truth is rarely pure and never simple.'

4 'Then there was the man who drowned crossing a stream with an average depth of six inches.'

5 'There are two kinds of statistics: the kind you look up and the kind you make up.'

6 'You know how dumb the average guy is? Well, by definition, half of them are even dumber than that.'

7 'Statistical thinking will one day be as necessary for efficient citizenship as the ability to read and write.'

8 'A statistician is a man who comes to the rescue of figures that cannot lie for themselves.'

9 'A single death is a tragedy, a million deaths is a statistic.'

10 'Everything should be made as simple as possible, but not simpler.'

CHAPTER 20

Probability

In this chapter you will learn how to:

- use theoretical and experimental probability
- make estimates for the results of a statistical experiment
- use conditional probability.

You will learn that:

- probabilities for a full set of mutually exclusive events add up to 1
- $P(A \text{ or } B) = P(A) + P(B)$ when A and B are mutually exclusive
- $P(A \text{ and } B) = P(A) \times P(B)$ when A and B are independent
- results of two trials may be displayed in a table or a tree diagram.

You will also be challenged to:

- investigate conditional probability.

Starter: Dice throws

One dice
Throw a single dice repeatedly – say 60 times. Record the scores obtained on each throw using a tally chart like this:

```
1 | |
2 |
3 | ||
4 |
5 | ||
6 | |
```

- Which score occurs most often? Is this what you expected?

Two dice
Now repeat the activity, but this time use *two* dice and record the total: the lowest score in your tally chart will be 2 and the highest will be 12.

- Which total score occurs most often? Is this what you expected?

20.1 Theoretical and experimental probability

If an experiment has a number of different outcomes, you can use **theoretical probability** to describe how likely the different outcomes are. Probability is a number between 0 and 1, with 0 indicating that something is impossible, and 1 that it is certain. This is a definition:

$$\text{Probability} = \frac{\text{number of favourable outcomes}}{\text{total number of (equally likely) possible outcomes}}$$

EXAMPLE

The whole numbers 1 to 10 are written on ten slips of paper, which are then folded and placed in a hat. One slip is removed at random.

Find the probability that it is:

a) 6   **b)** at least 8   **c)** 12

SOLUTION

a) There is only 1 way of getting the number 6:

$$P(6) = \tfrac{1}{10}$$

> $P(6)$ is a shorthand way of saying 'the probability of getting a 6'.

b) There are 3 ways of getting at least 8, namely 8, 9 or 10:

$$P(8, 9 \text{ or } 10) = \tfrac{3}{10}$$

c) There are 0 ways of getting 12:

$$P(12) = \tfrac{0}{10} = \underline{0}$$

When you conduct a probability experiment, or **trial**, then a particular result, A say, must either happen or not happen. So P(A does happen) + P(A does not happen) = 1. This leads to a very useful principle that:

$$P(A \text{ does not happen}) = 1 - P(A \text{ does happen})$$

EXAMPLE

The probability that it will snow on Christmas Day is 0.15.
Find the probability that it will not snow on Christmas Day.

SOLUTION

P(it will snow) + P(it will not snow) = 1

Therefore P(it will not snow) = 1 − P(it will snow)
$$= 1 - 0.15$$
$$= \underline{0.85}$$

Not all probability problems lend themselves to a theoretical approach. For example, if you throw a drawing pin in the air, and want to know the probability of it landing point up, there is no obvious theoretical method. Instead, you would do an experiment, and use the results to calculate an **experimental probability**.

EXAMPLE

A drawing pin is thrown in the air 20 times. It lands point up 13 times and point down 7 times.

a) Calculate the experimental probability that a single throw results in the pin landing point up.

b) The pin is then thrown 300 times. Work out an estimate for the number of times that the pin will land point down.

SOLUTION

a) The pin lands point up 13 times out of 20 trials.

$$P(\text{point up}) = \underline{\frac{13}{20}}$$

b) $\frac{7}{20}$ of the throws resulted in point down, so work out $\frac{7}{20}$ of 300:

$$300 \div 20 = 15, \text{ and then } 15 \times 7 = 105$$

Thus the estimated number of point down results in 300 trials $= \underline{105}$

Special care should be taken when finding probabilities from a two-way table.

EXAMPLE

The table shows the number of boys and girls in Years 7, 8 and 9 at a local school.

	Year 7	Year 8	Year 9	Total
Boys	72	66	70	208
Girls	48	44	60	152
Total	120	110	130	360

a) Find the probability that a randomly chosen member of Year 8 is a boy.

b) Find the probability that a randomly chosen boy is a member of Year 8.

c) Find the probability that a randomly chosen member of the school is a Year 8 boy.

SOLUTION

a) There are 110 students in Year 8. Of these, 66 are boys.

$$\text{Therefore } P(\text{boy}) = \frac{66}{110}$$
$$= \underline{\frac{3}{5}}$$

b) There are 208 boys. Of these, 66 are in Year 8.

$$\text{Therefore } P(\text{Year 8}) = \frac{66}{208}$$
$$= \underline{\frac{33}{104}}$$

c) There are 360 members of the school. Of these, 66 are Year 8 boys.

$$\text{Therefore } P(\text{Year 8 boy}) = \frac{66}{360}$$
$$= \underline{\frac{11}{60}}$$

EXERCISE 20.1

1 A bag contains 40 balls.
   There are 15 green balls and 9 yellow balls.
   The rest of the balls are red.
   A ball is chosen at random.
   Find the probability that this ball is:
   a) green
   b) red.

2 The colours of the rainbow are Red, Orange, Yellow, Green, Blue, Indigo and Violet.
   Crystal writes each of the colours of the rainbow on a slip of paper, and puts the seven slips in a bag.
   She then chooses a slip of paper at random.
   Find the probability that the colour written on it is:
   a) red
   b) not blue
   c) brown.

3 The probability that I win a game of chess when I play against my friend Boris is 0.22.
   Work out the probability that I do not win when I play him.

4 Tim throws 50 darts aimed at the bull on a dartboard.
   He hits the bull with 12 of his throws.
   a) Calculate Tim's experimental probability of hitting the bull.

   Tim throws more darts at the bull. He throws 400 darts in total.
   b) Estimate the number of times Tim hits the bull.

5 A bag contains 30 balls.
   10 of them are red and the rest are yellow or blue.
   There are three times as many yellow balls as blue balls.
   A ball is chosen at random.
   a) Work out the probability that it is red.
   b) Work out the probability that it is not red.
   c) Work out the probability that it is blue.

6 The table shows information about the
   numbers of books in a classroom library.

   a) Copy the table, and fill in the gaps.
   b) A science book is chosen at random.
      Work out the probability it is a hardback.
   c) A paperback is chosen at random.
      Work out the probability that it is a history book.

Book type	Science	History	Total
Hardback	10	18	
Paperback	20		
Total	30	50	80

7 Margaret is checking a book for spelling mistakes.
   She looks at a random sample of 50 pages.
   She finds spelling mistakes on four of the pages.
   a) Work out the experimental probability that a randomly chosen page contains no spelling mistakes.

   The book contains 350 pages altogether.
   b) Estimate the total number of pages that contain spelling mistakes.

8 A small school made a record of whether its pupils arrived on time or late yesterday.
It also noted what method of transport they used. The table shows the results.

Transport method	Walk	Bus	Other	Total
Late	3	21	5	29
On time	47	19	45	111
Total	50	40	50	140

a) Find the probability that a pupil who walked arrived late.
b) Find the probability that a pupil who arrived late came by bus.
c) The headmaster says these figures prove that pupils don't make enough of an effort to get to school on time. Do you agree or disagree? Explain your reasoning.

9 A café keeps records of how many drinks it sells during one day.

Type of drink sold	Tea	Coffee	Other	Total
Morning	78			110
Afternoon		48	20	
Total	100		20	200

a) Copy and complete the table.
b) Find the probability that a randomly chosen drink from the morning is coffee.
c) Find the probability that an afternoon drink is tea.
d) Next week the café expects to sell 1300 drinks. Estimate how many of these will be coffee.

10 In Class 3G at Mountview School there are 18 boys and 12 girls.
a) A pupil is chosen at random from Class 3G. Find the probability that it is a boy.
b) There are 720 pupils altogether at Mountview School. Estimate the total number of girls at the school.
c) Explain carefully why your estimate might not be very reliable.

11 ℰ = {the first 18 positive whole numbers}
A = {factors of 18}
B = {multiples of 3}
C = {even numbers}
a) Complete the Venn diagram to show the members of the sets A, B and C.

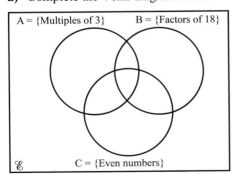

A bag contains 18 balls numbered 1 to 18.
A ball is taken at random from the bag.

b) Find the probability that the number on the ball is:
  i) a multiple of 3
  ii) an even number and a multiple of 3
  iii) both a factor of 18 and a multiple of 3
  iv) a member of set A ∪ B ∪ C.

12 A school science department has 100 students. The students study at least of the three sciences: Biology, Chemistry and Physics.
  20 students study all 3 sciences.
  10 study Physics and Chemistry.
  14 study Biology and Chemistry.
  59 study Physics.
  9 study Chemistry only.
  41 study exactly 2 sciences.
a) Draw a Venn diagram to show this information.
A science student is taken at random.
b) Find the probability that the student:
  i) studies Biology
  ii) studies exactly one science subject
  iii) studies Chemistry but not Physics.

20.2 Mutually exclusive outcomes

Supposing you roll a dice. The possible **outcomes**, or **events**, are scores of 1, 2, 3, 4, 5 and 6. These are said to be **mutually exclusive**, meaning that if one happens, the others do not.

Outcomes of experiments are not always mutually exclusive. For example, if you draw a card from a pack, one outcome is that it might be a heart, and another is that it might be a king. These are **not** mutually exclusive, since you can have a card that is both a heart and a king.

If the events A and B are mutually exclusive, then:

$$P(A \text{ or } B) = P(A) + P(B)$$

If you have a set of outcomes that cover all possible results of your experiment, then they are said to be **exhaustive**. The scores 1, 2, 3, 4, 5, 6 obtained by rolling a die are exhaustive, since no other score is possible.

If you have a set of outcomes that are mutually exclusive *and* exhaustive, then the corresponding probabilities must all add up to 1.

EXAMPLE

The breakfast menu at a works canteen is always one of four options. Some of these options are more likely to be on the menu than others. The table shows the options available on any day, together with three of the four probabilities.

Food	Sausages	Fish fingers	Bacon & eggs	Cereal
Probability	0.3	0.4	0.1	

a) Copy the table, and fill in the value of the missing probability.
b) Find the probability that the breakfast available on a randomly chosen day is:
   (i) cereal
   (ii) sausages or bacon & eggs
   (iii) not fish fingers.

SOLUTION

a)

Food	Sausages	Fish fingers	Bacon & eggs	Cereal
Probability	0.3	0.4	0.1	0.2

b) (i)   From the table, P(Cereal) = 0.2

   (ii) P(Sausages or Bacon & eggs) = 0.3 + 0.1
                                     = 0.4

   (iii) P(not Fish fingers) = 1 − 0.4
                             = 0.6

EXERCISE 20.2

1   A biased dice shows scores of 1, 2, 3, 4, 5, 6 with these probabilities:

Score	1	2	3	4	5	6
Probability	0.3	0.1	0.1	0.1	0.2	0.2

The dice is rolled once. Find the probability that the score obtained is:
a) 5   b) not 2   c) 3 or 4   d) an even number.

2   Tomorrow night Ginny is going to the cinema, **or** out for a pizza, **or** will stay in. She will do only one of those three things. The incomplete table shows some probabilities.

Activity	Cinema	Pizza	Stay in
Probability	0.25	0.45	

a) Copy and complete the table.
b) Which of the three things is Ginny most likely to do?
c) Find the probability that she **does not** go to the cinema.

3   Four types of bird visit my garden. The four types of bird are not all equally likely to be seen. The table shows the probability that a randomly observed bird is of a particular type.

Type of bird	Blackbird	Sparrow	Starling	Robin
Probability	0.35	0.25		0.1

a) Copy and complete the probability table.
b) Which type of bird is the most common in my garden?

A bird is observed at random. Find the probability that:
c) it is **not** a blackbird
d) it is a sparrow or a starling.

4 When Ricky plays computer chess he wins, draws or loses.
The probability that he wins is $\frac{1}{7}$ and the probability that he loses is $\frac{2}{7}$.
Find the probability that, in a randomly chosen game:
a) Ricky draws
b) Ricky draws or loses.

5 A biased spinner gives scores of 1, 2, 3 or 4.
The probability of getting 1 is 0.2. The probability of getting 2 is 0.3. The probability of getting 3 is 0.1.
a) Calculate the probability of getting an odd score.
b) Work out the probability of getting a score of 4.

6 Beth always eats one bowl of cereal for breakfast.
The probability that she chooses muesli is $\frac{1}{12}$.
The probability that she chooses porridge is $\frac{1}{4}$.
a) Work out the probability that she chooses muesli or porridge.
b) Work out the probability that she does **not** choose either of these cereals.

7 Alexei has a tin of crayons. Each crayon is either red or blue or yellow or green. The number of yellow crayons is the same as the number of green crayons.
Alexei chooses a crayon at random from the tin. The probability that he chooses a red crayon is 0.3.
The probability he chooses a blue crayon is 0.4.
a) Find the probability that the crayon is red or blue.
b) Find the probability that the crayon is yellow or green.
c) Find the probability that the crayon is green.

8 A garage tests cars to see if they are roadworthy.
The probability that a randomly chosen car has unsafe lights is 0.1.
The probability that a randomly chosen car has unsafe tyres is 0.15.

The probability that a randomly chosen car has unsafe lights or unsafe tyres must be 0.25.

Not necessarily – I think the correct figure could be lower than that.

Fred   Julie

a) Explain how you think Fred has obtained a probability of 0.25.
b) Explain why Julie thinks that Fred might not be correct.

9 A college has students who are in their first year, second year or third year. 36% of the students are in their first year, and 33% are in their second year. This information is shown in the probability table:

Year group	First year	Second year	Third year
Probability	0.36	0.33	

a) A student is chosen at random from the college. Work out the probability that it is a third year student.
b) 400 students from the college go to a Saturday night rock concert. Estimate the number of first year students who attend the concert.

10 A bag contains red, green, blue and yellow counters. In a probability experiment, one counter is chosen at random and removed from the bag. Its colour is noted, and it is returned to the bag. The table shows some probabilities for this experiment:

Colour	Red	Green	Blue	Yellow
Probability	0.28	0.44	0.08	

a) Find the probability of obtaining a yellow counter.
b) The experiment is carried out 250 times. Estimate the number of times a blue counter is obtained.
c) Rewrite the probability table to show the probabilities as fractions with the same denominator.
d) Altogether, the bag contains n counters. What is the lowest possible value of n?

20.3 Independent events

If two events are **independent**, then one does not interfere with the other.
For example, suppose you are tossing a coin and drawing a card from a pack.
The result of the coin (Heads or Tails) does not affect which card is selected.
You can solve simple problems on independent events by drawing up a table of equally likely outcomes.

EXAMPLE

Janine has two fair spinners. One of them has the numbers 1, 2, 3. The other has the numbers 3, 4, 5, 6. Janine spins both spinners, and adds the scores together.
a) Draw up a table to show all the possible total scores.
b) Work out the probability that Janine's total is 7.

SOLUTION

a)

+	3	4	5	6
1	4	5	6	⑦
2	5	6	⑦	8
3	6	⑦	8	9

Total scores of 7 occur in these three places.

b) There are 12 possible outcomes in the table. 3 of them correspond to a total of 7.

Therefore $P(\text{total of } 7) = \frac{3}{12}$
$$= \frac{1}{4}$$

If the events A and B are independent, then $P(A \text{ and } B) = P(A) \times P(B)$

This is a very useful method for working out the probability of a combined event, as in the next example.

EXAMPLE

When Laston drives home from work, he has to pass through two sets of traffic lights. The probability that the first set will be on red is 0.4. The probability that the second set will be on red is 0.3.

a) Calculate the probability that both sets will be on red.
b) Calculate the probability that neither set will be on red.

SOLUTION

a) P(both red) = P(first red and second red)
= P(first red) × P(second red)
= 0.4 × 0.3
= 0.12

b) P(neither red) = P(first not red and second not red)
= P(first not red) × P(second not red)
= (1 − 0.4) × (1 − 0.3)
= 0.6 × 0.7
= 0.42

EXERCISE 20.3

1 The probability that Brenda goes shopping this afternoon is 0.2. The probability that Millie goes shopping this afternoon is 0.45. Calculate the probability that:
a) Brenda and Millie both go shopping
b) Brenda goes shopping and Millie does not.

2 A fair spinner has four sides numbered 1, 2, 3 and 4. It is spun twice.
a) Draw up a two-way table to show the possible combinations of scores.
b) Find the probability that the total score is 7.
c) What is the most likely total score?

3 A bag contains two red balls and three green balls. A ball is chosen at random. It is then replaced, and a second ball is chosen.
a) Work out the probability that the first ball is red.
b) Work out the probability that both balls are red.

4 Tim has four green cards, with the numbers 1, 2, 3, 4 written on them. He has two red cards, with the numbers 5, 6 written on them. Tim chooses one red card and one green card at random.
a) Draw up a two-way table to show the possible combined events.
b) Work out the probability that the total of the two numbers on the cards is 8.
c) Work out the probability that the total of the two numbers on the cards is even.

5 Greg throws a red dice and a blue dice. Both dice are normal fair dice, labelled with the numbers 1, 2, 3, 4, 5, 6.
a) Write down the probability that the red dice shows a 4.
b) Work out the probability that the red dice shows a 4 but the blue dice does not.
c) Work out the probability that Greg throws a double 5.

6 Fahmi is playing a word game on a computer. The computer generates letters, from which she has to make a word. For each letter generated, the probability that it is a vowel is 0.3. Each letter is generated independently of the others.

a) Write down the probability that the first letter is *not* a vowel.

b) Work out the probability that the first two letters are both vowels.

Fahmi uses the computer to generate three letters.

c) Work out the probability that the first two letters are vowels but the third one is not.

7 Ian is a bowler in the school cricket team. Most of the time he bowls at his regular speed, but occasionally he tries to fool the batsman by bowling a slower ball. The probability that a randomly chosen ball is a slower one is $\frac{1}{12}$. Slower balls occur independently of each other.

Ian has two remaining balls to bowl.

a) Calculate the probability that the first ball is at his regular speed and the second one is a slower ball.

b) Calculate the probability that they are both regular speed balls.

Give your answers as exact fractions.

8 Each workday morning David selects a shirt and a tie at random from his wardrobe. His choice of shirt colour is made independently of his choice of tie colour. The table shows the probabilities of each colour being chosen.

Shirt colour	White	Blue	Grey	Pink
Probability	0.6	0.2	0.1	0.1

Tie colour	Red	Yellow	Green	Blue
Probability	0.4	0.3	0.25	0.05

a) Calculate the probability that David chooses a white shirt and a red tie.

b) Calculate the probability that David chooses a shirt and tie that are the same colour.

David's employer publishes a dress code document. It advises male workers not to wear a blue shirt with a green tie.

c) David works 300 days of the year. Estimate the number of these days on which he breaks the advice given in the dress code document.

20.4 Tree diagrams

In the previous two sections you have met and used these results:

Mutually exclusive outcomes: $P(A \text{ or } B) = P(A) + P(B)$

Independent events: $P(A \text{ and } B) = P(A) \times P(B)$

Some problems require using both of these principles together.
Tree diagrams are one way of doing this.

EXAMPLE

Oli and Dora are playing a game of Battleships. For each game, the probability that Oli wins is 0.4 and the probability that Dora wins is 0.6. They play two games.

a) Represent this information on a tree diagram.

b) Work out the probability that Oli wins both games.

c) Work out the probability that Dora wins exactly one game.

SOLUTION

a)

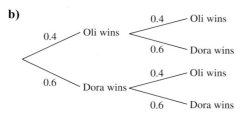

The first set of branches shows the possible results of the first game…

…while this second set shows the possible results of the second game.

b)

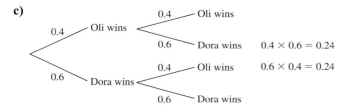

$0.4 \times 0.4 = 0.16$

$P(A \text{ and } B) = P(A) \times P(B)$ since the games are independent.

P(Oli wins both games) $= 0.4 \times 0.4$
$= \underline{0.16}$

c)

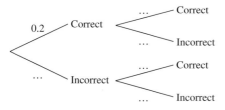

$0.4 \times 0.6 = 0.24$

$0.6 \times 0.4 = 0.24$

P(Dora wins exactly one game) $= 0.4 \times 0.6 + 0.6 \times 0.4$
$= 0.24 + 0.24$
$= \underline{0.48}$

EXERCISE 20.4

1 Fred is trying to guess the last two answers to a multiple-choice test. For each questions there is a probability of 0.2 that he is able to guess the answer correctly.

a) Copy and complete the tree diagram to show the possible outcomes to Fred's guesses.

```
                    ...  ┌─ Correct
         0.2   Correct <
        /           ... └─ Incorrect
       /
       \           ...  ┌─ Correct
         ...  Incorrect <
                    ... └─ Incorrect
```

b) Work out the probability that Fred guesses the correct answer to both questions.

c) Work out the probability that Fred guesses one answer correctly and one incorrectly.

2 The probability that Ravi beats Leon at tennis is $\frac{2}{3}$. They play two matches.

a) Draw a tree diagram to represent the possible outcomes of the two matches.

b) Work out the probability that Ravi wins both matches.

c) Work out the probability that Ravi wins the first match and Leon wins the second.

d) Work out the probability that Ravi wins just one of the two matches.

3 Joan and Simi both go shopping. The probability that Joan will buy a Lottery ticket is 0.3, and the probability that Simi will buy a Lottery ticket is (independently) 0.4.

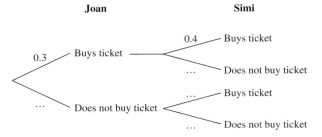

a) Copy and complete the tree diagram.

b) Find the probability that Joan and Simi buy one Lottery ticket in total.

c) Find the probability that neither Joan nor Simi buys a Lottery ticket.

4 Jesse has 20 CDs, 10 of which are classical. Camilla has 32 CDs, 24 of which are classical. Jesse and Camilla choose one CD each at random from their own collections.

a) Write down the probability that Jesse chooses a classical CD.

b) Write down the probability that Camilla chooses a classical CD.

c) Draw a tree diagram to illustrate their choices.

d) Hence find the probability that one of them chooses a classical CD and the other does not.

5 Andy is an archer. The probability that with any arrow he hits the centre of the target (the 'gold') is 0.3. The result of shooting any arrow is independent of previous shots.

Andy fires three arrows at the target.

a) Illustrate the possible results on a tree diagram.

b) Work out the probability that all three arrows hit the gold.

c) Work out the probability that exactly one of the three arrows hits the gold.

20.5 Conditional probability

In the last section you used tree diagrams to solve simple probability problems with two trials. The two trials were independent of each other, so the probabilities on the second set of branches did not depend on the first set.

Sometimes you will need to adjust the second set of probabilities, depending on the outcome from the first set. It sounds complicated, but the next example will show you how this is done.

EXAMPLE

A school choir is made up of 6 girls and 4 boys. The school's music teacher chooses, at random, two members of the choir to look after the sheet music.

a) Draw a tree diagram to illustrate this information.

b) Use your tree diagram to work out the probability that the teacher chooses two boys.

SOLUTION

a)

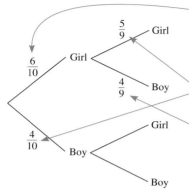

To begin with, there are six girls out of 10, so this probability is $\frac{6}{10}$.

There are four boys, so this probability is $\frac{4}{10}$.

By this stage, one girl has now been selected. So there are five remaining girls out of nine to select, giving a probability of $\frac{5}{9}$.

There are still four boys available, so this is $\frac{4}{9}$.

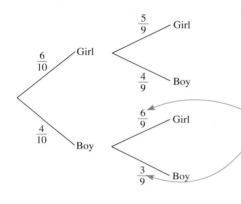

Here, one boy has been selected instead. So there are only three remaining boys out of nine, giving a probability of $\frac{3}{9}$, and six remaining girls, giving a probability of $\frac{3}{9}$.

b) From the tree diagram:

$$\text{Probability (two boys)} = \frac{4}{10} \times \frac{3}{9}$$
$$= \frac{12}{90}$$
$$= \frac{2}{15}$$

This topic is sometimes called **conditional probability**, since the probabilities on the second branch are conditional on what happened on the first branch. It is also often called **sampling without replacement**, since it describes situations when two items are chosen one after the other, without replacing the first one.

EXAMPLE

My bookcase contains ten novels and six science books. I choose two books at random. Find the probability that they are both the same type of book.

SOLUTION

The tree diagram shows the various possibilities.

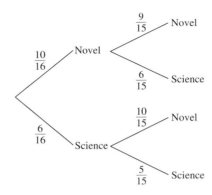

From the tree diagram:

Probability (two novels) $= \dfrac{10}{16} \times \dfrac{9}{15} = \dfrac{90}{240}$

Probability (two science) $= \dfrac{6}{16} \times \dfrac{5}{15} = \dfrac{30}{240}$

Therefore, probability (both the same) $= \dfrac{90}{240} + \dfrac{30}{240} = \dfrac{120}{240} = \underline{\dfrac{1}{2}}$

> Both novels OR both science.
> OR means ADD the probabilities.

EXERCISE 20.5

1   A bag contains ten coloured discs. Four of the discs are red and six of the discs are black.
    Asif is going to take two discs at random from the bag, **without** replacement.

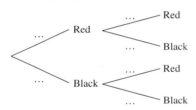

a) Copy and complete the tree diagram.
b) Work out the probability that Asif will take two black discs.
c) Work out the probability that Asif takes two discs of the same colour.   [Edexcel]

2   In a class of 12 students, nine use black pens and three use blue pens. Two students are chosen at random.
    a) Illustrate the situation with a tree diagram.
    b) Calculate the probability that both students use black pens.
    c) Calculate the probability that both students use the same colour pen.

3   A bag contains ten red balls and five black balls. Two balls are chosen, without replacement.
    a) Draw a tree diagram to show this information.
    b) Calculate the probability that the two balls are different colours.

4   In a class there are ten girls and eight boys. Three children are chosen at random from the class. Work out
    the probability that all three children are girls.

5   A box contains 12 red pencils, five green pencils and three blue pencils. Two pencils are chosen, without
    replacement.
    a) Draw a tree diagram to show this information.
    b) Calculate the probability that the two pencils are the same colour.

6 In a freezer there are 12 iced lollies. There are six orange lollies, four cola lollies and two strawberry lollies. Two lollies are chosen at random.
   a) Draw a tree diagram to show this information.
   b) Find the probability that both lollies are cola.
   c) Work out the probability that the two lollies are **different** flavours.

7 16 students attend a 'calculator allowed' mathematics exam. Ten of the students have solar-powered calculators, and the other six have battery-powered calculators.
   a) Two of the 16 students are chosen at random. Find the probability that they both have battery-powered calculators.
   b) Three of the 16 students are chosen at random. Find the probability that they all have solar-powered calculators.

REVIEW EXERCISE 20

1 Shreena has a bag of 20 sweets. Ten of the sweets are red. Three of the sweets are black. The rest of the sweets are white. Shreena chooses one sweet at random.
   What is the probability that Shreena will choose:
   a) a red sweet   **b)** a white sweet?   [Edexcel]

2 A bag contains coloured beads. A bead is selected at random.
   The probability of choosing a red bead is $\frac{5}{8}$.
   Write down the probability of choosing a bead that is **not** red from the bag.   [Edexcel]

3 A bag contains counters which are white or green or red or yellow.
   The probability of taking a counter of a particular colour at random is:

Colour	White	Green	Red	Yellow
Probability	0.15	0.25		0.4

Laura is going to take a counter at random and then put it back in the bag.
   a) **(i)** Work out the probability that Laura will take a red counter.
      (ii) Write down the probability that Laura will take a blue counter.

Laura is going to take a counter from the bag at random 100 times.
Each time she will put the counter back in the bag.
   b) Work out an estimate for the number of times that Laura will take a yellow counter.   [Edexcel]

4 Asif has a box of 25 pens. 12 of the pens are blue. 8 of the pens are black. The rest of the pens are red.
   Asif chooses one pen at random from the box. What is the probability that Asif will choose:
   a) a blue pen,   **b)** a red pen?   [Edexcel]

5 Sharon has 12 computer discs. Five of the discs are red. Seven of the discs are black.
   She keeps all the discs in a box.
   Sharon removes one disc at random. She records its colour and replaces it in the box.
   Sharon removes a second disc at random, and again records its colour.
   a) Copy and complete the tree diagram.

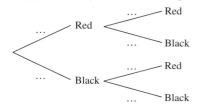

b) Calculate the probability that the two discs removed:

(i) will both be red

(ii) will be different colours.

[Edexcel]

6 Here is a five-sided spinner.

Its sides are labelled 1, 2, 3, 4, 5.

Alan spins the spinner and throws a coin.

One possible outcome is (3, Heads).

a) List all the possible outcomes.

The spinner is biased. The probability that the spinner
will land on each of the numbers 1 to 4 is given in the table.

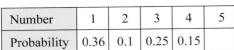

Number	1	2	3	4	5
Probability	0.36	0.1	0.25	0.15	

Alan spins the spinner once.

b) (i) Work out the probability that the spinner will land on 5.

(ii) Write down the probability that the spinner will land on 6.

(iii) Write down the number that the spinner is most likely to land on.

(iv) Work out the probability that the spinner will land on an even number.

Alan spins the spinner and throws a fair coin.

c) Work out the probability that the spinner will land on 3 and the coin will show Heads. [Edexcel]

7 Jason has 10 cups.

6 of the cups are Star Battle cups. 4 of the cups are Top Pops cups.

On Monday Jason picks at random one cup from the 10 cups.

On Tuesday he also picks at random one cup from the same 10 cups.

a) Copy and complete the probability tree diagram.

Monday　　　　**Tuesday**

```
                         0.6      Star Battle
              Star    /
      0.6  /  Battle  
         /           \  ...     Top Pops
       /
      \           ...     Star Battle
        \   Top  /
   ...    \ Pops 
            \   ...    Top Pops
```

b) Work out the probability that Jason will pick a Star Battle cup on both Monday and Tuesday.

c) Work out the probability that Jason will pick one of each type of cup. [Edexcel]

8 Tony carries out a survey about the words in a book. He chooses a page at random.

He then counts the number of letters in each of the first hundred words on the page.

The table shows Tony's results.

Number of letters in a word	1	2	3	4	5	6	7	8
Frequency	6	9	31	24	16	9	4	1

A word is chosen at random from the hundred words.

a) What is the probability that the word will have 5 letters?
The book has 25 000 words.

b) Estimate the number of 5-letter words in the book.

The book has 125 pages with a total of 25 000 words.
The words on each of the first 75 pages are counted. The mean is 192.

c) Calculate the mean number of words per page for the remaining 50 pages.   [Edexcel]

9 Jack has two fair dice.
One of the dice has 6 faces numbered from 1 to 6.
The other dice has 4 faces numbered from 1 to 4.
Jack is going to throw the two dice.
He will add the scores on the two dice to get a total.

Work out the probability that he will get:
a) a total of 7   **b)** a total of less than 5.   [Edexcel]

10 Chris is going to roll a biased dice. The probability he will get a six is 0.09.
a) Work out the probability that he will **not** get a six.
Chris is going to roll the dice 30 times.
b) Work out an estimate for the number of sixes he will get.
Tina is going to roll the same biased dice **twice**.
c) Work out the probability that she will get
(i) **two** sixes   **(ii)** **exactly one** six.   [Edexcel]

11 Helen and Joan are going to take a swimming test.
The probability that Helen will pass the swimming test is 0.95.
The probability that Joan will pass the swimming test is 0.8.
The two events are independent.

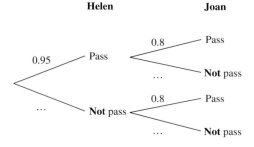

a) Copy and complete the tree diagram.
b) Work out the probability that both Helen and Joan will pass the swimming test.
c) Work out the probability that one of them will pass the swimming test and the other one
will **not** pass the swimming test.   [Edexcel]

12 A game is played with two spinners.
You multiply the two numbers on which
the spinners land to get the score.

Spinner A    Spinner B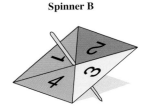

This score is $2 \times 4 = 8$

a) Copy and complete the table to show all the possible
scores. One score has been done for you.

Spinner B

	×	1	2	3	4
Spinner A	1				
	2				8
	3				

b) Work out the probability of getting a score of 6.
c) Work out the probability of getting a score that is an odd number.   [Edexcel]

13 Robin has 20 socks in a drawer. Twelve of the socks are red. Six of the socks are blue. Two of the socks
are white. He picks two socks at random from the drawer.
Calculate the probability that he chooses two socks of the same colour.   [Edexcel]

14 In a bag there are ten counters. Four of the counters are red and the rest of the counters are blue.
Ann and Betty are going to play a game.
Ann is going to remove two counters at random from the bag. She will not put them back.
If both counters are the same colour, Ann will win the game.

a) Calculate the probability that Ann will win the game.

If the counters are different colours, it will be Betty's turn.
Betty will remove one counter at random from the 8 counters still in the bag.
If the counter is red, Betty will win the game. If the counter is blue, the result will be a draw.

b) Calculate the probability that the result will be a draw.   [Edexcel]

15 The probability that Betty will be late for school tomorrow is 0.05.
The probability that Colin will be late for school tomorrow is 0.06.
The probability that both Betty and Colin will be late for school tomorrow is 0.011.
Fred says that the events 'Betty will be late tomorrow' and 'Colin will be late tomorrow' are
independent.
Justify whether Fred is correct or not.   [Edexcel]

16 A bag contains three black beads, five red beads and two green beads.
Gianna takes a bead at random from the bag, records its colour and replaces it.
She does this two more times.
Work out the probability that, of the three beads Gianna takes, exactly two are the same colour. [Edexcel]

Key points

1 Probability is used to describe the likelihood of an uncertain event. Probabilities always lie between 0 and 1; 0 indicates that an event cannot happen, while 1 indicates that it must happen.

2 If an experiment has a number of equally likely outcomes, then probability is defined as:

$$\text{Probability} = \frac{\text{number of favourable outcomes}}{\text{total number of (equally likely) possible outcomes}}$$

3 Not all probabilities can be computed in a theoretical way. This applies, for example, to the probability of a drawing pin landing point up. In such a case, you would need to carry out a large number of trials, and then use the formula:

$$\text{Probability} = \frac{\text{number of trials leading to a favourable outcome}}{\text{total number of trials conducted}}$$

4 A probability obtained from a theoretical calculation is called a theoretical probability, whereas a probability computed from the results of many trials is called an experimental probability.

5 The probability of an event not happening may be computed using this result:

$$P(A \text{ does not happen}) = 1 - P(A \text{ does happen})$$

6 If two events A and B are mutually exclusive, then they cannot both occur at the same time. The combined probability of A or B may be found by using this expression:

$$P(A \text{ or } B) = P(A) + P(B)$$

7 If two events A and B are independent, then the probability of the second is not dependent on the outcome of the first trial. The combined probability of A and B may be found by using this expression:

$$P(A \text{ and } B) = P(A) \times P(B)$$

8 Many problems involving more than one trial may be formulated using tree diagrams. Probabilities are multiplied along the branches to obtain the final combined probability.

9 When using tree diagrams to solve probability problems, watch out for situations where an object is selected and then not replaced before a second one is selected. Tree diagrams should be drawn, with differing probabilities on the second set of branches, according to what has happened on the first set of branches.

   In the examination, read questions such as this carefully, to see whether the first item has been replaced or not before the second is selected.

Internet Challenge 20

Cars and goats

Imagine you are playing a TV game show. You are shown three doors. Behind one door there is a brand new car. Behind each of the other two doors is a goat. Your goal is to try to win the car by guessing which door it is behind.

The doors are labelled 1, 2 and 3. You select a door, and the game show host then opens one of the two doors you did not select, to reveal a goat.

Now it is decision time! If you wish, you may switch from your original decision and pick the other door instead.

Does it matter? Is the probability of winning the car going to change if you switch doors?

You may like to try some probability experiments to help you decide what happens.

Then, you could use the internet to find out more about this problem, which is based on a real television show that was broadcast in America during the 1970s. The show was called *Let's Make a Deal*. There was much discussion in the media at the time, and some professional mathematics lecturers were drawn into giving the wrong answer!

CHAPTER 21

Direct and inverse proportion

In this chapter you will learn how to:

- construct formulae and solve problems using direct proportion
- construct formulae and solve problems using inverse proportion
- understand graphical representations of proportion
- use compound measures such as density, speed and pressure.

You will also be challenged to:

- investigate planets and their orbits.

Starter: A sense of proportion

Puzzle 1

In a biology experiment, a bean shoot grows by 35 mm in 5 days. How much would it be expected to grow in 7 days?

Puzzle 2

For a high altitude expedition to the Himalayas, a group of eight mountaineers plan to take bottled oxygen to breathe while they are sleeping. Their oxygen supply is enough to last them for 12 nights. But before they start, two are ill and leave the expedition. How many nights will the oxygen last the remaining mountaineers?

Puzzle 3

Nine lumberjacks can chop 6 piles of logs in 20 minutes. How long would it take for 18 lumberjacks to chop 15 piles of logs?

Puzzle 4

A farmer sees 10 crows perched on a fence. He shoots one. How many remain?

Puzzle 5

It takes 10 monkeys 10 minutes to eat 10 bananas. How long does it take 1 monkey to eat 1 banana?

21.1 Direct proportion

Suppose two variables are related in such a way that one of them is a constant multiple of the other, for example $y = 3x$.

Then y is said to be **directly proportional** to x.

The constant multiplier, in this case 3, is the **constant of proportionality**.

Direct proportion can be indicated by the symbol $\propto$, so $y \propto x$ is simply a short way of writing 'y is directly proportional to x'. Algebraically, you would write $y = kx$, where k represents the constant of proportionality.

EXAMPLE

Two quantities, x and y, are such that $y \propto x$.
a) Write an algebraic formula for y in terms of x.
b) When $x = 4$, $y = 10$. Find the constant of proportionality.
c) Using your formula, work out **(i)** the value of y when $x = 6$ and **(ii)** the value of x when $y = 8$.

SOLUTION

a) $y = kx$

b) Using $x = 4$ and $y = 10$:

$$10 = 4k$$
$$k = \frac{10}{4}$$
$$k = 2.5$$

Thus $y = 2.5x$

c) **(i)** When $x = 6$:
$$y = 2.5x$$
$$= 2.5 \times 6$$
$$= 15$$

(ii) When $y = 8$:
$$8 = 2.5x$$
$$x = \frac{8}{2.5}$$
$$= 3.2$$

Sometimes one variable is directly proportional to a power or square root of another. This situation can be developed in the same way, though the relationship between the original variables is in this case no longer linear.

EXAMPLE

y is directly proportional to the square of x. When x is 10, y is 300.
a) Obtain a formula for y in terms of x.
b) Use your formula to find the value of y when x is 20.
c) Use your formula to find the value of x when y is 675.

SOLUTION

a) y is directly proportional to the square of x, that is, $y = kx^2$

When x is 10, y is 300, so:

$$300 = k \times 10^2$$
$$300 = 100k$$
$$k = \frac{300}{100} = 3$$

Therefore $y = 3x^2$

> Substitute $x = 10$ and $y = 300$ into the equation $y = kx^2$ in order to find the value of k.

b) When $x = 20$:

$y = 3x^2$

$\quad = 3 \times 20^2$

$\quad = 3 \times 400$

$\quad = \underline{1200}$

c) When $y = 675$:

$675 = 3x^2$

$x^2 = \dfrac{675}{3}$

$\quad = 225$

$x = \pm\sqrt{225}$

$\quad = \underline{15 \text{ or } -15}$

Some questions will ask you to formulate the proportional equation from information given in words. You might also find that the constant of proportionality is a fractional amount: in that case, use a fraction rather than a decimal approximation. The next example demonstrates how this works.

EXAMPLE

The weight, W kg, of a spherical garden ornament is directly proportional to the cube of its diameter, d cm. An ornament of diameter 20 cm weighs 2 kg.
a) Find a formula for W in terms of d.
b) Find the weight of an ornament of diameter 15 cm.
c) Tony struggles to lift an ornament weighing 30 kg. Work out the diameter of this ornament. Give your answer correct to the nearest centimetre.

SOLUTION

a) Since W is directly proportional to the cube of d:

$W = kd^3$

When $d = 20$, $W = 2$

$2 = k \times 20^3$

$2 = 8000k$

$k = \dfrac{2}{8000}$

$k = \dfrac{1}{4000}$

so $\underline{W = \dfrac{d^3}{4000}}$

b) When $d = 15$:

$W = \dfrac{15^3}{4000}$

$\quad = \dfrac{3375}{4000}$

$\quad = 0.84375$

So weight of a 15 cm ornament:

$\quad = \underline{0.84 \text{ kg}} \text{ (2 d.p.)}$

c) When $W = 30$:

$30 = \dfrac{d^3}{4000}$

$d^3 = 30 \times 4000$

$\quad = 120\,000$

$d = \sqrt[3]{120000}$

$\quad = 49.324\,241\,48$

So diameter of the ornament:

$\quad = \underline{49 \text{ cm}} \text{ (to nearest cm)}$

EXERCISE 21.1

1 y is directly proportional to x, and $y = 21$ when $x = 7$. Find the value of y when $x = 13$.

2 y is directly proportional to x, and $y = 15$ when $x = 6$. Find x when $y = 40$.

3 Each of the tables below shows a set of matching x and y values, where y is directly proportional to x.
Find a formula for y in terms of x, and work out the missing values in each case.

a)

x	1	2	3
y	2	4	

b)

x	1	3		30
y		12	18	120

c)

x	12	18		102
y		3	7	

d)

x	2	8	
y		5	35

4 y is directly proportional to x, and it is known that $y = 10$ when $x = 15$.
   a) Obtain an equation for y in terms of x.
   b) Use your equation to find the values of:
      (i) y, when $x = 60$ **(ii)** x, when $y = 25$.

5 y is directly proportional to x^2, and it is known that $y = 20$ when $x = 10$.
   a) Obtain an equation for y in terms of x.
   b) Use your equation to find the values of:
      (i) y, when $x = 30$ **(ii)** x, when $y = 125$.

6 y is directly proportional to the square of x, and $y = 16$ when $x = 2$. Find y when $x = 3$.

7 y is directly proportional to the cube of x, and $y = 50$ when $x = 5$. Find y when $x = 8$.

8 y is directly proportional to $\sqrt{x}$. When $x = 25$, $y = 20$.
   a) Express y in terms of x.
   b) Hence, or otherwise,
      i) calculate the value of y when $x = 36$
      ii) calculate the value of x when $y = 12$.

9 F is directly proportional to the square root of s. When $s = \frac{1}{4}$, $F = \frac{1}{4}$.
   a) Express F in terms of s.
   b) Hence, or otherwise,
      i) calculate the value of F when $s = 16$
      ii) calculate the value of s when $F = 1$.

10 Dave is working out different ways of travelling round Europe during his gap year. He is making maps showing the cities he might visit. The time, T minutes, that it takes Dave to draw a map is directly proportional to the square of the number of cities, c, he puts on the map. A map with 8 cities takes 10 minutes to draw.
   a) Find a formula for T in terms of c.
   b) Work out how long it would take to draw a map with 12 cities.
   c) Dave eventually spent an hour and a half making his map. How many cities did he decide to include?

11 My computer has a program that can work out the decimal value of π to a large numbers of digits. You can specify the number of digits required. The time it takes is directly proportional to the square of the number of digits specified.

The computer can work out π to 5000 significant figures in exactly half a second.
a) Find a formula for the number of digits, n, that the computer can work out in t seconds.
b) Use the formula to find out how long it would take to calculate π to one million digits.
c) How many digits can the computer work out in 10 minutes?

12 Square carpet tiles are sold in three sizes – small, medium and large. The cost of a carpet tile is directly proportional to the square of the diagonal dimension of the tile.

Small tiles cost £2.70 each and have a diagonal of length 15 cm.
a) A medium tile has diagonal of length 20 cm. Work out its cost.
b) A large tile costs £10.80. Work out the length of its diagonal dimension.

21.2 Inverse proportion

Consider a rectangle whose area is fixed at 20 cm². Here are some possible dimensions for the rectangle:

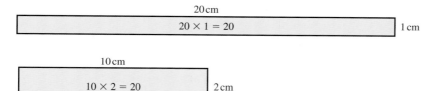

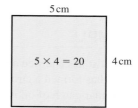

Notice that, as one dimension goes down in size, the other one increases. The **product** of the two dimensions remains **constant**, that is, $xy = 20$, where x and y are the length and breadth of the rectangle respectively.

You could also write this as $y = \dfrac{20}{x}$. This type of relation is called **inverse proportion**.

You could write $y \propto \dfrac{1}{x}$.

More formally, you would write $y = \dfrac{k}{x}$,

where k is the constant of proportionality.

EXAMPLE

Two quantities x and y are such that y is inversely proportional to x.
When $x = 20$, $y = 6$.
a) Find a formula connecting x and y.
b) Work out the value of y when $x = 40$.
c) What is the value of x when y is 2?

SOLUTION

a) Let $y = \frac{k}{x}$. Substitute $x = 20$, $y = 6$ into this to obtain:

$$6 = \frac{k}{20}$$

$$k = 20 \times 6$$

$$k = 120$$

Thus the required formula is $y = \dfrac{120}{x}$

b) When $x = 40$:

$$y = \frac{120}{x}$$

$$= \frac{120}{40}$$

$$y = 3$$

c) When y is 2:

$$y = \frac{120}{x}$$

$$2 = \frac{120}{x}$$

$$2x = 120$$

$$x = \frac{120}{2}$$

$$x = 60$$

As with direct proportion, some inverse proportion problems can involve powers such as squares.

EXAMPLE

A scientist has squirted a droplet of oil on to the surface of a container of water. The oil has formed a circular patch, and is slowly expanding. The thickness of the circular patch of oil is inversely proportional to the square of its radius.

When the patch has radius 3 cm, its thickness is 60 microns
(1 micron $= 1 \times 10^{-6}$ m). At any time, the radius is r cm and the thickness is d microns.
a) Write down a formula connecting d and r.
b) Find the thickness of the patch when the radius is 6 cm.
c) The scientist hopes to end up with an oil patch that is exactly 1 micron thick. What will the radius of the circular patch become if she can achieve this aim?

SOLUTION

a) Since d is inversely proportional to r squared, we have $d = \dfrac{k}{r^2}$

With $r = 3$ and $d = 60$:

$$d = \frac{k}{r^2}$$

$$60 = \frac{k}{9}$$

$$k = 9 \times 60$$

$$k = 540$$

Therefore the required formula is $d = \dfrac{540}{r^2}$

b) When $r = 6$,

$$d = \frac{540}{6^2}$$

$$= \frac{540}{36}$$

$$= \underline{15 \text{ microns}}$$

c) When $d = 1$:

$$1 = \frac{540}{r^2}$$

$$r^2 = 540$$

$$r = \sqrt{540}$$

$$= \underline{23.2 \text{ cm}} \text{ (correct to 3 s.f.)}$$

EXERCISE 21.2

1 y is inversely proportional to x, and $y = 20$ when $x = 5$. Find the value of y when $x = 4$.

2 y is inversely proportional to x, and $y = 12$ when $x = 1$. Find x when $y = 4$.

3   Each of the tables below shows a set of matching x and y values, where y is inversely proportional to x.
Find a formula for y in terms of x, and work out the missing values in each case.

a)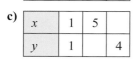

x	1		30
y	60	12	

b)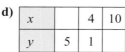

x		10	40
y	45		9

c)

x	1	5	
y	1		4

d)

x		4	10
y	5	1	

4 r is inversely proportional to t, and $r = 7$ when $t = 42$.
a) Obtain an equation for r in terms of t.
b) Use your equation to find the values of **(i)** t, when $r = 14$, **(ii)** r, when $t = 49$.

5 p is inversely proportional to s, and it is known that $p = 12$ when $s = 15$.
a) Obtain an equation for p in terms of s.
b) Use your equation to find the values of **(i)** p, when $s = 9$, **(ii)** s, when $p = 60$.

6 y is inversely proportional to the square root of x, and $y = 9$ when $x = 16$. Find y when $x = 8$.

7 y is inversely proportional to the cube of x, and $y = 1$ when $x = 4$. Find x when $y = 8$.

8   Light intensity follows an inverse square law, that is, the intensity of light is inversely proportional to the square of the distance to the source of the light.

A lamp is at a distance of 10 metres from a light detector, and it produces an intensity reading of 300 units.
a) Find the intensity reading for a similar lamp, at a distance of 16 metres.
b) How distant would the lamp need to be to in order to produce an intensity reading of 500 units?

9   A bowling machine is able to simulate the action of a fast bowler at cricket. The machine can project cricket balls at a batsman. The velocity, v km/h, at which a ball is projected is inversely proportional to the mass, m kg, of the ball.

A regular cricket ball has a mass of 0.156 kg. The machine can project it at 54 km/h.
a) Write a formula to express v in terms of m.
b) A lightweight ball has a mass of 0.144 kg. How fast can the machine project the lightweight ball?

10 During the run-up to an election, a statistician is trying to forecast what percentage of the votes will be cast for each party. She takes a sample of voters, and uses their responses to make her forecast. The statistician knows that her forecast will only be accurate to within a certain amount, known as the standard error. The size of the standard error is inversely proportional to the square root of the number of voters in the sample.

When a sample of 1067 voters is taken, the standard error is 3%.
The statistician wants to improve her accuracy.
Work out the number of voters to be sampled if the standard error is to be reduced to:
a) 2%
b) 1%.

21.3 Graphical representation of direct and inverse proportion

Look at the three graphs below.
Do you think any of them indicate that x and y are in direct proportion?

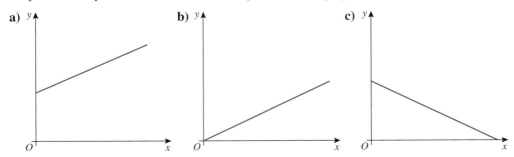

a) **b)** **c)**

All three graphs are linear, but only **b)** shows direct proportion, since the other two do not pass through the origin.

If y is **directly proportional** to x, then the graph of y against x *must* be a **straight line passing through the origin**.

Now look at the three graphs below.

Do you think any of them indicates that x and y are in inverse proportion?

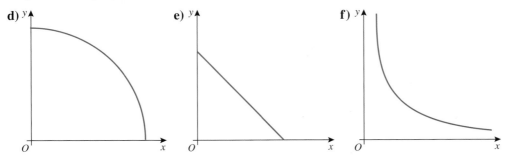

d) **e)** **f)**

All three graphs show that y decreases as x increases, but only **f)** does so in a way that matches the graph of $y = \dfrac{1}{x}$.

This distinctively shaped graph is called a **hyperbola**.

Note that **e)** shows a constant rate of decrease of y as x increases, but this is **not** the same as inverse proportion.

EXAMPLE

Look at the graphs and equations below. Only one of them is an example of y being directly proportional to x. Decide which one it is, giving a graphical explanation for your choice.

a) $y = 3x^2$

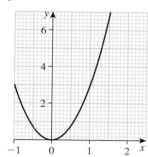

b) $y = 10 - x$

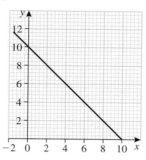

c) $y = 2x$

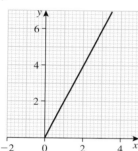

d) $y = x + 2$

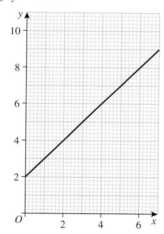

SOLUTION

For direct proportion, the graph of the equation must be a straight line through the origin.

a) $y = 3x^2$ ✗ This passes through the origin but is not a straight line, due to the square term.

b) $y = 10 - x$ ✗ This is a straight line but does not pass through the origin – it has a y intercept of 10.

c) $y = 2x$ ✓ $y = 2x$ is of the form $y = kx$, so it indicates direct proportion. Its graph is a straight line through the origin.

d) $y = x + 2$ ✗ This is a straight line but does not pass through the origin – it has a y intercept of 2.

Thus $\underline{y = 2x}$ is the case where y is proportional to x

EXERCISE 21.3

Copy these six sketch-graphs and equations.
Alongside each, write either 'Direct Proportion', 'Inverse Proportion' or 'Neither'.

1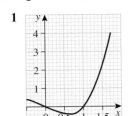

2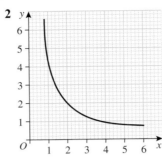

3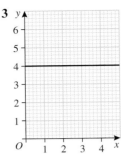

4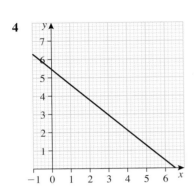

5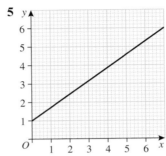

6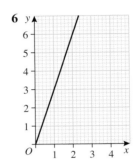

21.4 Compound measures

A compound measure is made up from two (or more) other measures.
Speed is an example of a compound measure. It is a measure of how fast an object is travelling. Speed usually has units of kilometres per hour (km/h) or metres per second (m/s).
You need to learn the formula

$$\text{average speed} = \frac{\text{total distance}}{\text{total time}}$$

> For a fixed distance: speed is inversely proportional to the time taken.

EXAMPLE

Sophie flew from Dubai to Tokyo.
Her flight covered a distance of 7954 km.
The flight time was 10 hours 15 minutes.
Work out the average speed for the flight.

SOLUTION

Using the formula:

$$\text{average speed} = \frac{\text{total distance}}{\text{total time}}$$

$$= \frac{7954}{10.25}$$

$$= 776 \text{ km/h}$$

> Take care when using decimals to write time.
> 15 minutes $= \frac{15}{60}$ hours $= 0.25$ hours
> So 10 hours 15 minutes $= 10.25$ hours

The average speed is the speed at which she would travel, if she travelled at the same speed for the whole flight.

EXAMPLE

Tim runs a distance of 100 m at an average speed of 6.25 m/s

a) Find the time he takes to run 100 m.

b) What is his average speed in kilometres per hour?

SOLUTION

a) Rearranging the formula for speed gives:

$$\text{total time} = \frac{\text{total distance}}{\text{average speed}}$$

$$\text{So time} = \frac{100}{6.25} = 16 \text{ seconds}$$

b) 6.25 m/s means in 1 second Tim runs 6.25 m
so in 60 seconds (1 minute) Tim runs 375 m
and in 60 minutes (1 hour) Tim runs 22 500 m
22 500 m = 22.5 km
So Tim's speed is 22.5 km/h.

Density is a measure of how much mass a certain volume of a material has. Density usually has units of kilograms per cubic metre (kg/m³) or grams per cubic centimetre (g/cm³).

> In everyday English the word 'weight' is used to mean 'mass'.

For example, a material with a density of:

- 200 kg/m³ means 1 m³ of the material has a mass of 200 kg
- 5 g/cm³ means 1 cm³ of the material has a mass of 5 g.

You need to learn the formula

$$\text{density} = \frac{\text{mass}}{\text{volume}}$$

> For a particular material: mass is directly proportional to volume.

EXAMPLE

The diagram shows a steel cylinder. It has a radius of 10 centimetres and is 2 centimetres thick.

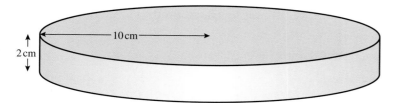

a) Work out the volume of the cylinder.

The steel has a density of 7.6 grams per cubic centimetre.

b) Work out the mass of the cylinder.

SOLUTION

a) The cylinder has $r = 10$ and $h = 2$.

$$\text{Volume} = \pi r^2 h$$
$$= \pi \times 10^2 \times 2$$
$$= 628.3185\ 307$$
$$= \underline{629\ \text{cm}^3}\ (3\ \text{s.f.})$$

b) Mass of cylinder $= 628.3185\ 307 \times 7.6$
$$= 4775.220\ 833$$
$$= \underline{4780\ \text{grams}}\ (3\ \text{s.f.})$$

Another example of a compound measure is pressure. The units of pressure are Newtons per square metre (N/m²) or Pascals (Pa).

$$\text{pressure} = \frac{\text{force}}{\text{area}}$$

You will be given this formula if you need it in the exam.

Exercise 21.4

1 Ali drove for 5 hours at an average speed of 70 km/h.
   Work out the distance that Ali drove.

2 Stefan cycled for $1\frac{1}{2}$ hours at an average speed of 10 km/h.
   Work out the distance that Stefan cycled.

3 Tamsin walks 16.5 km in 2 hours 45 minutes.
   Work out her average speed in km/h.

4 A tortoise moves 150 cm in $2\frac{1}{2}$ minutes.
   Work out its average speed.
   Give your answer in m/s.

5 Hans drives 40 km at an average speed of 30 km/h.
   Find the time for his journey. Give your answer in hours and minutes.

6 Bradley cycles for 50 seconds at an average speed of 15 km/h.
   How far does Bradley cycle? Give your answer correct to the nearest metre.

7 In 2009, Usain Bolt ran 100 m in 9.58 seconds.
   Calculate his average speed in
   i) metres per second
   ii) kilometres per hour.
   Give your answers correct to 3 significant figures.

8 The density of silver is 10.49 g/cm³.
   Giving your answers correct to 3 significant figures, find the volume of silver used to make:
   a) a silver ring of mass 25 g
   b) a silver spoon of mass 125 g

   A silver coin has a volume of 1.91 cm³
   c) Find the mass of the coin.

9 An aluminium hubcap has a volume of $2510\,cm^3$.
The mass of the hubcap is $6.8\,kg$
Giving your answers correct to 2 significant figures, calculate the density of aluminium in
a) grams per cubic centimetre
b) kilograms per cubic metre.

10 The formula for pressure is:
$$\text{pressure} = \frac{\text{force}}{\text{area}}$$
A particular force exerts a pressure of 20 Pascals over an area of $0.1\,m^2$.
Find the pressure exerted by the same force when it is applied over an area of $0.25\,m^2$.

11 An ice hockey puck is in the shape
of a cylinder with a radius of 3.8 cm,
and a thickness of 2.5 cm.

It is made out of rubber with a
density of 1.5 grams per cm^3.

Work out the mass of the ice hockey
puck. Give your answer correct to
3 significant figures.

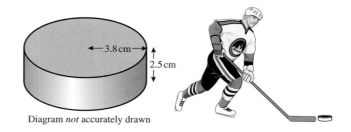

Diagram *not* accurately drawn

12 Population density is given by:
$$\text{population density} = \frac{\text{population size}}{\text{area}}$$
a) Mauritius has a population of 1.3 million and an area of $2000\,km^2$
Calculate the population density of Mauritius.
b) The population density of Taiwan is 650 people/km^2.
The population of Taiwan is 23.4 million.
What is the area, in km^2, of Taiwan?
c) Manila has a population of 1.6 million people and an area of $40\,km^2$.
Paris has a population of 2.3 million people and an area of $100\,km^2$.
How many more people are there per square kilometre in Manila than in Paris?

REVIEW EXERCISE 21

1 y is directly proportional to x, and $y = 21$ when $x = 7$. Find the value of y when $x = 13$.

2 y is inversely proportional to x, and $y = 8$ when $x = 4$. Find the value of y when $x = 16$.

3 y is directly proportional to x^2, and $y = 1$ when $x = 5$. Find the value of y when $x = 15$.

4 y is inversely proportional to x^3, and $y = 40$ when $x = 2$. Find the value of y when $x = 1$.

5 Hooke's Law says that the tension, T, in a stretched string is directly proportional to its extension, x.
A certain string has a tension of 20 units when its extension is 30 units.
a) Write down a formula to express T in terms of x.
b) Find the tension T when the extension is 36 units.
c) Work out the extension x when the tension is 48 units.

6 The periodic time, T seconds, that it takes a pendulum to complete one swing is directly proportional to the square root of the pendulum's length, l cm. A pendulum of length 25 cm has a periodic time of 1 second.
   a) Write a formula for T in terms of l.
   b) Find the periodic time for a pendulum of length 35 cm.

7 A weight is hung at the end of a beam of length L.
   This causes the end of the beam to drop a distance d. d is directly proportional to the cube of L.
   $d = 20$ when $L = 150$.

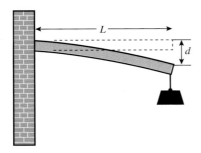

   a) Find a formula for d in terms of L.
   b) Calculate the value of L when $d = 15$.   [Edexcel]

8 y is inversely proportional to x^2. $y = 3$ when $x = 4$.
   a) Write y in terms of x.
   b) Calculate the value of y when $x = 5$.   [Edexcel]

9 y is directly proportional to x^2. When $x = 2$, $y = 36$.
   a) Express y in terms of x.

   z is inversely proportional to x. When $x = 3$, $z = 2$.
   b) Show that $z = cy^n$, where c and n are numbers and $c > 0$.
      (You must find the values of c and n.)   [Edexcel]

10 y is inversely proportional to x. When $x = 3$, $y = 24$.
   a) Write a formula for y in terms of x.

   Hence, or otherwise,
   b) **(i)** calculate the value of y when $x = 6$
      (ii) calculate the value of x when $y = 4.8$.   [Edexcel]

11 d is directly proportional to the square of t. $d = 80$ when $t = 4$.
   a) Express d in terms of t.
   b) Work out the value of d when $t = 7$.
   c) Work out the positive value of t when $d = 45$.   [Edexcel]

12 The force, F, between two magnets is inversely proportional to the square of the distance, x, between them. When $x = 3$, $F = 4$.
   a) Find an expression for F in terms of x.
   b) Calculate F when $x = 2$.
   c) Calculate x when $F = 64$.   [Edexcel]

13 A car travelling at a speed of V metres per second has a stopping distance of d metres.

The straight-line graph of $\dfrac{d}{V}$ against V has been drawn on the grid.

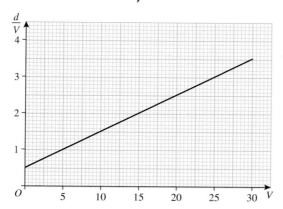

The car travels at a speed of 18 m s^{-1}.

a) Use the straight-line graph to find the stopping distance of the car.

b) By first finding the equation of the line, obtain a formula for d in terms of V.

14 The shutter speed, S, of a camera varies inversely as the square of the aperture setting, f.
When $f = 8$, $S = 125$.

a) Find a formula for S in terms of f.

b) Hence, or otherwise, calculate the value of S when $f = 4$. **[Edexcel]**

15 The heaviest stick of rock ever made was in the shape of a cylinder. The cylinder had a length of 503 cm and a radius of 21.6 cm.

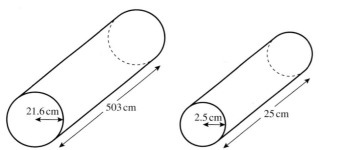

Diagram *not* accurately drawn

a) Work out the volume of the cylinder. Give your answer correct to 3 significant figures.

A small stick of rock, in the shape of a cylinder, has a length of 25 cm and a radius of 2.5 cm. It is made using the same recipe as the heaviest stick of rock. The weight of the heaviest stick of rock ever made was 413.6 kg.

b) Calculate the weight of the small stick of rock. Give your answer, in grams, correct to 3 significant figures. *[Edexcel]*

Key points

1   When y is directly proportional to x, we can write $y = kx$ where k is a constant.

The graph of y against x will then be a straight line through the origin:

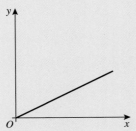

2   When y is inversely proportional to x, we can write $y = \dfrac{k}{x}$ where k is a constant.

The graph of y against x will then be a curve (more particularly, a hyperbola):

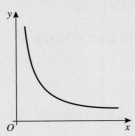

IGCSE questions will often tell you that y is directly (or inversely) proportional to some power of x, and will give you an x value and a corresponding y value. You should set up an equation using k (e.g. $y = kx^2$) and then use the given x and y values to determine the value of k. This formula can then be used to answer the rest of the question.

3   Average speed $= \dfrac{\text{total distance}}{\text{total time}}$

Make sure you use consistent units when working with speed. For example, convert a time given in minutes to hours when the speed is in km/h.

4   Density $= \dfrac{\text{mass}}{\text{volume}}$

The units of density are g/cm^3 or kg/m^3.

Internet Challenge 21

The planets and their orbits

The table shows information about the orbits of the nine planets and dwarf planets in our Solar System.

The time T years it takes for a planet to complete one orbit around the Sun is directly proportional to the 1.5th power of its mean distance d from the Sun. For simplicity, the distances have been scaled so that the Earth–Sun distance is 1 unit. The orbital period of the Earth is 1 year.

Planet or dwarf planet	Mean distance, d, from the Sun (Earth–Sun = 1 unit)	Orbital period, T (years)
Mercury	0.387	
Venus	0.723	
Earth	1	1
Mars	1.524	
Jupiter	5.203	
Saturn	9.529	
Uranus	19.19	
Neptune	30.06	
Pluto	39.53	

1 Use the fact that $T \propto d^{1.5}$ to show that Mars has an orbital period of 1.88 years.

2 Copy and complete the table to include the orbital periods for all the outer planets, namely Mars to Pluto.

3 In a similar way, work out the orbital periods for the inner planets, Mercury and Venus. These should each be less than 1 year, so give your answer in days.

4 Now use the internet to check that you have worked out these orbital periods correctly.

5 The $T \propto d^{1.5}$ law was one of three published by the astronomer Johannes Kepler. Use the internet to find out about Kepler's other laws.

6 In 2003, a remote object named Sedna was discovered far beyond Pluto. Sedna's mean distance from the Sun is approximately 510 times the Sun–Earth distance. Use Kepler's laws to calculate the orbital period for Sedna. Give your answer to the nearest 100 years.

7 Is Sedna really a planet? Is Pluto? Use the internet to help you decide.

CHAPTER 22

Quadratic equations, curves and inequalities

In this chapter you will learn how to:
- solve quadratic equations by factorising
- solve quadratics by the general formula
- complete the square
- set up and solve problems using quadratics
- solve quadratic inequalities.

You will also be challenged to:
- investigate conic sections.

Starter: **Solutions of equations**

Here are some equations, and some suggested solutions.

Substitute the suggested values into each equation to discover which are correct.

Equation	Suggested solutions
$5x + 3 = 18$	$x = 1, x = 2, x = 3, x = 4, x = 5$
$5x^2 + 4 = 9$	$x = -1, x = 0, x = 1, x = 2, x = 3$
$x^2 = 7x - 10$	$x = 1, x = 2, x = 3, x = 4, x = 5$
$12x - 3 = 45$	$x = 1, x = 2, x = 3, x = 4$
$x^2 = 1$	$x = -2, x = -1, x = 0, x = 1, x = 2$
$x + 4 = 10 - x$	$x = 0, x = 1, x = 2, x = 3, x = 4$
$x(x + 1) = 2$	$x = -2, x = -1, x = 0, x = 1, x = 2$
$x^2 = 36$	$x = -6, x = -3, x = 0, x = 3, x = 6$
$4x^2 = 100$	$x = -5, x = -3, x = 0, x = 1, x = 5$
$x^3 - 6x^2 + 11x - 6 = 0$	$x = 1, x = 2, x = 3, x = 4, x = 5$

Here is an extract from an old mathematics book:

> *Linear equations like* $3x + 5 = 21$ *have only one solution. Equations containing an* x^2 *term often have two solutions, however, and equations containing* x^3 *terms may have as many as three solutions.*

Do your results support this extract?

22.1 Solving quadratic equations – factorising

An equation like $x^2 + 4x + 3 = 0$ is called a **quadratic equation**. Quadratic equations must contain a square term, (such as the x^2 in this example) with no higher power of x, such as x^3. You may be able to spot a solution of a quadratic equation by inspection (i.e. by guesswork), but this is not a reliable method because quadratics may have two solutions. **Factorising** is a method of making sure that all of the solutions to a quadratic equation are found.

EXAMPLE

Solve the equation $x^2 + 4x + 3 = 0$

SOLUTION

$x^2 + 4x + 3 = 0$
$(x + 1)(x + 3) = 0$
$x + 1 = 0$ or $x + 3 = 0$
So, $x = -1$ or $x = -3$

> If $(x + 1)(x + 3) = 0$ then one of the brackets must be equal to 0.

Factorisation can be more difficult, especially if the coefficient of x^2 is greater than 1.

EXAMPLE

Solve the equation $2x^2 - 9x - 5 = 0$

SOLUTION

$2x^2 - 9x - 5 = 0$
$(2x + 1)(x - 5) = 0$
$2x + 1 = 0$ or $x - 5 = 0$
So, $x = -\frac{1}{2}$ or $x = 5$

Some quadratics contain only two terms, not three. If the constant term at the end is missing, then all you need to do is take out a common factor of x.

EXAMPLE

Solve the equation $10x^2 - 4x = 0$

SOLUTION

$10x^2 - 4x = 0$
$2x(5x - 2) = 0$
$2x = 0$ or $5x - 2 = 0$
So, $x = 0$ or $x = \frac{2}{5}$

If, instead, the middle term is missing, then you can simply solve to find x^2. Then take the square root of both sides to find x. Remember to allow for both positive and negative answers.

EXAMPLE

Solve the equation $5x^2 - 80 = 0$

SOLUTION

$5x^2 - 80 = 0$
$\qquad 5x^2 = 80$
$\qquad\quad x^2 = \dfrac{80}{5}$
$\qquad\quad x^2 = 16$

Square rooting both sides gives:
$x = 4$ or $x = -4$

> Alternatively, by factorising:
> $5x^2 - 80 = 0$
> $5(x^2 - 16) = 0$
> $5(x - 4)(x + 4) = 0$
> and so $x = 4$ or $x = -4$

EXERCISE 22.1

Solve each of these quadratic equations by using the factorisation method.

1 $x^2 + 3x + 2 = 0$
2 $x^2 + 6x + 5 = 0$
3 $x^2 + 7x - 8 = 0$

4 $x^2 + x - 2 = 0$
5 $x^2 + 2x - 8 = 0$
6 $x^2 + 4x - 12 = 0$

7 $x^2 - 7x + 12 = 0$
8 $x^2 - 8x + 15 = 0$
9 $x^2 - 2x - 8 = 0$

10 $x^2 - 4x + 4 = 0$
11 $2x^2 + 3x + 1 = 0$
12 $2x^2 + 5x - 3 = 0$

13 $3x^2 + 7x + 2 = 0$
14 $2x^2 + x - 3 = 0$
15 $3x^2 + 8x + 4 = 0$

16 $2x^2 - 9x + 9 = 0$
17 $3x^2 + 8x + 5 = 0$
18 $2x^2 - 9x + 10 = 0$

19 $5x^2 + 26x + 5 = 0$
20 $4x^2 + 4x + 1 = 0$

Here are some more difficult quadratic equations. Solve them by the factorisation method.

21 $6x^2 + x - 1 = 0$
22 $5x^2 - x = 0$
23 $4x^2 - 1 = 0$

24 $3x^2 - 3x = 0$
25 $12x^2 - 7x + 1 = 0$
26 $10x^2 - x = 0$

27 $8x^2 - 10x + 3 = 0$
28 $8x^2 - 11x + 3 = 0$
29 $4x^2 + 12x + 9 = 0$

30 $4x^2 - 9 = 0$

Rearrange these quadratic equations so that the right-hand side is zero. Then solve them, by factorisation.

31 $x^2 - 6x = 7$
32 $x^2 + 40 = 13x$
33 $x^2 + 20x = 7x - 30$

34 $x^2 + 10x = 3x + 44$
35 $2x^2 = 11x + 6$
36 $8 - 23x = 3x^2$

37 $2 = x + 3x^2$
38 $4x^2 = 8x - 3$
39 $6x^2 + 6x = x + 6$

40 $5x^2 + 30 = x^2 + 55$

22.2 Completing the square

When you cannot solve a quadratic equation by factorisation, you can solve it using the method of 'completing the square'.

Quadratic expressions in the form:

$$x^2 + 2ax + a^2 = (x + a)^2 \quad \text{and} \quad x^2 - 2ax + a^2 = (x - a)^2$$

are called perfect squares.

> The constant term is a square number: a^2.

> The coefficient of x is $2a$.

You can adjust a quadratic expression to write it in the form:

$$(x + a)^2 + b$$

> Note: a and b can be positive or negative.

This is called completing the square.

> A *perfect square* + a *constant*.

EXAMPLE

a) Factorise $x^2 + 12x + 36$

b) Hence write:
   i) $x^2 + 12x + 40$
   ii) $x^2 + 12x - 1$
   in the form $(x + a)^2 + b$.

SOLUTION

a) $x^2 + 12x + 36 = (x + 6)^2$

b) **i)** $x^2 + 12x + 40$
     $= x^2 + 12x + 36 + 4$ $\boxed{40 = 36 + 4}$
     $= (x + 6)^2 + 4$

  ii) $x^2 + 12x - 1$
     $= x^2 + 12x + 36 - 37$ $\boxed{-1 = 36 - 37}$
     $= (x + 6)^2 - 37$

Here is a method you can use to write a quadratic expression in the form $(x + a)^2 + b$.

EXAMPLE

Write each of the following in the form $(x + a)^2 + b$.

a) $x^2 - 10x + 16$

b) $x^2 - 7x + 3$

SOLUTION

a) $x^2 - 10x + 16$
   Take the coefficient of x -10 This number goes in the bracket.
   Halve it -5
   Square it 25 Add and subtract this square.

So $x^2 - 10x + 16$

$= \underbrace{x^2 - 10x + 25}_{} - 25 + 16$

$= (x - 5)^2 - 25 + 16$

$= (x - 5)^2 \qquad - 9$

$x^2 - 10x + 25$ is a perfect square.

So $x^2 - 10x + 16 = (x - 5)^2 - 9$ in completed square form.

b) $x^2 - 7x + 3$

Take the coefficient of x -7

This number goes in the bracket.

Halve it $\qquad -\dfrac{7}{2}$

Square it $\qquad \dfrac{49}{4}$

Add and subtract this square.

So $\qquad\qquad x^2 - 7x + 3$

$= \qquad \underbrace{x^2 - 7x + \dfrac{49}{4}}_{} - \dfrac{49}{4} + 3$

$x^2 - 7x + \dfrac{49}{4}$ is a perfect square.

$= \qquad (x - \dfrac{7}{2})^2 - \dfrac{49}{4} + 3$

$= \qquad (x - \dfrac{7}{2})^2 - \dfrac{37}{4}$

So $x^2 - 10x + 16 = (x - \dfrac{7}{2})^2 - \dfrac{37}{4}$ in completed square form.

You can use completing the square to solve a quadratic equation.

EXAMPLE

Use the completed square from the last example to solve
$x^2 - 10x + 16 = 0$

SOLUTION

Add 9 to both sides.

$(x - 5)^2 - 9 = 0$

Square root.

$(x - 5)^2 = 9$

$x - 5 = 3 \quad$ or $\quad x - 5 = -3$

Don't forget the negative square root!

So the solution is $\underline{x = 8}$ or $\qquad \underline{x = 2}$

You can also complete the square when the coefficient of x^2 is not equal to 1.

EXAMPLE

a) Write $3x^2 - 12x + 8$ in the form $a(x + b)^2 + c$

b) Hence, or otherwise, solve $3x^2 - 9x + 8 = 0$
Give your solutions correct to 3 significant figures.

SOLUTION

a) $3x^2 - 12x + 8$ in the form $a(x + b)^2 + c$

$3x^2 - 12x + 8$

$= 3[x^2 - 4x] + 8$

$= 3[x^2 - 4x + 4 - 4] + 8$

$= 3[(x - 2)^2 - 4] + 8$

$= 3(x - 2)^2 - 12 + 8$

$= 3(x - 2)^2 - 4$

So $3x^2 - 12x + 8 = \underline{3(x - 2)^2 - 4}$

> Take out a factor of 3 from the first 2 terms ...

> ... now complete the square on $x^2 - 4x$
> Coefficient of x: -4
> Halve it: $\qquad -2$
> Square it: $\qquad +4$

> Remove square brackets

b) $3(x - 2)^2 - 4 = 0$

$3(x - 2)^2 = 4$

$(x - 2)^2 = \dfrac{4}{3}$

> Add 4 to both sides and then divide by 3.

> square root.

$x - 2 = \sqrt{\dfrac{4}{3}}$ or $x - 2 = -\sqrt{\dfrac{4}{3}}$

> Don't forget the negative square root!

So $x = 2 + \sqrt{\dfrac{4}{3}} = \underline{3.15}$ or $x = 2 - \sqrt{\dfrac{4}{3}} = \underline{0.845}$

EXERCISE 22.2

1 a) Factorise $x^2 + 14x + 49$.

  b) Hence write:

    i) $x^2 + 14x + 60$ **ii)** $x^2 + 14x + 55$

    iii) $x^2 + 14x + 40$ **iv)** $x^2 + 14x - 10$

    in the form $(x + a)^2 + b$.

2 a) Factorise $x^2 - 6x + 9$.

  b) Hence write:

    i) $x^2 - 6x + 10$ **ii)** $x^2 - 6x + 15$

    iii) $x^2 - 6x + 3$ **iv)** $x^2 - 6x + 8$

    in the form $(x + a)^2 + b$.

3 Write each of these in the form $(x + a)^2 + b$.

  a) $x^2 + 6x - 15$ **b)** $x^2 + 6x - 20$ **c)** $x^2 + 2x + 6$

  d) $x^2 + 10x - 9$ **e)** $x^2 - 4x + 1$ **f)** $x^2 - 2x - 5$

  g) $x^2 + x + 1$ **h)** $x^2 - x + 1$ **i)** $x^2 + 7x - 10$

  j) $x^2 + 3x - 4$ **k)** $x^2 + 5x + 8$ **l)** $x^2 - 3x + 6$

4 Write each of these in the form $a(x + b)^2 + c$.

  a) $2x^2 + 4x + 5$ **b)** $2x^2 + 8x - 3$ **c)** $2x^2 + 6x + 9$

  d) $3x^2 + 12x - 5$ **e)** $3x^2 + 9x + 1$ **f)** $4x^2 + 10x + 5$

5 For each of these equations:

  i) write in completed square form

  ii) hence solve the equation, giving your solutions correct to 3 significant figures.

  a) $x^2 + 4x - 7 = 0$ **b)** $x^2 + 8x - 12 = 0$ **c)** $x^2 + 10x + 8 = 0$

  d) $2x^2 + 4x - 3 = 0$ **e)** $2x^2 + 6x - 5 = 0$ **f)** $3x^2 + 6x + 2 = 0$

22.3 Solving quadratic equations – formula

A quadratic equation contains three **coefficients**. For example:

$$x^2 + 4x + 3 = 0$$

has an x^2 coefficient of 1, an x coefficient of 4 and a constant term of 3.

$2x^2 - 4x - 1 = 0$ has an x^2 coefficient of 2, an x coefficient of -4 and a constant term of -1.

Similarly, $4x^2 - 1 = 0$ has an x^2 coefficient of 4, an x coefficient of 0 and a constant term of -1.

There is a formula that can be used to find solutions to a quadratic equation. If $ax^2 + bx + c = 0$ is a quadratic equation, then the solutions are given by the formula:

$$x = \frac{-b \pm \sqrt{b^2 - 4ac}}{2a}$$

The sign $\pm$ is read as 'plus or minus'.

You obtain one of the solutions of the quadratic by using $x = \dfrac{-b + \sqrt{b^2 - 4ac}}{2a}$

and the other one by using $x = \dfrac{-b - \sqrt{b^2 - 4ac}}{2a}$.

The formula method can be applied to a much wider range of quadratic equations than the factorising method. You would normally use the formula if the equation cannot be factorised in an obvious way. The quadratic formula will be given to you in an IGCSE exam, on the formula sheet.

EXAMPLE

Solve the equation $2x^2 - 4x - 1 = 0$. Give your answers to 3 decimal places.

SOLUTION

There is no obvious factorisation, so use the formula.
Inspecting the equation, $a = 2$, $b = -4$ and $c = -1$.
Then substituting these values into the formula:

$$x = \frac{-b \pm \sqrt{b^2 - 4ac}}{2a}$$

gives

$$x = \frac{-(-4) \pm \sqrt{(-4)^2 - 4(2)(-1)}}{2(2)}$$

$$= \frac{4 \pm \sqrt{16 + 8}}{4}$$

$$= \frac{4 \pm \sqrt{24}}{4}$$

$$= 2.224\,744\,871 \text{ or } -0.224\,744\,871$$

$$= \underline{2.225} \text{ or } \underline{-0.225} \text{ (3 d.p.)}$$

If you are asked to solve a quadratic equation in an IGCSE exam, and the number that you calculate under the square root sign is negative, for example $\sqrt{-25}$, then you know you must have made an error.

EXERCISE 22.3

Solve these equations using the quadratic equation formula. Give your answers correct to 3 decimal places.

1 $x^2 + 5x + 2 = 0$

2 $x^2 + 10x + 7 = 0$

3 $2x^2 - 14x + 13 = 0$

4 $2x^2 + 11x - 5 = 0$

5 $x^2 - 7x + 1 = 0$

6 $3x^2 + 2x - 3 = 0$

7 $x^2 + 5x - 1 = 0$

8 $2x^2 - 3x - 4 = 0$

9 $5x^2 - x - 1 = 0$

10 $2x^2 + 9x - 2 = 0$

Rearrange the equations below so that they are in the form $ax^2 + bx + c = 0$. Then solve them using the formula method. Give your answers correct to 3 significant figures.

11 $x^2 + 5x = 7$

12 $2x^2 = 3x + 1$

13 $3x^2 = 5 + 4x$

14 $x^2 + x = 2 - 9x$

15 $11x = 1 - 2x^2$

16 $3x^2 = 12x + 1$

17 $2x = 5x^2 - 4$

18 $21x + 1 = 7x^2$

19 $20x + 4 = 3x - 6x^2$

20 $9x^2 = 2 + x$

22.4 Problems leading to quadratic equations

At IGCSE you may be expected to set up a problem that leads to a solution involving a quadratic equation. You will then need to solve the quadratic equation to complete the problem.

EXAMPLE

A rectangular flower bed measures $2x + 5$ metres by $x + 3$ metres.
It has an area of 45 square metres.
a) Draw a sketch to show this information.
b) Show that x must satisfy the equation $2x^2 + 11x - 30 = 0$.
c) Solve this equation, to find the value of x. Hence find the dimensions of the flower bed.

SOLUTION

a)

2x + 5

x + 3

b)
$$(2x + 5)(x + 3) = 45$$
$$2x^2 + 5x + 6x + 15 = 45$$
$$2x^2 + 11x + 15 = 45$$
$$2x^2 + 11x - 30 = 0$$

c) Factorising $2x^2 + 11x - 30 = 0$ gives:
$$(2x + 15)(x - 2) = 0$$
So, $2x + 15 = 0$ or $x - 2 = 0$

Therefore $x = -7\frac{1}{2}$ or $x = 2$

But $x = -7\frac{1}{2}$ will lead to negative dimensions for the flower bed, so it must be rejected.

Therefore $x = 2$

We know that the dimensions of the flower bed are:

$2x + 5$ metres by $x + 3$ metres

Substituting $x = 2$ gives dimensions of 9 metres by 5 metres.

EXERCISE 22.4

1 Two whole numbers x and $x + 7$ are multiplied together. The result is 144.
   a) Write down an equation in x.
   b) Show that this equation can be expressed as $x^2 + 7x - 144 = 0$
   c) Solve the equation, to find the values of the two whole numbers (there are two possible sets of answers, and you should give both).

2 A rectangular playing field is x metres wide and $2x - 5$ metres long. Its area is 3000 m^2.
   a) Write down an equation in x.
   b) Show that this equation can be expressed as $2x^2 - 5x = 3000$
   c) Solve the equation, to find the value of x. Hence find the dimensions of the playing field.

3 Hannah and Jamal each thought of a positive whole number. Jamal's number was 3 more than Hannah's number. Let Hannah's number be represented by x.
   a) Their two numbers multiply together to make 180. Write down an equation in x.
   b) Show that this equation can be expressed as $x^2 + 3x - 180 = 0$
   c) Solve the equation, and hence find the numbers that Hannah and Jamal thought of.

4 A square measures x cm along each side, and a rectangle measures x cm by $2x + 1$ cm. The total area of the square and the rectangle is 114 cm^2.
   a) Write down an equation in x.
   b) Show that this equation can be expressed as $3x^2 + x - 114 = 0$
   c) Solve the equation, to find the value of x.

5 A rectangle measures $3x + 1$ cm by $2x + 5$ cm. Two squares, each of side x cm, are removed from it. The remaining shape has an area of 55 cm^2.
   a) Express this information as an equation in x.
   b) Show that this equation can be expressed as $4x^2 + 17x - 50 = 0$
   c) Solve your equation, and hence find the dimensions of the rectangle.

6 A rectangle measures x cm by $2x + 3$ cm. A second rectangle measures $x + 3$ cm by $x + 4$ cm.

a) Write down expressions for the areas of the two rectangles.

Both rectangles have the same area.

b) Write an equation in x.

c) Solve this equation. Hence determine the dimensions of each rectangle.

7 The first term of an arithmetic series is 7 and the common difference is 3.

The sum of the first n terms is 710.

Find the value of n.

8 The first term of an arithmetic series is 20 and the fifth term is 18.

The sum of the first n terms is 392.

Find the possible values of n.

22.5 Quadratic curves

A function is a way of expressing a relationship between two variables.

$y = x^2 + 3x + 1$ is an example of a quadratic function.

> You will learn more about functions in Chapter 24.

You can use a table of values to help you draw the graph of a quadratic function.

EXAMPLE

Draw up a table of values of the function $y = x^2 + 2x$
for $x = -3, -2, -1, 0, 1, 2, 3$.

SOLUTION

x	-3	-2	-1	0	1	2	3
y							

> Begin by listing the x values you are going to work with.

x	-3	-2	-1	0	1	2	3
y	3						

> Now work out $(-3)^2 + 2 \times (23) = 9 - 6 = 3$
> Write the result in the table.

x	-3	-2	-1	0	1	2	3
y	3	0	-1	0	3	8	15

> Continue until the table is complete.

Once you have made a table of values, the points can be plotted on graph paper to make a graph of the function. You should join the points using a *smooth* curve. Use a pencil, so that you can easily change your graph if it is not quite right at the first attempt.

EXAMPLE

The table shows values of the function $y = x^2 + 2x$
for $x = -4, -3, -2, -1, 0, 1, 2, 3$.

x	-4	-3	-2	-1	0	1	2	3
y	8	3	0	-1	0	3	8	15

a) Draw a set of axes so that x can range from -4 to 3 and y from -5 to 20, and plot these points on your axes.
b) Join your points with a smooth curve.
c) Use your graph to find: (i) the value of y when $x = 2.5$
                              (ii) the values of x when $y = 5$.

SOLUTION

a)

Draw a smooth curve, not a set of straight-line segments.

Take extra care here, where the graph curves rapidly.

Plot the points carefully, using a neat cross or dot.
Use pencil so that errors can be corrected.

b)

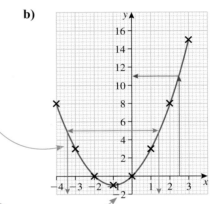

c) (i)   When $x = 2.5$, $y = 11.3$ (1 d.p.)
   (ii) When $y = 5$, $x = -3.4$ or 1.4

Note that whenever you plot the graph of a quadratic function, the curve will always have a distinctive bowl shape: it is a **parabola**.

When the coefficient of x^2 is positive the parabola looks like

When the coefficient of x^2 is negative the parabola looks like

Exercise 22.5

1 Complete the table of values for the function $y = x^2 - 5$.
One value has been filled in for you.

x	-3	-2	-1	0	1	2	3
y		-1					

2 Complete the table of values for the function $y = 2x^2 + 3$.
Two values have been filled in.

x	-2	-1	0	1	2	3
y		5			11	

3 Complete the table of values for the function $y = 3x^2 + x$.

x	-2	-1	0	1	2	3
y						

4 Complete the table of values for the function $y = x^2 - 4x$.

x	-1	0	1	2	3	4	5
y							

5 The table shows values of the function $y = x^2 - 3$ for $x = -3, -2, -1, 0, 1, 2, 3$.

x	-3	-2	-1	0	1	2	3
y	-6	1	-2	-3	-2	1	6

    a) Draw a set of axes so that x can range from -3 to 3 and y from -5 to 10, and plot these points on your axes.

    b) Join your points with a smooth curve.

    c) Use your graph to find the value of y when $x = 1.5$.

    d) Find the coordinates of the lowest point on the curve.

6 The table shows some values of the function $y = x^2 + 4x$.

x	-3	-2	-1	0	1	2	3
y			-3			12	

    a) Copy and complete the table.

    b) Draw a set of coordinate axes on squared paper, so that x can run from -3 to 3 and y from -5 to 22. Plot these points on your graph, and join them with a smooth curve.

    c) Give the coordinates of the lowest point on the graph.

7 The table shows some values of the function $y = 2x^2 - 3$.

x	-3	-2	-1	0	1	2	3
y		5			-1		

a) Copy and complete the table.
b) Draw a set of coordinate axes on squared paper, so that x can run from -3 to 3 and y from -5 to 20. Plot these points on your graph, and join them with a smooth curve.
c) Use your graph to find all the solutions to the equation $2x^2 - 3 = 0$.

8 The table shows some values of the function $y = 8 - x^2$.

x	-3	-2	-1	0	1	2	3
y		4		8			-1

a) Copy and complete the table.
b) Draw a set of coordinate axes on squared paper, so that x can run from -3 to 3 and y from -5 to 10. Plot these points on your graph, and join them with a smooth curve.
c) State the coordinates of the point on the curve where y takes its maximum value.
d) Use your graph to find the two solutions to the equation $8 - x^2 = 0$.

9 The table shows some x values for the function $y = x^2 - x - 6$.

x	-3	-2	-1	0	1	2	3	4
y	6			-6			0	

a) Copy and complete the table.
b) Draw a set of coordinate axes on squared paper, so that x can run from -3 to 4 and y from -8 to 8. Plot these points on your graph, and join them with a smooth curve.
c) Write down the solutions to the equation $x^2 - x - 6 = 0$.
d) Give the coordinates of the minimum point on the curve.

22.6 Solving quadratic inequalities

When you solve a quadratic inequality you should:
- find the critical values by solving the inequality as though it were an equation
- find the correct regions by using a sketch of the curve or by testing points.

EXAMPLE

Solve:
a) $x^2 + x - 2 < 0$
b) $x^2 + 6x - 8 \geqslant 0$
Illustrate your solutions on a number line.

SOLUTION

a) Identify critical values:

$$x^2 + x - 2 = 0$$

$$(x+2)(x-1) = 0$$

Factorise x.

Solve equation.

So $\quad x = -2 \quad$ or $\quad x = 1$

Look at a sketch of the curve:

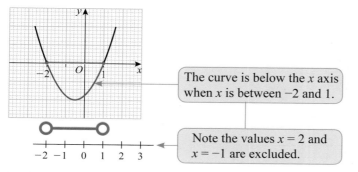

The curve is below the x axis when x is between -2 and 1.

Note the values $x = 2$ and $x = -1$ are excluded.

So the solution is $-2 < x < 1$

b) Identify critical values:

$$x^2 + 6x - 8 = 0$$

$$(x+4)(x-2) = 0$$

So $\quad x = -4 \quad$ or $x = 2$

Look at a sketch of the curve:

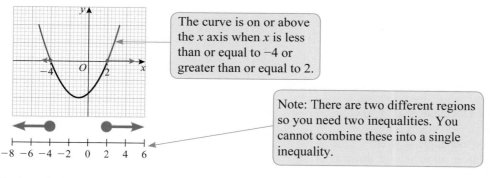

The curve is on or above the x axis when x is less than or equal to -4 or greater than or equal to 2.

Note: There are two different regions so you need two inequalities. You cannot combine these into a single inequality.

So the solution is $x \le -4$ or $x \ge 2$.

Exercise 22.6

Solve the following inequalities.

1 $\quad x^2 - 5x + 6 < 0$

2 $\quad x^2 - 5x + 6 \ge 0$

3 $\quad x^2 + 7x + 12 \ge 0$

4 $\quad x^2 + 2x - 15 \le 0$

5 $\quad 2x^2 - 3x - 2 < 0$

6 $\quad 3x^2 + x - 2 \ge 0$

7 $\quad (2-x)(x+3) > 0$

8 $\quad -2x^2 - 5x + 3 < 0$

REVIEW EXERCISE 22

1 a) Factorise $x^2 - 6x + 8$.
   b) Solve the equation $x^2 - 6x + 8 = 0$.   [Edexcel]

2 The table shows some values of the function $y = x^2 - 5x + 1$.

x	-1	0	1	2	3	4	5	6
y			-3				1	

   a) Copy and complete the table.
   b) On a copy of the grid below, draw the graph of $y = x^2 - 5x + 1$.

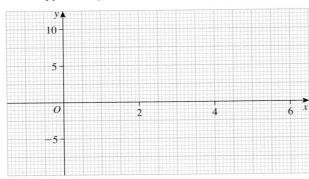

   c) Use your graph to estimate the two values of x for which $x^2 - 5x + 1 = 0$.
   d) Write down the coordinates of the minimum point on the curve.

3 Solve the equation $(2x - 3)^2 = 100$.   [Edexcel]

4 Find the solutions of the equation $x^2 - 4x - 1 = 0$.
   Give your solutions correct to three decimal places.   [Edexcel]

5 $(x + 3)(x - 2) = 1$.
   a) Show that $x^2 + x - 7 = 0$.
   b) Solve the equation $x^2 + x - 7 = 0$.
   Give your answers correct to 3 significant figures.   [Edexcel]

6 The length of a rectangle is $(x + 4)$ cm.
   The width is $(x - 3)$ cm.
   The area of the rectangle is 78 cm².
   a) Use this information to write down
      an equation in terms of x.
   b) (i) Show that your equation in part **a)**
      can be written as $x^2 + x - 90 = 0$.
   (ii) Find the values of x which are
      solutions of the equation $x^2 + x - 90 = 0$.
   (iii) Write down the length and the width of the rectangle.   [Edexcel]

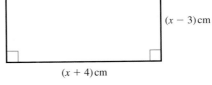

Diagram *not*
accurately drawn

$(x - 3)$ cm

$(x + 4)$ cm

7 AT is a tangent to a circle, centre O. OT = x cm, AT = $(x + 5)$ cm and OA = $(x + 8)$ cm.

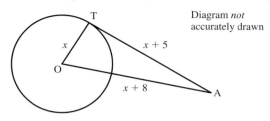

Diagram *not* accurately drawn

a) Show that $x^2 - 6x - 39 = 0$.
b) Solve the equation $x^2 - 6x - 39 = 0$ to find the radius of the circle.
Give your answer correct to 3 significant figures.

[Edexcel]

8 The diagram shows a prism.
The cross-section of the prism is a right-angled triangle.
The lengths of the sides of the triangle are $3x$ cm, $4x$ cm and $5x$ cm.
The total length of all the edges of the prism is E cm.
a) Show that the length, L cm, of the prism is given
by the formula $L = \frac{1}{3}(E - 24x)$.

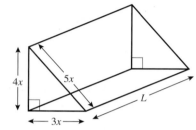

The surface area, A cm^2, of the prism is given by
the formula $A = 12x^2 + 12Lx$. $E = 98$ cm and $A = 448$ cm.
b) Substitute these values into the formulae of L and A to show that x satisfies the equation
$3x^2 - 14x + 16 = 0$. Make the stages in your working clear.
c) Solve the equation $3x^2 - 14x + 16 = 0$.

[Edexcel]

9 The diagram shows a trapezium.
The measurements on the diagram
are in centimetres.
The lengths of the parallel sides are
x cm and 20 cm. The height of the
trapezium is $2x$ cm.
The area of the trapezium is 400 cm^2.
a) Show that $x^2 + 20x = 400$.
b) Find the value of x. Give your answer correct to 3 decimal places.

Diagram *not* accurately drawn

[Edexcel]

10 a) Write $x^2 + 12x + 8$ in the form $(x + b)^2 + c$.
b) Hence, or otherwise, solve $x^2 - 12x + 8 = 0$.
Give your solutions correct to 3 significant figures.

11 a) Write $2x^2 + 6x - 7$ in the form $a(x + b)^2 + c$.
b) Hence, or otherwise, solve $2x^2 + 6x - 7 = 0$.
Give your solutions correct to 3 significant figures.

12 Solve: **a)** $x^2 - 9x + 20 < 0$ **b)** $2x^2 - x - 3 \geqslant 0$.

Illustrate your solutions on a number line.

13 a) (i) Factorise $2x^2 - 35x + 98$.
    (ii) Solve the equation $2x^2 - 35x + 98 = 0$.

A bag contains $(n + 7)$ tennis balls. n of the balls are yellow. The other seven balls are white.
John will take a ball at random from the bag. He will look at its colour and then put it back in the bag.
b) (i) Write down an expression, in terms of n, for the probability that John will take a white ball.
      Bill states that the probability that John will take a white ball is $\frac{2}{5}$.
    (ii) Prove that Bill's statement cannot be correct.

After John has put the ball back into the bag, Mary will then take at random a ball from the bag.
She will note its colour.
c) Given that the probability that John and Mary will take balls with different colours is $\frac{4}{9}$,
    prove that $2n^2 - 35n + 98 = 0$.
d) Using your answer to part **a) ii)**, or otherwise, calculate the probability that John and
    Mary will both take white balls.
                                                             [Edexcel]

Key points

1 Quadratic equations contain a term in x^2, and are often written in the form
$ax^2 + bx + c = 0$. Sometimes a solution may seem obvious, but you should
always use formal methods to solve the equation fully since quadratics can have
two solutions.

2 If the factors of a quadratic are easy to spot, then the factorising method is best.
Otherwise, use the quadratic equation formula:

$$x = \frac{-b \pm \sqrt{b^2 - 4ac}}{2a}$$

3 If an exam question asks you to solve a quadratic correct to 3 significant figures,
this is a clue that the quadratic formula will be required.

4 You can solve a quadratic equation by completing the square. This means you
write the quadratic in the form $a(x+b)^2+c$.

5 The graph of a quadratic function is a parabola.
When the coefficient of x^2 is positive the parabola looks like

When the coefficient of x^2 is negative the parabola looks like

6 To solve a quadratic inequality first identify the critical values by solving the
inequality as though it were an equation.
Then use a sketch of the curve to help you identify the correct regions.

Internet Challenge 22

Conic sections

When a quadratic expression is graphed, the result is a distinctive curve called a parabola.

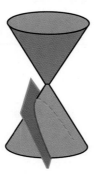

1   The diagram above shows a plane slice through a cone. The slice is parallel to one edge of the cone. What shape is the curve (marked in red) where the plane cuts the cone?

2   The parabola is one of four conic sections. Use the internet to find out the names of the other three.

3   Draw up a set of coordinate axes on squared paper. Then draw some line segments like this:
    Join the point (10, 0) to the origin (0, 0)
    Join (9, 0) to (0, 1)
    Join (8, 0) to (0, 2), etc.

   You should see a curve forming inside these lines. Is it a conic section? If so, which one?

4   When a body such as a planet or a comet moves through the Solar System, it traces out a path known as its orbit. The Earth's orbit, for example, is an ellipse. What shapes are the orbits followed by other bodies in the Solar System?

5   Use the internet to find a method for drawing an ellipse using a string and two drawing pins. Then use the method to draw some ellipses. Is this a good method?

CHAPTER 23

Advanced algebra

In this chapter you will learn how to:

- manipulate and simplify surds
- rationalise the demoninator
- add and subtract algebraic fractions, and solve equations containing them
- simplify algebraic fractions by cancelling common factors
- solve simultaneously one linear and one quadratic equation
- change the subject of an equation when the symbol occurs twice
- use algebra in proofs.

You will also be challenged to:

- investigate well-known mathematical formulae.

Starter: How many shapes?

Count the squares and rectangles in this 2 by 2 grid. You should be able to find nine.

Now count the number of squares and rectangles in these grids. Try to work systematically.

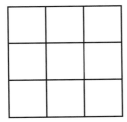   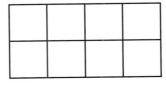

It is suggested that the number of squares or rectangles contained within a grid measuring m by n is given by an expression of the form:

$$\frac{m(m + 1)n(n + 1)}{k}$$

where k is a fixed number.

Assuming this expression is correct, use your results to work out the value of k.

Now try to prove that this expression is correct, using algebra.

23.1 Working with surds

Some quantities in mathematics can only be written exactly using a square root symbol.

For example, if $x^2 = 5$, then the exact value of x is $\sqrt{5}$ (or $-\sqrt{5}$).

Quantities like these, written using roots, are called **surds**.
Note that $\sqrt{5} \times \sqrt{5} = 5$.

Some surds can be simplified by writing them in terms of simpler surds. You should look for roots of perfect squares (4, 9, 16, 25, etc.) to help achieve this.

EXAMPLE

Simplify $\sqrt{48}$

SOLUTION

Since $48 = 16 \times 3$:
$$\sqrt{48} = \sqrt{16 \times 3}$$
$$= \sqrt{16} \times \sqrt{3}$$
$$= \underline{4\sqrt{3}}$$

EXAMPLE

Write $\sqrt{48} + \sqrt{27}$ as a single surd term.

SOLUTION

$$\sqrt{48} + \sqrt{27} = \sqrt{16 \times 3} + \sqrt{9 \times 3}$$
$$= \sqrt{16} \times \sqrt{3} + \sqrt{9} \times \sqrt{3}$$
$$= 4\sqrt{3} + 3\sqrt{3}$$
$$= \underline{7\sqrt{3}}$$

If you need to multiply two surd expressions together, just follow the ordinary rules for multiplying algebraic expressions.

EXAMPLE

A rectangle has a length of $(2 + \sqrt{5})$ cm and a width of $(3 - \sqrt{5})$ cm.
a) Show this information on a sketch.
b) Find the perimeter of the rectangle.
c) Find the area of the rectangle.

SOLUTION

a)

$(2 + \sqrt{5})$

$(3 - \sqrt{5})$

b) Perimeter $= (2 + \sqrt{5}) + (3 - \sqrt{5}) + (2 + \sqrt{5}) + (3 - \sqrt{5})$

$\qquad = \underline{10 \text{ cm}}$

c) Area $= (2 + \sqrt{5})(3 - \sqrt{5})$

$\qquad = 2 \times 3 - 2 \times \sqrt{5} + \sqrt{5} \times 3 - \sqrt{5} \times \sqrt{5}$

$\qquad = 6 - 2\sqrt{5} + 3\sqrt{5} - 5$

$\qquad = \underline{1 + \sqrt{5} \text{ cm}^2}$

Sometimes a fraction has a single surd in the denominator (the bottom line). Such a fraction is not in its simplest form. You can simplify it by multiplying the top and the bottom of the fraction by this surd, so that the bottom becomes a simple whole number. This process is called **rationalising the denominator**.

EXAMPLE

Write $\dfrac{3 + 2\sqrt{2}}{\sqrt{2}}$ in a form that does not have surds in the denominator.

SOLUTION

$$\frac{3 + 2\sqrt{2}}{\sqrt{2}} = \frac{(3 + 2\sqrt{2})}{\sqrt{2}} \times \frac{\sqrt{2}}{\sqrt{2}}$$

$$= \frac{3\sqrt{2} + 2\sqrt{2}\sqrt{2}}{\sqrt{2}\sqrt{2}}$$

$$= \frac{3\sqrt{2} + 4}{2}$$

EXAMPLE

a) Expand $(5 - \sqrt{2})(5 + \sqrt{2})$

b) Hence rationalise the denominator of $\dfrac{3}{(5 - \sqrt{2})}$

> This means find an equivalent fraction with a rational number on the bottom line.

SOLUTION

a) $(5 - \sqrt{2})(5 + \sqrt{2}) = 5 \times 5 + 5 \times \sqrt{2} - \sqrt{2} \times 5 - \sqrt{2} \times \sqrt{2}$

> Multiply each term in the second bracket by 5 and then by $-\sqrt{2}$

$= 25 + 5\sqrt{2} - 5\sqrt{2} - \sqrt{2} \times \sqrt{2}$

$= 25 - 2$

> Remember $\sqrt{2} \times \sqrt{2} = 2$

$= 23$

Notice that the middle terms, $5\sqrt{2}$ and $-5\sqrt{2}$, cancelled out.
This is because the expression $(5 + \sqrt{2})(5 - \sqrt{2})$ is of the form $(a + b)(a - b)$
and $(a + b)(a - b) = a^2 - b^2$ which is the difference of two squares.

b) $\dfrac{3}{(5 - \sqrt{2})} = \dfrac{3}{(5 - \sqrt{2})} \times \dfrac{(5 + \sqrt{2})}{(5 + \sqrt{2})}$

$= \dfrac{3(5 + \sqrt{2})}{23}$

> This fraction is equal to 1 as the top line and bottom line are the same...

> ...and multiplying by 1 doesn't change the value of the fraction!

You can use the difference of two squares to help you rationalise the denominator.

For fractions in the form $\dfrac{1}{(a + \sqrt{b})}$ multiply the top and bottom lines by $(a - \sqrt{b})$

to give:

$$\dfrac{1}{(a + \sqrt{b})} \times \dfrac{(a - \sqrt{b})}{(a - \sqrt{b})} = \dfrac{(a - \sqrt{b})}{a^2 - b}$$

or

$$\dfrac{1}{(a - \sqrt{b})} \times \dfrac{(a + \sqrt{b})}{(a + \sqrt{b})} = \dfrac{(a + \sqrt{b})}{a^2 - b}$$

> Remember $\sqrt{b} \times \sqrt{b} = b$

When you solve a quadratic equation by using the formula, your result will contain a square root sign. Instead of using a calculator to work out the answer to 3 or 4 significant figures, you could instead be asked to give an exact answer using surds.

EXAMPLE

Solve the equation $x^2 - 10x + 3 = 0$ using the quadratic equation formula.
Leave your answer in surd form.

SOLUTION

For the equation $x^2 - 10x + 3 = 0$ we have $a = 1, b = -10, c = 3$.

$$x = \frac{-b \pm \sqrt{b^2 - 4ac}}{2a}$$

$$= \frac{-(-10) \pm \sqrt{(-10)^2 - 4 \times 1 \times 3}}{2 \times 1}$$

$$= \frac{10 \pm \sqrt{100 - 12}}{2} = \frac{10 \pm \sqrt{88}}{2}$$

$$= \frac{10 \pm 2\sqrt{22}}{2} = 5 \pm \sqrt{22}$$

EXERCISE 23.1

Simplify the following surds.

1 $\sqrt{18}$ **2** $\sqrt{32}$ **3** $\sqrt{50}$ **4** $\sqrt{45}$ **5** $\sqrt{150}$ **6** $\sqrt{24}$ **7** $\sqrt{99}$ **8** $\sqrt{108}$

Write each of these as a single surd term.

9 $\sqrt{48} + \sqrt{12}$ **10** $\sqrt{8} + \sqrt{50}$ **11** $\sqrt{75} - \sqrt{12}$

12 $\sqrt{18} + \sqrt{32}$ **13** $\sqrt{11} + \sqrt{99}$ **14** $\sqrt{98} - \sqrt{18}$

15 Simplify $\sqrt{3}(4 + 2\sqrt{12})$ **16** Simplify $(6 + \sqrt{2})(1 + \sqrt{2})$ **17** Simplify $(5 - \sqrt{3})(5 + \sqrt{3})$

18 Simplify $(4 + \sqrt{5})(3 + 2\sqrt{5})$ **19** Simplify $\dfrac{3 + 2\sqrt{5}}{\sqrt{5}}$ **20** Simplify $\dfrac{2\sqrt{7} - 7}{\sqrt{7}}$

21 a) Expand and simplify $(1 + \sqrt{2})(1 - \sqrt{2})$ **b)** Hence, or otherwise, simplify $\dfrac{3}{(1 + \sqrt{2})}$

22 Show that $\dfrac{3}{(2 + \sqrt{5})} = 3\sqrt{5} - 6$

Show your working clearly.

23 Express $\dfrac{9}{(6 - 3\sqrt{5})}$ in the form $a(b + \sqrt{5})$ where a and b are integers.

24 Rationalise the denominators.

    a) $\dfrac{19}{(8 - 3\sqrt{5})}$ **b)** $\dfrac{26}{(5 + 2\sqrt{3})}$

25 A rectangle measures $(4 + \sqrt{28})$ cm long by $(5 - \sqrt{7})$ cm wide.
    a) Write the length of the rectangle in its simplest form.
    b) Work out the perimeter of the rectangle. Give your answer as an exact surd, in its simplest form.
    c) Work out the area of the rectangle. Give your answer as an exact surd, in its simplest form.

26 A rectangle has an area of 15 cm². The length of the rectangle is $(2 + \sqrt{7})$ cm.
Calculate the exact perimeter of the rectangle.
Give your answer in the form $a\sqrt{b} + c$ where a, b and c are whole numbers to be found.

Solve each of the following quadratic equations, using the quadratic formula.
Leave your answers in surd form.

27 $x^2 + 4x - 7 = 0$

28 $x^2 + x - 1 = 0$

29 $x^2 + 3x - 1 = 0$

30 $2x^2 + 8x + 3 = 0$

31 $x^2 - 5x + 2 = 0$

32 $x^2 + 6x - 2 = 0$

23.2 Algebraic fractions

Algebraic fractions should be treated in just the same way as numerical
fractions. In order to add (or subtract) two fractions, you need to write them
with the same denominator.

EXAMPLE

Write as a single fraction:

$$\frac{x+1}{4} + \frac{3x+2}{6}$$

SOLUTION

The fractions have denominators of 4 and 6. These can be written with a
common denominator of 12:

$$\frac{x+1}{4} + \frac{3x+2}{6} = \frac{3 \times (x+1)}{3 \times 4} + \frac{2 \times (3x+2)}{2 \times 6}$$

$$= \frac{3x+3}{12} + \frac{6x+4}{12}$$

$$= \frac{3x+3+6x+4}{12}$$

$$= \frac{9x+7}{12}$$

EXAMPLE

Write as a single fraction $\dfrac{3}{x+1} - \dfrac{2}{x+5}$

SOLUTION

The fractions have denominators of $(x + 1)$ and $(x + 5)$. These can be written with
a common denominator of $(x + 1)(x + 5)$. The top and bottom of the first fraction
must be multiplied by $(x + 5)$, and similarly $(x + 1)$ for the second fraction:

$$\frac{3}{x+1} - \frac{2}{x+5} = \frac{3}{(x+1)} \times \frac{(x+5)}{(x+5)} - \frac{2}{(x+5)} \times \frac{(x+1)}{(x+1)}$$

451

$$= \frac{3(x+5)}{(x+1)(x+5)} - \frac{2(x+1)}{(x+1)(x+5)}$$

$$= \frac{3(x+5) - 2(x+1)}{(x+1)(x+5)}$$

$$= \frac{3x+15-2x-2}{(x+1)(x+5)}$$

Note carefully how the subtraction affects the signs with the second bracket here.

$$= \frac{x+13}{(x+1)(x+5)}$$

Sometimes you may meet an equation containing algebraic fractions. You could simplify the equation to end up with a single fraction on each side, and then cross-multiply.

An alternative method is to multiply both sides by a factor large enough to clear the fractions away. The next example shows you both approaches.

EXAMPLE

Solve the equation:

$$\frac{5x-9}{3} + \frac{2x+1}{12} = \frac{3}{4}$$

SOLUTION

Method 1

$$\frac{5x-9}{3} + \frac{2x+1}{12} = \frac{3}{4}$$

Express the two fractions on the left-hand side so that they both have denominator 12. Then you can add them together.

$$\frac{4 \times (5x-9)}{4 \times 3} + \frac{2x+1}{12} = \frac{3}{4}$$

$$\frac{20x-36}{12} + \frac{2x+1}{12} = \frac{3}{4}$$

$$\frac{20x - 36 + 2x + 1}{12} = \frac{3}{4}$$

$$\frac{22x-35}{12} = \frac{3}{4}$$

$$4 \times (22x - 35) = 3 \times 12 \qquad \text{Cross-multiply at this stage.}$$

$$4(22x - 35) = 36$$

$$22x - 35 = 9$$

$$22x = 9 + 35$$

$$22x = 44$$

$$x = 2$$

Method 2

$$\frac{5x-9}{3} + \frac{2x+1}{12} = \frac{3}{4}$$

$$\frac{12 \times (5x-9)}{3} + \frac{12 \times (2x+1)}{12} = \frac{12 \times 3}{4}$$

$$\frac{\cancel{12}^{4} \times (5x-9)}{\cancel{3}} + \frac{\cancel{12} \times (2x+1)}{\cancel{12}} = \frac{\cancel{12}^{3} \times 3}{\cancel{4}}$$

$$4(5x-9) + (2x+1) = 3 \times 3$$

$$20x - 36 + 2x + 1 = 9$$

$$22x - 35 = 9$$

$$22x = 44$$

$$\underline{x = 2}$$

> Here, all three fractions are made 12 times larger. This clears the fractions away entirely.

EXERCISE 23.2

Express these as a single fraction.

1 $\dfrac{x}{3} + \dfrac{3+1}{5}$

2 $\dfrac{x+2}{8} + \dfrac{x-1}{6}$

3 $\dfrac{x+1}{5} + \dfrac{3x}{10}$

4 $\dfrac{3x+2}{2} - \dfrac{2x+3}{3}$

5 $\dfrac{x}{4} + \dfrac{2x+1}{5}$

6 $\dfrac{3x}{4} + \dfrac{x+1}{6}$

7 $\dfrac{5}{x} - \dfrac{3}{x+1}$

8 $\dfrac{2}{x+1} + \dfrac{3}{x+2}$

9 $\dfrac{2}{x+3} + \dfrac{1}{2x+1}$

10 $\dfrac{2}{x+4} + \dfrac{1}{x+3}$

11 $\dfrac{3}{x-2} - \dfrac{2}{x+5}$

12 $\dfrac{5}{(x+1)(x+2)} + \dfrac{4}{x+2}$

Solve these equations involving algebraic fractions.

13 $\dfrac{x-2}{5} + \dfrac{x}{10} = \dfrac{1}{2}$

14 $\dfrac{x+2}{8} + \dfrac{x-1}{4} = \dfrac{3}{4}$

15 $\dfrac{x+3}{5} + \dfrac{x+4}{15} = \dfrac{1}{3}$

16 $\dfrac{x+3}{5} + \dfrac{x+4}{15} = \dfrac{1}{3}$

17 $\dfrac{1}{12} + \dfrac{x+1}{6} = \dfrac{x}{4}$

18 $\dfrac{1}{x} + \dfrac{2}{3x} = \dfrac{1}{3}$

19 $\dfrac{1}{x-2} + \dfrac{1}{x} = \dfrac{3}{4}$

20 $\dfrac{1}{x} + \dfrac{1}{2x+1} = \dfrac{7}{10}$

23.3 Cancelling common factors in rational expressions

Some algebraic fractions can be simplified by factorising the top and/or the bottom. You can then cancel any common factors.

EXAMPLE

Simplify the expression:

$$\frac{8x+4}{6x+4}$$

SOLUTION

$$\frac{8x+4}{6x+4} = \frac{4(2x+1)}{2(3x+2)}$$

$$= \frac{{}^{2}\cancel{4}(2x+1)}{{}^{1}\cancel{2}(3x+2)}$$

$$= \frac{2(2x+1)}{3x+2}$$

EXAMPLE

Simplify the expression:

$$\frac{x^2 + 8x}{2x}$$

SOLUTION

$$\frac{x^2+8x}{2x} = \frac{x(x+8)}{2x}$$

$$= \frac{{}^{1}\cancel{x}(x+8)}{2\cancel{x}{}^{1}}$$

$$= \frac{(x+8)}{2}$$

$$= \frac{x+8}{2}$$

You may be asked to simplify a fraction where both the top and the bottom are quadratic expressions. In such cases, factorise the quadratics first. Then look to see if there is a common factor on the top and bottom that can be cancelled out.

EXAMPLE

Simplify the expression:

$$\frac{x^2 + 7x + 10}{x^2 + x - 2}$$

SOLUTION

$$\frac{x^2 + 7x + 10}{x^2 + x - 2} = \frac{(x+5)(x+2)}{(x+2)(x-1)}$$

$$= \frac{(x+5)\cancel{(x+2)}}{\cancel{(x+2)}(x-1)}$$

$$= \frac{x+5}{x-1}$$

You might have more than one factor available for cancelling, as in the next example.

EXAMPLE

Simplify the expression:

$$\frac{8(x-3)^5}{4(x-3)^2}$$

SOLUTION

$$\frac{8(x-3)^5}{4(x-3)^2} = \frac{\overset{2}{\cancel{8}}(x-3)^{\cancel{5}\,3}}{\underset{1}{\cancel{4}}(x-3)^{\cancel{2}\,0}}$$

$$= \frac{2(x-3)^3}{1}$$

$$= \underline{2(x-3)^3}$$

EXERCISE 23.3

Simplify these algebraic fractions.

1 $\dfrac{2x+6}{4x+2}$

2 $\dfrac{12x+20}{8x+2}$

3 $\dfrac{20x+30}{15}$

4 $\dfrac{2x^2+10}{4x}$

5 $\dfrac{5x+10}{5x}$

6 $\dfrac{15(x+3)^6}{3(x+3)^2}$

7 $\dfrac{x^2+10x}{5x}$

8 $\dfrac{9x+6}{3x}$

9 $\dfrac{x^2-5x}{x^2}$

10 $\dfrac{24(2x+1)^3}{6(2x+1)^5}$

11 $\dfrac{x^2+8x}{2x}$

12 $\dfrac{x^2-5x}{2x-10}$

Simplify these expressions fully.

13 $\dfrac{x^2+x}{x^2+3x+2}$

14 $\dfrac{x^2-x-6}{x^2+4x-21}$

15 $\dfrac{x^2+3x+2}{x^2+5x+4}$

16 $\dfrac{x^2+7x+10}{x^2+5x+6}$

17 $\dfrac{x^2+6x+9}{x^2+7x+12}$

18 $\dfrac{x^2+3x-4}{x^2+x-2}$

19 $\dfrac{x^2+5x}{x^2-3x}$

20 $\dfrac{x+4}{x^2-16}$

23.4 Simultaneous equations, one linear and one quadratic

In Chapter 7 you solved simultaneous equations using the elimination method. This approach can also be used when one of the equations is quadratic. Since quadratics often have two solutions, you should be prepared to find two different solutions to the simultaneous equations.

EXAMPLE

Solve the equations:
$y = x + 1$
$y = x^2 - 1$

SOLUTION

Since both equations are of the form $y = ...$, then the two right-hand sides must be equal.

$x^2 - 1 = x + 1$
$x^2 - x - 2 = 0$
$(x - 2)(x + 1) = 0$
$x = 2$ or $x = -1$

> Check: If $x = 2$ and $y = 3$, then
> $y = x + 1 \rightarrow 3 = 2 + 1$ ✓

If $x = 2$:	If $x = -1$:
$y = x + 1$	$y = x + 1$
$\quad = 2 + 1$	$\quad = -1 + 1$
$\quad = 3$	$\quad = 0$

> Check: If $x = -1$ and $y = 0$, then
> $y = x + 1 \rightarrow 0 = -1 + 1$ ✓

Thus the solutions are $\underline{x = 2, y = 3}$ or $\underline{x = -1, y = 0}$

If both x and y appear in square form in the second equation, the elimination is done by substituting.

EXAMPLE

Solve the equations:
$\quad\quad y = x - 5$
$x^2 + y^2 = 17$

SOLUTION

$\quad\quad y = x - 5$
$x^2 + y^2 = 17$

Replace y with $x - 5$ in the second equation:
$\quad\quad x^2 + (x - 5)^2 = 17$
$x^2 + x^2 - 10x + 25 = 17$
$\quad 2x^2 - 10x + 25 = 17$
$\quad 2x^2 - 10x + 8 = 0$
$\quad\quad x^2 - 5x + 4 = 0$
$\quad\quad (x - 4)(x - 1) = 0$

> Check: If $x = 4$ and $y = -1$, then
> $y = x - 5 \rightarrow -1 = 4 - 5$ ✓
> $x^2 + y^2 = 17 \rightarrow (4)^2 + (-1)^2 = 17$ ✓

Thus $x = 4$ or $x = 1$.
If $x = 4$, then $y = x - 5 = 4 - 5 = -1$.
If $x = 1$, then $y = x - 5 = 1 - 5 = -4$.

> Check: if $x = 1$ and $y = -4$,
> then $y = x - 5 \rightarrow -4 = 1 - 5$ ✓
> $x^2 + y^2 = 17 \rightarrow (1)^2 + (-4)^2 = 17$ ✓

Thus the solutions are $\underline{x = 4, y = -1}$ or $\underline{x = 1, y = -4}$

EXERCISE 23.4

Solve these simultaneous equations.

1 $y = x$
$y = x^2 - 2$

2 $y = x + 7$
$y = x^2 + 1$

3 $y = 4x + 7$
$y = 2x^2 + 1$

4 $y = 11x - 2$
$y = 5x^2$

5 $x^2 = y - 1$
$y = 4x + 1$

6 $x = y + 2$
$y = x^2 - 4$

Solve these simultaneous equations.

7 $y = x - 2$
$x^2 + y^2 = 10$

8 $y = 2x - 2$
$x^2 + y^2 = 8$

9 $x = y + 4$
$x^2 + y^2 = 10$

10 $y + 7 = x$
$x^2 + y^2 = 37$

Solve these simultaneous equations.

11 $y = 2x + 3$
$y = x^2 - 12$

12 $x - y = 5$
$y = x^2 - 35$

13 $y = 2x$
$y = x^2 - x + 2$

14 $y = x^2 - 3x - 1$
$y = 2x - 7$

15 $y = 2x - 7$
$x^2 + y^2 = 34$

16 $y - x = 1$
$x^2 + y^2 = 5$

23.5 Changing the subject of an equation where the symbol occurs twice

Sometimes you need to change the subject of an equation where the required symbol appears twice. It is necessary to collect all the terms containing that symbol on to one side of the equation, and then take the symbol out as a common factor.

EXAMPLE

Make x the subject of the equation $3x + 5 = y - ax$

SOLUTION

$$3x + 5 = y - ax$$
$$3x + 5 + ax = y$$
$$3x + ax = y - 5$$
$$x(3 + a) = y - 5$$
$$x = \frac{y - 5}{3 + a}$$

Sometimes you might need to clear away a fraction first.

EXAMPLE

Make x the subject of the equation:

$$10 = \frac{ax + 12}{bx + 1}$$

SOLUTION

$$10 = \frac{ax + 12}{bx + 1}$$
$$10(bx + 1) = ax + 12$$
$$10bx + 10 = ax + 12$$
$$10bx - ax = 12 - 10$$
$$x(10b - a) = 2$$
$$x = \frac{2}{10b - a}$$

EXERCISE 23.5

1 Make x the subject of the equation
$3x - 5 = mx$

2 Make x the subject of the equation
$ax + b = cx + d$

3 Make x the subject of the equation
$2x = k(2 + x)$

4 Make y the subject of the equation
$$d = \frac{y + 1}{y + 2}$$

5 Make t the subject of the equation
$$\frac{t + a}{t + b} = c$$

6 Make x the subject of the equation
$3x - n = kx + 2$

7 Make x the subject of the equation
$$5x + a = \frac{x}{b}$$

8 Make x the subject of the equation
$$\frac{ax - 3}{x} = 2$$

9 Make x the subject of the equation
$$k = \frac{x}{x + a}$$

10 Make u the subject of the equation
$$\frac{1}{u} + \frac{1}{v} = \frac{1}{f}$$

[Hint: Multiply through by uvf first.]

23.6 Algebraic proofs

In this section, a number of general results about properties of numbers will be proved using algebra.

EXAMPLE

Prove that the sum of the squares of two consecutive integers is always odd.

PROOF

Let the two consecutive integers be n and $n + 1$.

The sum of the squares of these numbers is:

$$n^2 + (n + 1)^2 = n^2 + n^2 + 2n + 1$$
$$= 2n^2 + 2n + 1$$
$$= 2(n^2 + n) + 1$$
$$= \text{an even number} + 1$$
$$= \underline{\text{an odd number}}$$

> **Consecutive integers** are integers that are next to each other on a number line, e.g. 9 and 10. Algebraically we write consecutive integers as n and $n + 1$.

In the last example, n and $n + 1$ were used to represent consecutive integers. Some problems will ask about even or odd numbers. $2n$ can be used to represent an even number, and $2n + 1$ can be used to represent an odd number (where n is an integer). If you are using two unrelated even or odd numbers, you must use different variables for each.

> **Consecutive** numbers follow each other in number order:
> 3, 4, 5, $n, n + 1, n + 2$
> $2 \times$ any integer is **even**:
> 2×3 $2n$
> $2 \times$ (any integer) $+ 1$ is **odd**:
> $2 \times 3 + 1$ $2n + 1$

EXAMPLE

Prove that the product of an even number and an odd number is always even.

PROOF

Let the even number be $2n$ and the odd number be $2m + 1$.
Then the product of these two numbers is:

$$2n \times (2m + 1) = 4mn + 2n$$
$$= 2(2mn + n)$$
$$= 2(2mn + n)$$
$$= 2k \quad \text{(where } k = 2mn + n)$$
$$= \text{an even number}$$

EXERCISE 23.6

1 Prove that the sum of two consecutive integers is always odd.

2 Prove that the product of any two even numbers is always even.

3 Prove that the product of any two odd numbers is always odd.

4 Prove that the sum of three consecutive integers is always a multiple of three.

5 Prove that the difference between the squares of any two odd integers is always divisible by four.

6 The diagram shows a square measuring $(a + b)$ along each side. A smaller square, of side c, is inscribed inside the larger square.
  a) Show that the total area of the four triangles is $2ab$.
  b) Obtain expressions for the total area of the shape in two ways:
    (i) by adding together the areas of the four triangles and the inner square
    (ii) by expanding $(a + b)^2$.
  c) Use your results from part **b)** to prove that $c^2 = a^2 + b^2$.
  d) What well-known theorem have you just proved?

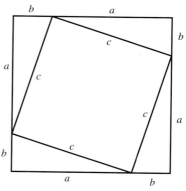

7 **a)** Show that $(100x + 1)(100x - 1) = 10\,000x^2 - 1$.
  b) Hence show that 89 999 is not prime.

8 By writing the nth term of the sequence 1, 3, 5, 7, ... as $(2n - 1)$, or otherwise, show that the difference between the squares of any two consecutive odd numbers is a multiple of 8.   [Edexcel]

REVIEW EXERCISE 23

1 a) Show that $(3+\sqrt{5})(3-\sqrt{5}) = 4$.

   b) Hence, or otherwise, show that $\dfrac{8}{(3-\sqrt{5})} = 6+2\sqrt{5}$.
   Show your working clearly.

2 Express $\dfrac{12}{(6-2\sqrt{3})}$ in the form $a+\sqrt{3}$ where a is a positive integer.

3 a) Find the value of $\sqrt{5} \times \sqrt{20}$

   $\sqrt{5} + \sqrt{20} = k\sqrt{5}$, where k is an integer.
   b) Find the value of k.

   c) Find the value of $\dfrac{\sqrt{5} + \sqrt{45}}{\sqrt{20}}$ [Edexcel]

4 Work out $\dfrac{(5 + \sqrt{3})(5 - \sqrt{3})}{\sqrt{22}}$. Give your answer in its simplest form. [Edexcel]

5 a) Find the value of:
   (i) m when $\sqrt{128} = 2^m$
   (ii) n when $(\sqrt{8} - \sqrt{2})^2 = 2^n$

   A rectangle has a length of 2^t cm and a width of $(\sqrt{8} - \sqrt{2})$ cm.
   The area of the rectangle is $\sqrt{128}$ cm^2.
   b) Find t. [Edexcel]

6 a) Find the value of $16^{\frac{1}{2}}$.
   b) Given that $\sqrt{40} = k\sqrt{10}$, find the value of k.

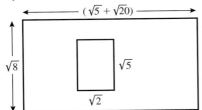

Diagram *not* accurately drawn

   A large rectangular piece of card is $(\sqrt{5} + \sqrt{20})$ cm long and $\sqrt{8}$ cm wide.
   A small rectangle $\sqrt{2}$ cm long and $\sqrt{5}$ cm wide is cut out of the piece of card.
   c) Express the area of the card that is left as a percentage of the area of the large rectangle. [Edexcel]

7 a) Express $\dfrac{1}{x-2} + \dfrac{2}{x+4}$ as a single algebraic fraction.

   b) Hence, or otherwise, solve $\dfrac{1}{x-2} + \dfrac{2}{x+4} = \dfrac{1}{3}$ [Edexcel]

8 Solve $\dfrac{2}{x} + \dfrac{3}{2x} = \dfrac{1}{3}$ [Edexcel]

9 a) Write down an expression, in terms of n, for the nth multiple of 5.
   b) Hence, or otherwise:
   (i) prove that the sum of two consecutive multiples of 5 is always an odd number,
   (ii) prove that the product of two consecutive multiples of 5 is always an even number. [Edexcel]

10 Prove that $(n + 1)^2 - (n - 1)^2$ is a multiple of 4, for all positive integer values of n. [Edexcel]

11 a) Show that $(2a - 1)^2 - (2b - 1)^2 = 4(a - b)(a + b - 1)$.
  b) Prove that the difference between the squares of any two odd numbers is a multiple of 8.
  (You may assume that any odd number can be written in the form $2r - 1$, where r is
  an integer.) [Edexcel]

12 John says, 'For all prime numbers, n, the value of $n^2 + 3$ is always an even number.'
  Give an example to show that John is **not** correct. [Edexcel]

13 a) Factorise $2x^2 + 7x + 5$

  b) Write as a single fraction in its simplest form: $\dfrac{3}{x + 1} + \dfrac{5x}{2x^2 + 7x + 5}$ [Edexcel]

14 Simplify fully:
  a) $2(3x + 4) - 3(4x - 5)$
  b) $(2xy^3)^5$
  c) $\dfrac{n^2 - 1}{n + 1} \times \dfrac{2}{n - 2}$ [Edexcel]

You may use a calculator in the remaining questions.

15 a) Solve $\dfrac{40 - x}{3} = 4 + x$ **b)** Simplify fully $\dfrac{4x^2 - 6x}{4x^2 - 9}$ [Edexcel]

16 Rearrange $4y = k(2 - 3y)$ to write y in terms of k. [Edexcel]

17 Make x the subject of the formula $y = \dfrac{x}{a - x}$ [Edexcel]

18 Solve the simultaneous equations:
$$y = 3x - 1$$
$$x^2 + y^2 = 29$$

19 Bill said that the line $y = 6$ cuts the curve $x^2 + y^2 = 25$ at two points.
  a) By eliminating y, show that Bill is incorrect.
  b) By eliminating y, find the solutions to the simultaneous equations:
$$x^2 + y^2 = 25$$
$$y = 2x - 2$$
[Edexcel]

Key points

1 A surd is an expression containing a root, like $\sqrt{45}$. You can often simplify a surd without using a calculator, by looking for a perfect square inside the root; for example:

$$\sqrt{45} = \sqrt{9 \times 5} = \sqrt{9} \times \sqrt{5} = 3\sqrt{5}$$

2 You can rationalise a fraction like $\dfrac{6 + \sqrt{3}}{\sqrt{3}}$ to clear the surd from the bottom; for example:

$$\frac{6 + \sqrt{3}}{\sqrt{3}} = \frac{(6 + \sqrt{3}) \times \sqrt{3}}{\sqrt{3} \times \sqrt{3}} = \frac{6\sqrt{3} + 3}{3} = 2\sqrt{3} + 1$$

3 Algebraic fractions can be added or subtracted in a similar way to ordinary numerical fractions. You must rewrite the fractions to have the same bottom (common denominator) first.

4 Algebraic fractions involving quadratic expressions can sometimes be simplified by cancelling common factors. You may need to factorise the top and bottom separately first.

5 Some IGCSE problems on simultaneous equations may give you one linear and one quadratic equation. Use an elimination method to obtain a quadratic equation, and remember to look for both solutions to the quadratic. Give each pair of answers at the end, e.g.

$x = 2$ and $y = 3$ or $x = -1$ and $y = 0$

6 Some IGCSE problems on changing the subject of an equation will have the new symbol occurring twice. You must isolate the terms containing this symbol on one side of the equation, then extract the symbol as a common factor.

7 Algebraic proofs are often about even and odd numbers. You can write any even number in the form $2m$, whilst any odd number has the form $2n + 1$, where m and n are integers.

8 You may also meet questions about consecutive integers, that is, integers that are next to each other on a number line. You can represent two consecutive integers as n and $n + 1$. Two consecutive odd numbers would be $2n - 1$ and $2n + 1$.

Internet Challenge 23

Famous formulae

Here are some famous mathematical formulae. Use the internet to help you find out what each one represents. You should know some of them already.

1 $c^2 = a^2 + b^2$

2 $C = 2\pi r$

3 $A = \dfrac{(a + b)h}{2}$

4 $V = IR$

5 $V = \frac{1}{3}\pi r^2 h$

6 $x = \dfrac{-b \pm \sqrt{b^2 - 4ac}}{2a}$

7 $E = mc^2$

8 $A = 4\pi r^2$

9 $s = ut + \frac{1}{2}at^2$

10 $T = 2\pi\sqrt{\dfrac{l}{g}}$

11 $F - E + V = 1$

12 $C = \frac{5}{9}(F - 32)$

13 $E = \frac{1}{2}mv^2$

14 $E = mgh$

15 $\dfrac{1}{u} + \dfrac{1}{v} = \dfrac{1}{f}$

16 $\dfrac{1}{R} = \dfrac{1}{R_1} + \dfrac{1}{R_2}$

17 $I = \dfrac{PRT}{100}$

18 $A = \frac{1}{2}\,ab\sin C$

19 $F = \dfrac{Gm_1 m_2}{d^2}$

20 $W = Fd$

CHAPTER 24

Functions and function notation

In this chapter you will learn how to:

- use function notation to describe simple functions (mappings)
- find the range of a function, for a given domain
- find the inverse of a given function
- work with composite functions.

You will also be challenged to:

- investigate the Greek alphabet.

Starter: **Number crunchers**

Here are some instructions for a number cruncher machine:

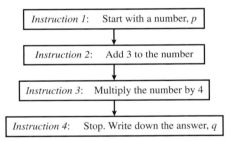

Instruction 1:   Start with a number, p

Instruction 2:   Add 3 to the number

Instruction 3:   Multiply the number by 4

Instruction 4:   Stop. Write down the answer, q

a) Debbie starts with the number $p = 2$. Work through the number cruncher to show that she should get an answer $q = 20$. Show all the steps of your working.

b) Arthur can't remember what number he started with, but he did get the answer $q = 32$. What number did he start with?

c) Alison got instructions 2 and 3 the wrong way round. She ended up with $q = 35$. What number should she have ended up with?

24.1 Introducing functions and function notation

Consider the following 'number machine':

Input → Multiply by 2 → Add 5 → Output

For any given value of the input number, the instructions tell you to multiply that number by 2 first, and then add 5, to obtain the output number. If the input number is x, then the output number is $2x + 5$.

This number machine is an example of a **function**, that is, a process that takes one number and turns it into (maps it to) another one. In this case, x is **mapped** to $2x + 5$.

Functions are often given names such as f, g, h and so on. The rule for the above function could be written as:

$$f(x) = 2x + 5$$

Sometimes an arrow is used instead, like this:

$$f : x \rightarrow 2x + 5$$

EXAMPLE

a) Write this number machine as a function using the notation $f : x \rightarrow \ldots$

Input → Multiply by 4 → Add 1 → Output

b) Write this number machine as a function using the notation $g(x) = \ldots$

Input → Subtract 2 → Multiply by 4 → Output

c) Work out the values of:

   (i) $g(10)$ **(ii)** $g(-2)$

SOLUTION

a) If the input is x, then this becomes $4x$, then 1 is added, to obtain $4x + 1$, so

$$\underline{f : x \rightarrow 4x + 1}$$

b) If the input is x, then this becomes $x - 2$, then this result is multiplied by 4, to obtain $4(x - 2)$, so

$$\underline{g(x) = 4(x - 2)}$$

c) (i) $g(10) = 4(10 - 2)$

$$= 4 \times 8$$

$$= \underline{32}$$

   (ii) $g(-2) = 4(-2 - 2)$

$$= 4 \times -4$$

$$= \underline{-16}$$

EXERCISE 24.1

1 a) Write this number machine as a function using the notation f : $x \rightarrow$ …

b) Find the values of: **(i)** f(3)   **(ii)** f(10)

2 a) Write this number machine as a function using the notation g(x) = …

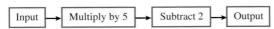

b) Find the values of: **(i)** g(7)   **(ii)** g(−1)

3 The function f(x) is given by f(x) = 2x + 7.
Find the values of:

a) f(1)   **b)** f(3)   **c)** f(10)

4 The function g(x) is given by g : $x \rightarrow x^2 + 3$.
Find the values of:

a) g(2)   **b)** g(5)   **c)** g(−2)

5 The function h(x) is given by h(x) = $\dfrac{3x+1}{2}$.
Find the values of:

a) h(5)   **b)** h(−1)   **c)** h(6)

24.2 Domain and range

Look again at the number machine from the previous section:

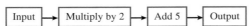

Supposing the input numbers were to be 1, 2, 3 and 4. Then the corresponding output numbers would be 7, 9, 11, 13. We could represent this on a **mapping diagram**:

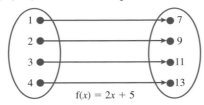

f(x) = 2x + 5

The set of numbers that provide the *input* values is called the **domain** of the function, in this case, {1, 2, 3, 4}. The corresponding *output* numbers form the **range** set, in this case {7, 9, 11, 13}.

Examination questions will often tell you the domain set for a function, and ask you to find the range. Sometimes the domain will be a small discrete set of numbers (for example, {1, 2, 3, 4}), other times it may be an infinite set, such as all possible numbers on the number line, including fractions and decimals. These are known as the **real numbers**.

EXAMPLE

a) The function $f(x) = 6x - 2$ has domain $\{0, 1, 2, 3, 4\}$. Find the range of $f(x)$.

b) The function $g(x) = x^2 + 3$ has as its domain all the real numbers. Find the range of $g(x)$.

SOLUTION

a) Taking each of the values 0, 1, 2, 3, 4 and multiplying by 6, then subtracting 2, we obtain $-2, 4, 10, 16$ and 22 so the range of $f(x)$ is $\{-2, 4, 10, 16, 22\}$.

b) The real numbers take positive and negative values, but when squared must give a positive value (or zero). Thus x^2 takes all real values from 0 upwards, so the range of $x^2 + 3$ takes all real values greater than or equal to 3.

Some functions might lead to arithmetic that cannot be carried out, such as division by zero, or finding the square root of a negative number. In such cases it is necessary to **restrict the domain**, so that the function is properly defined, that is, some values might need to be **excluded** from the domain of the function.

EXAMPLE

a) The function $f(x)$ is defined as $f : x \to \dfrac{3}{x - 2}$, where x is a real number.

Which real number must be excluded from the domain of the function $f(x)$?

b) The function $g(x)$ is defined as $g : x \to \sqrt{x - 4}$ where x is a real number.
Which real numbers must be excluded from the domain of the function $g(x)$?

SOLUTION

a) To avoid division by zero, $x - 2$ must not be zero, so $x = 2$ must be excluded from the domain of $f(x)$.

b) To avoid square rooting a negative number, $x - 4$ must be 0 or positive, so x must be at least 4. Thus all real numbers less than 4 must be excluded from the domain of $g(x)$.

EXERCISE 24.2

1 The function $f(x) = 2x + 1$ has domain $\{1, 2, 3, 4\}$. Find the range of $f(x)$.

2 The function $g(x) = \dfrac{12}{x}$ has domain $\{1, 2, 3, 4\}$. Find the range of $g(x)$.

3 The function $h(x) = 5 - x$ has domain $\{0, 1, 2, 3, 4, 5\}$. Find the range of $h(x)$.

4 The function $f(x) = 10x + 5$ has domain $\{-1, 0, 1, 2\}$. Find the range of $f(x)$.

5 The function $g(x) = 2(x + 3)$ has domain $\{0, 2, 4, 6, 8\}$. Find the range of $g(x)$.

6 The function $h(x) = x^2 + 1$ has domain $\{-1, 0, 1, 2, 3, 4\}$. Find the range of $h(x)$.

7 State the value of x that must be excluded from the domain of the function $f(x) = \dfrac{16}{x+1}$.

8 State the values of x that must be excluded from the domain of the function $g(x) = \sqrt{x-3}$.

9 The function $f(x) = x^2 + 6$ has domain all real numbers. Find the range of $f(x)$.

10 The function $g(x) = 2x + 1$ has domain $\{x: 0 \leqslant x \leqslant 20\}$. Find the range of $g(x)$.

24.3 Inverse functions

Here is our number machine again:

Input → Multiply by 2 → Add 5 → Output

This diagram corresponds to the function $f(x) = 2x + 5$.

The **inverse** function $f^{-1}(x)$ is the function that undoes the effect of the function $f(x)$, in other words, it returns each member of the range set back to its corresponding number in the domain.

In this particular case, the inverse function is $f^{-1}(x) = \dfrac{x-5}{2}$.

There are two good methods for finding inverse functions, namely the *reverse flow diagram* method and *algebraic rearrangement*. Each method is illustrated in the examples below.

EXAMPLE

The function $f(x)$ is defined as $f(x) = 4x + 1$, where x is a real number.
Find the inverse function $f^{-1}(x)$.

SOLUTION

Here is a number machine, or flow diagram, for the function $f(x)$:

Input → Multiply by 4 → Add 1 → Output

The reverse flow diagram is made by reversing the arrows and replacing each box with its own inverse process:

Output ← Divide by 4 ← Subtract 1 ← Input

To complete the solution, read out the instructions from right to left:

$$f^{-1}(x) = \dfrac{x-1}{4}$$

EXAMPLE

The function f(x) is defined as f(x) = $3x^2 + 7$, where x is a positive real number. Find the inverse function f^{-1}(x).

SOLUTION

Begin by writing x = f(y), so, in this case

$$x = 3y^2 + 7$$

Now rearrange, to make y the subject:

$$3y^2 + 7 = x$$
$$3y^2 = x - 7$$
$$y^2 = \frac{x - 7}{3}$$
$$y = \sqrt{\frac{x - 7}{3}}$$

Thus the inverse function is

$$f^{-1}(x) = \sqrt{\frac{x - 7}{3}}$$

EXERCISE 24.3

1 The function f(x) is defined as f(x) = $4x + 3$.
   a) Find f(5).
   b) Find f^{-1}(35).

2 The function g(x) is defined as g : $x \rightarrow 2(x + 5)$.
   Find the value of x for which g(x) = 24.

3 Given that f(x) = $2x + 11$ find f^{-1}(x). Write your answer in the form f^{-1}(x) = ...

4 Given that h : $x \rightarrow \frac{1}{2}x + 3$ find h^{-1}(x). Write your answer in the form h^{-1} : $x \rightarrow$ ...

5 The functions p and q are defined as follows:

   $$p : x \rightarrow 3x - 1 \qquad q : x \rightarrow x^2 + 1$$

   a) Find: (i) p(4)   (ii) q(5)
   b) Find: (i) p^{-1}(5)   (ii) q^{-1}(50)

6 Find the inverse of each of these functions:
   a) f(x) = $5x + 7$ b) g(x) = $\frac{1}{x} + 3$ c) h(x) = $\frac{3}{2 - x}$

7 The function p is defined as p : $x \rightarrow 2x + 7$
   a) Find the value of p(4).
   b) Given that p(a) = 1, work out the value of a.

8 The function f is defined as $f : x \to x^2 + 1$ for all positive values of x.
   a) Find the value of p(2).
   b) Given that $p(n) = 50$, work out the value of n.
   c) Find the inverse function. Write your answer in the form $f^{-1}(x) = \ldots$

9 The function g is defined as $g : x \to 4x + 3$ for all positive values of x.
   a) Find the value of g(3).
   b) Solve the equation $g(x) = 9$.
   c) Find the inverse function. Write your answer in the form $g^{-1}(x) = \ldots$

10 The function h is defined as $h(x) = \dfrac{x - 5}{2}$.
   Solve the equation $h^{-1}(x) = 6x$.

24.4 Composite functions

Suppose you have two different functions, for example $f(x) = 2x + 5$ and $g(x) = x^2$. The **composite function** $fg(x)$ is the result of applying one of the functions to an input value of x, then applying the other function to the result. You always process the function nearest the (x) bracket first, so $fg(x)$ means 'apply g to x first, then apply f to the result'.

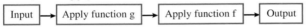

It is very important to take great care with the order in which the functions are applied since, in most cases, $fg(x)$ and $gf(x)$ are not the same.

EXAMPLE

The functions f and g are defined over the domain of all the real numbers as $f(x) = 2x + 5$ and $g(x) = x^2$.
a) Find the values of: (i) fg(3)   (ii) gf(3)
b) Express the composite function fg in the form $fg(x) = \ldots$
c) Express the composite function gf in the form $gf(x) = \ldots$

SOLUTION

a) (i) fg(3) = f(9)
            $= 2 \times 9 + 5$
            $= 18 + 5$
            $= \underline{23}$

   (ii) gf(3) = g(11)
            $= 11^2$
            $= \underline{121}$

b) $fg(x) = f(x^2)$
            $= 2 \times x^2 + 5$
            $= \underline{2x^2 + 5}$

c) $gf(x) = g(2x + 5)$
            $= \underline{(2x + 5)^2}$

EXERCISE 24.4

1 The functions f and g are defined over the domain of all the real numbers as $f(x) = 3x + 2$ and $g(x) = 5x$.
  a) Find the values of:
     (i) fg(4)   **(ii)** gf(4)
  b) Express the composite function fg in the form $fg(x) = \ldots$

2 The functions f and g are defined over the domain of all the real numbers as $f(x) = x^2 + 1$ and $g(x) = 2x$.
  a) Find the values of:
     (i) fg(2)   **(ii)** gf(5)
  b) Express the composite function gf in the form $gf : x \rightarrow \ldots$

3 The functions p and q are defined over the domain of all the real numbers as $p(x) = 4x + 1$ and $q(x) = x + 1$.
  a) Express the composite function pq in the form $pq(x) = \ldots$
  b) Find the value of a if $pq(a) = 7$.

4 $f : x \rightarrow 3x + 1$ $g : x \rightarrow 2x - 3$
  Solve the equation $fg(x) = g(x)$.

5 Functions p and q are defined as $p : x \rightarrow 2 + \sqrt{x}$ $q : x \rightarrow 5x + 4$
  Find the values of:
  a) p(16)   **b)** q(−2)   **c)** $p^{-1}q(1)$

6 $f : x \rightarrow x^2$ $g : x \rightarrow 2x - 9$
  Solve the equation $gf(x) = f(x)$.

7 $f : x \rightarrow 3x + 1$ $g : x \rightarrow x - 2$
  Solve the equation $fg(x) = g^{-1}(x)$.

8 Functions p and q are defined as $p : x \rightarrow \dfrac{2x + 1}{5}$ $q : x \rightarrow \dfrac{5x - 1}{2}$
  a) Find the composite function pq in the form $pq(x) = \ldots$
  b) Describe the relationship between the functions p and q.

REVIEW EXERCISE 24

1 Here are three functions:

  $f(x) = 5 + 2x$ $g(x) = \sqrt{10 - x}$ $h(x) = \dfrac{1}{x + 3}$

  a) Find:
     (i) f(1.5)   **(ii)** g(6)   **(iii)** h(7)
  b) **(i)** Given that $f(a) = 2$, find a.
     (ii) Given that $g(b) = 5$, find b.
     (iii) Given that $h(c) = 1$, find c.

2 The functions $f(x)$ and $g(x)$ are defined as follows:
  $f : x \rightarrow \dfrac{1}{x - 2}$ $g : x \rightarrow x^2$
  a) Find:
     (i) fg(3)   **(ii)** gf(1.5)
  b) Express the composite function fg in the form $fg(x) = \ldots$
  c) Which value must be excluded from the domain of $f(x)$?

3 The function f is defined as $f(x) = 3x + 1$.
  a) Find:
  (i) $f(4)$ **(ii)** $f^{-1}(7)$
  b) Express the function $ff(x)$ in the form $ax + b$, stating the values of a and b.

4 Three functions are defined with their domains as follows:
 $f(x) = 2x + 15$ with domain $\{x: x \text{ is any real number}\}$
 $g(x) = x^2$ with domain $\{x: x \text{ is any real number}\}$
 $h(x) = \sin(x°)$ with domain $\{x: 0 \leqslant x \leqslant 90\}$
  a) Find the range of each function.
  b) Find the values of x for which $f(x) = g(x)$.
  c) Find the value of x for which $h(x) = 0.5$.

5 The functions g and h are defined, with their domains, as
 $g(x) = 2x + 3$ with domain $\{x: 0 \leqslant x \leqslant 6\}$
 $h(x) = 4x - 1$ with domain $\{x: m \leqslant x \leqslant n\}$
  a) Find the range of g.
  b) The functions g and h have the same range. Find the values of m and n.

6 $p(x) = \dfrac{x+1}{x-2}$ $q(x) = \dfrac{2x+1}{x-1}$
  a) Find the values of:
  (i) $pq(4)$ **(ii)** $pq(2)$
  b) Express the composite function pq in the form $pq(x) = \dots$
  c) Describe the relationship between the functions p and q.

7 $f: x \to \dfrac{1}{x}$ $g: x \to 4x - 1$
  a) Find the values of:
  (i) $f(0.2)$ **(ii)** $fg\left(\tfrac{1}{2}\right)$
  b) Express the inverse function f^{-1} in the form $f^{-1}: x \to \dots$
  c) **(i)** Express the composite function fg in the form $fg: x \to \dots$
  (ii) Which value of x must be excluded from the domain of fg?

8 Three functions p, q, r are defined as follows:
 $p(x) = 5x$ with domain $\{x: x \text{ is a real number such that } 0 \leqslant x \leqslant 18\}$
 $q(x) = x^2 - 6$ with domain $\{x: x \text{ is any real number}\}$
 $r(x) = \cos(x°)$ with domain $\{x: 0 \leqslant x \leqslant 90\}$
  a) Find the value of $rp(12)$.
  b) State the range of the function r.
  c) Solve the equation $p(x) = q(x)$.

9 $f: x \to 2x - 1$ $g: x \to \dfrac{3}{x}, x \neq 0$
  a) Find the values of:
  (i) $f(3)$ **(ii)** $fg(6)$
  b) Express the inverse function f^{-1} in the form $f^{-1}: x \to \dots$
  c) **(i)** Express the composite function gf in the form $gf: x \to \dots$
  (ii) What value of x must be excluded from the domain of gf?   [Edexcel]

10 $f: (x) = x^2$ $g: (x) = x - 6$
 Solve the equation $fg(x) = g^{-1}(x)$.   [Edexcel]

Key points

1. A function is a rule for processing an input to obtain an output value. Functions may be written in one of two forms:

$$f(x) = 2x + 5 \qquad f : x \rightarrow 2x + 5$$

2. The set of values that provide the input values for a function forms the domain; the corresponding set of output values forms the range. For example, if the function $f(x) = 2x + 5$ is defined over the domain $\{1, 2, 3\}$ then the range will be $\{7, 9, 11\}$.

3. Sometimes a value must be excluded from the domain; this is to prevent problems when evaluating the function. For example, the function $f : x \rightarrow \dfrac{5}{x - 2}$ must have the value $x = 2$ excluded from its domain, otherwise a division by zero would occur.

4. For a function $f(x)$ the inverse function $f^{-1}(x)$ restores each member of the range back to its original value in the domain: if $f(x) = y$, then $f^{-1}(y) = x$. Inverse functions may be found by algebraic rearrangement or by the reverse flow diagrams; both methods are described fully in section 24.3.

5. A composite function such as $fg(x)$ is the result of applying the single functions f and g in succession. Note that the function nearest the (x) bracket is carried out first, so that $fg(x)$ means apply function g first, then apply function f to the result.

6. In most cases $fg(x)$ and $gf(x)$ are not the same.

Internet Challenge 24

The Greek Alphabet

The table below gives the Greek alphabet, in order. Unscramble the jumbled words in the statements alongside each one. Note: the table gives the capital (A) form first, then the lower case (α).

A α	alpha	Denotes the brightest star in a ACEILLNNOOSTT.
B β	beta	Used to describe a trial version of a piece of CEMOPRTU AEFORSTW.
Γ γ	gamma	The gamma function is a generalised form of the AACFILORT function.
Δ δ	delta	Sometimes used to denote the value of $b^2 - 4ac$ in the AACDIQRTU AEINOQTU formula.
E ε	epsilon	The fifth caste of society in Huxley's ABERV ENW DLORW.
Z ζ	zeta	Riemann's zeta function has widespread applications in BEMNRU EHORTY.
H η	eta	Eta functions have been defined by the mathematicians DDDEEIKN and CDEHIILRT.
Θ θ	theta	Symbol used in geometry to denote an KNNNOUW AEGLN.
I ι	iota	Iota is one of the EENSV ELOSVW in the Greek alphabet.
K κ	kappa	English mathematician JNHO WARBRO (1630 to 1677) published work on the tangents to the kappa curve, predating the formal development of calculus.
Λ λ	lambda	Used by physicists to denote the AEEGHLNTVW of radiation.
M μ	mu	Denotes the SI prefix micro- , ENO HIILLMNOT of.
N ν	nu	Used to denote the CEEFNQRUY of a wave.
Ξ ξ	xi	In the system of Greek numerals, xi has a value of ISTXY.
O o	omicron	In the system of Greek numerals, omicron has a value of EENSTVY.
Π π	pi	The ratio of CCCEEEFIMNRRU to ADEEIMRT in a circle.
P ρ	rho	Used in ACIISSSTTT as a measure of correlation between two variables.
Σ σ	sigma	Capital sigma denotes the process of AIOMMNSTU, or adding up.
T τ	tau	Can be used as a symbol for EOQRTU, a measure of rotational force.
Υ υ	upsilon	The star upsilon Andromedae is believed to have a AAELNPRTY EMSSTY.
Φ φ	phi	Used to denote the LLNU EST in set theory.
X χ	chi	Statisticians used the chi-squared test to measure DEGNOOSS FO IFT.
Ψ ψ	psi	This letter has given us English words such as CGHLOOPSYY.
Ω ω	omega	In astronomy the Omega Nebula is also known as the ANSW or EEHHOORSS.

CHAPTER 25

Further trigonometry

In this chapter you will **learn how to**:

- solve trigonometry problems using the sine and cosine rules
- find the area of a triangle using $\frac{1}{2}ab \sin C$
- find the area of a segment of a circle
- calculate distances and angles in 3-D.

You will also be **challenged to**:

- investigate Heron's formula.

Starter: How tall is the church?

A mathematical vicar wants to work out the height of the spire on his parish church. He notices that from a point A the spire has an angle of elevation of 32°. When he walks 25 metres nearer the spire, to point B, the angle increases to 65°.

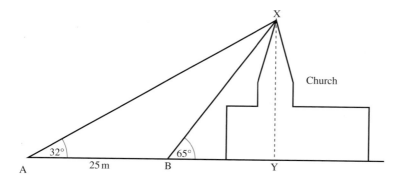

Make a scale drawing, and measure the height of the spire.
A scale of 1 cm = 5 metres should work well.

In this chapter you will meet methods for calculating sides and angles in a non-right-angled triangle. You will meet this question again in the Review exercise, and you will be able to calculate the answer, to see how accurate your scale drawing is.

25.1 The sine rule

In your earlier work on trigonometry you used the sine ratio in a right-angled triangle. It is also possible to use the sine ratio in a non-right-angled triangle. The triangle ABC below has no right angles. The line CX is the perpendicular height, that is, angle CXB = 90°.

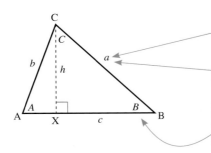

The lengths of the sides are labelled a, b, c using italic small letters.

Notice that side a is opposite angle A, and so on.

The angles are labelled A, B, C using italic capital letters.

The triangle AXC is right angled, so you can use ordinary trigonometry to obtain:

$$h = b \times \sin A$$

The triangle BXC is also right angled, so you can use trigonometry in this triangle, to obtain:

$$h = a \times \sin B$$

Since both these equations refer to the same height h, then the expressions on the right-hand sides must be equal:

$$a \times \sin B = b \times \sin A$$

This may be rearranged into the following form:

$$\frac{a}{\sin A} = \frac{b}{\sin B}$$

By drawing a perpendicular from A on to side BC, or from B on to side AC, it is possible to obtain a similar result including side c and sin C. Putting this altogether, the result is the **sine rule**:

$$\frac{a}{\sin A} = \frac{b}{\sin B} = \frac{c}{\sin C}$$

EXAMPLE

Find the missing lengths, x cm and y cm, in this triangle. Give your answers to 3 significant figures.

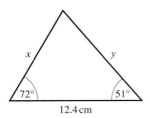

SOLUTION

Labelling the sides and
angles we have:

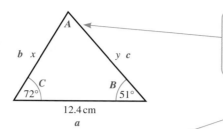

Since $A + B + C = 180°$,
then $A = 180° - (51° + 72°)$
$= 57°$

The sine rule gives: $\dfrac{a}{\sin A} = \dfrac{b}{\sin B} = \dfrac{c}{\sin C}$

For x: $\dfrac{x}{\sin 51°} = \dfrac{12.4}{\sin 57°}$

$x = \dfrac{\sin 51° \times 12.4}{\sin 57°}$

$= 11.490\ 339\ 94$

$= \underline{11.5\text{ cm}}$ (3 s.f.)

For y: $\dfrac{y}{\sin 72°} = \dfrac{12.4}{\sin 57°}$

$y = \dfrac{\sin 72° \times 12.4}{\sin 57°}$

$= 14.061\ 660\ 51$

$= \underline{14.1\text{ cm}}$ (3 s.f.)

You can also use the sine rule to find an unknown angle in a triangle.

EXAMPLE

Find the size of the angle
marked x, in the triangle.

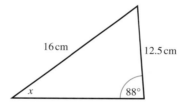

SOLUTION

By the sine rule, we have:

$\dfrac{16}{\sin 88°} = \dfrac{12.5}{\sin x}$

Cross multiplying:

$16 \times \sin x = 12.5 \times \sin 88°$

$\sin x = \dfrac{12.5 \times \sin 88°}{16}$

$= 0.780774084$

$x = 51.331\ 504\ 51$

$= \underline{51.3°}$ (nearest 0.1°)

EXERCISE 25.1

Find the lengths of the sides represented by letters. Give your answers to 3 significant figures.

1

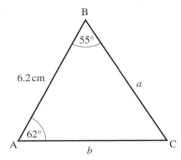

2

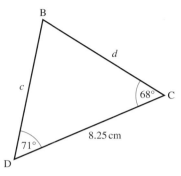

3

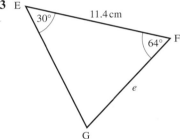

4

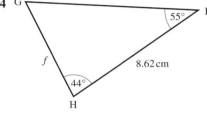

5

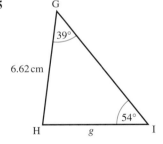

6

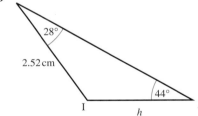

7

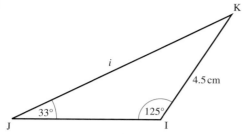

8

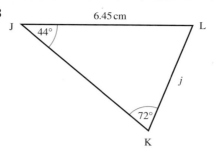

Find the angles represented by letters. Give your answers to the nearest 0.1°.

9

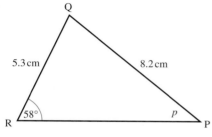

10

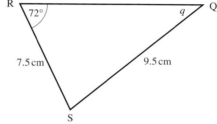

11

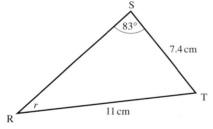

12

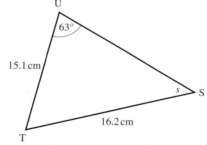

25.2 The cosine rule

The diagram shows a general triangle ABC, labelled so that the sides of length a, b and c are opposite the angles A, B and C.

For all triangles labelled in this way, it may be proved that:

$$c^2 = a^2 + b^2 - 2ab \cos C$$

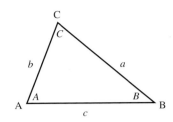

This is the **cosine rule**. At IGCSE you will need to use the cosine rule to find unknown sides or angles, but you do not need to know how to prove it.

Notice that if the angle C were to be a right angle, then $\cos C$ becomes $\cos 90°$, which is zero, and the cosine rule reduces to $c^2 = a^2 + b^2$. Thus you can think of the cosine rule as being a version of Pythagoras' Theorem that may be used in triangles when they do not have right angles.

The cosine rule contains one angle and all three sides. The version above has been written from the point of view of angle C. There are two other, equivalent forms based on each of the other two angles. So, for any triangle ABC, the cosine rule states that:

$$a^2 = b^2 + c^2 - 2bc \cos A$$
$$b^2 = a^2 + c^2 - 2ac \cos B$$
$$c^2 = a^2 + b^2 - 2ab \cos C$$

> Notice that the side forming the subject of the equation corresponds with the angle in the cosine expression, for example: $a^2 = b^2 + c^2 - 2bc \cos A$

EXAMPLE

In triangle ABC, AB = 10 cm, BC = 12 cm and angle ABC = 55°.
a) Make a sketch of the triangle, and mark this information on the sketch.
b) Calculate the length of AC. Give your answer correct to 3 significant figures.

SOLUTION

a)

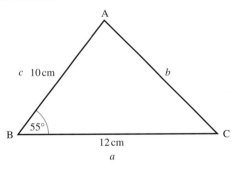

b) Using the cosine rule in the form $b^2 = a^2 + c^2 - 2ac \cos B$, we have:

$$b^2 = 12^2 + 10^2 - 2 \times 12 \times 10 \times \cos 55°$$
$$= 144 + 100 - 240 \cos 55°$$
$$= 106.341\ 655\ 3$$
$$b = \sqrt{106.341\ 655\ 3}$$
$$= 10.312\ 209\ 04$$

Thus AC = <u>10.3 cm</u> (3 s.f.)

You can also use the cosine rule to find an unknown angle. If the unknown angle lies between 0° and 90°, its cosine will be *positive*, while if it is between 90° and 180° its cosine will be *negative*. This means that the unknown angle can always be found uniquely.

EXAMPLE

Find the angles x and y indicated on these sketch diagrams.
Give your answers to the nearest 0.1°.

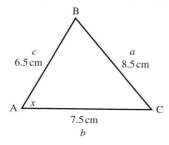
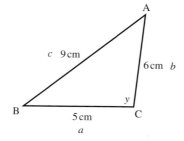

SOLUTION

To find x, use the cosine rule in the form $a^2 = b^2 + c^2 - 2bc \cos A$:

$$8.5^2 = 7.5^2 + 6.5^2 - 2 \times 7.5 \times 6.5 \times \cos x$$

Rearranging:

$$2 \times 7.5 \times 6.5 \times \cos x = 7.5^2 + 6.5^2 - 8.5^2$$

$$\cos x = \frac{7.5^2 + 6.5^2 - 8.5^2}{2 \times 7.5 \times 6.5}$$

$$= \frac{26.25}{97.5}$$

$$= 0.269\ 230\ 769$$

$$x = \cos^{-1}(0.269\ 230\ 769)$$

$$= 74.381\ 501\ 72$$

$$= \underline{74.4°} \text{ (nearest 0.1°)}$$

To find y, use the cosine rule in the form $c^2 = a^2 + b^2 - 2ab \cos C$:

$$9^2 = 5^2 + 6^2 - 2 \times 5 \times 6 \times \cos y$$

Rearranging:

$$2 \times 5 \times 6 \times \cos y = 5^2 + 6^2 - 9^2$$

$$\cos y = \frac{5^2 + 6^2 - 9^2}{2 \times 5 \times 6}$$

$$= \frac{-20}{60}$$

$$= -0.333\ 333\ 333$$

$$y = \cos^{-1}(-0.333\ 333\ 333)$$

$$= 109.471\ 220\ 6$$

$$= \underline{109.5°} \text{ (nearest 0.1°)}$$

> The negative sign here indicates that the angle is more than 90°; in other words, it is obtuse.

$$a^2 = b^2 + c^2 - 2bc \cos A \qquad \text{rearranges to give} \qquad \cos A = \frac{b^2 + c^2 - a^2}{2bc}$$

$$b^2 = a^2 + c^2 - 2ac \cos B \qquad \text{rearranges to give} \qquad \cos B = \frac{a^2 + c^2 - b^2}{2ac}$$

$$c^2 = a^2 + b^2 - 2ab \cos C \qquad \text{rearranges to give} \qquad \cos C = \frac{a^2 + b^2 - c^2}{2ab}$$

EXERCISE 25.2A

Use the cosine rule to find the unknown sides indicated by letters in these triangles.

1
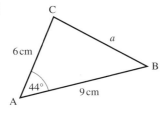

2
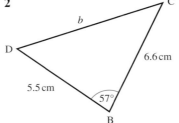

3
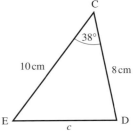

4
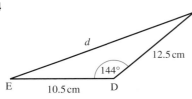

5
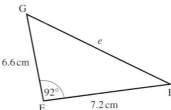

6
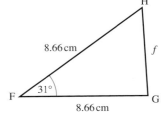

Use the cosine rule to find the unknown angles indicated by letters in these triangles.

7
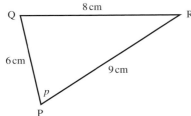

8
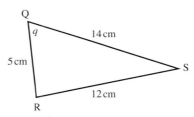

9
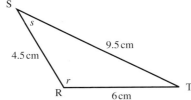

10
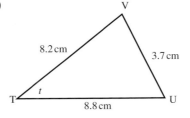

EXERCISE 25.2B

These are mixed questions on sine and cosine rules.

Find the missing sides represented by letters. Give your answers correct to 3 significant figures.

1
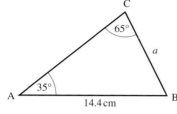

2
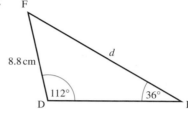

3
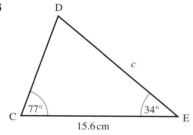

4
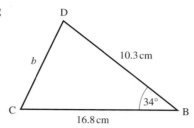

5
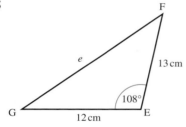

6
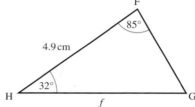

Find the unknown angles indicated by letters. Give your answers to the nearest 0.1°.

7
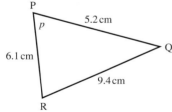

8
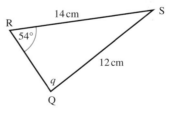

9
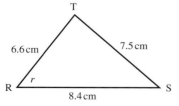

10
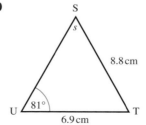

483

25.3 Area of a triangle using $\frac{1}{2}ab$ sin C, and segments of circles

Suppose you are given two sides of a non-right-angled triangle, say a and b, and the *included* angle between them, C. This information can be used to find the area of the triangle.

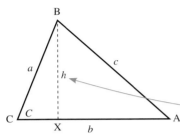

The area of the triangle is:

$$\text{Area} = \frac{1}{2} \times \text{base} \times \text{height}$$
$$= \frac{1}{2} \times AC \times BX$$
$$= \frac{1}{2} \times b \times (a \sin C)$$
$$\text{Area} = \frac{1}{2}ab \sin C$$

Since triangle CXB is right angled at X, then BX = h = a sin C

As with the cosine rule, you may wish to learn this formula in three versions, according to which angle you have been given. In each case, in order to use the formula, you must have SAS, meaning 'two Sides and the included Angle'.

$$\text{Area of triangle} = \frac{1}{2}ab \sin C = \frac{1}{2}bc \sin A = \frac{1}{2}ac \sin B$$

EXAMPLE

Find the area of this triangle, correct to 3 significant figures.

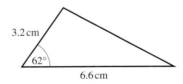

SOLUTION

$$\text{Area of triangle} = \frac{1}{2}ab \sin C$$
$$= \frac{1}{2} \times 3.2 \times 6.6 \times \sin 62°$$
$$= 9.323\,926\,581$$
$$= 9.32 \text{ cm}^2 \,(3 \text{ s.f.})$$

You might need to use the formula for the area of a triangle in reverse.

EXAMPLE

The diagram shows a triangle PQR in which PQ = 8 cm and QR = 10 cm. The triangle has an area of 20 cm². Angle PQR is acute.

a) Calculate the size of angle PQR.
b) Calculate the length of PR.
Give your answer to 3 significant figures.

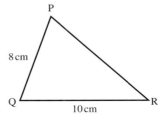

SOLUTION

a) Since the area of the triangle is 20 cm²:

$$'\tfrac{1}{2} ab \sin C' = \text{area of triangle}$$
$$\tfrac{1}{2} \times 8 \times 10 \times \sin Q = 20$$
$$40 \sin Q = 20$$
$$\sin Q = \frac{20}{40}$$
$$\sin Q = 20$$
$$Q = \sin^{-1}(0.5)$$
$$Q = 30°$$

Thus angle $\underline{PQR = 30°}$.

b) The length of PR may now be found, using the cosine rule:

$$PR^2 = 8^2 + 10^2 - 2 \times 8 \times 10 \times \cos 30$$
$$= 64 + 100 - 160 \cos 30°$$
$$= 25.435\ 935\ 39$$
$$PR = \sqrt{25.435\ 935\ 39}$$
$$= \underline{5.04\ \text{cm}}\ (3\ \text{s.f.})$$

Finally, you can use $\frac{1}{2} ab \sin C$ to help find the area of a segment of a circle. The next example illustrates the method.

EXAMPLE

Find the area of the segment shaded in the diagram below.
Give your answer to 3 significant figures.

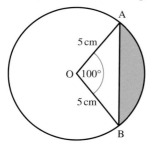

SOLUTION

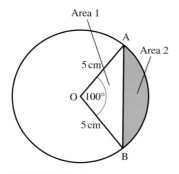

Area 1 + Area 2 form a sector of angle 100°.
The area of the sector is:

$$\frac{100}{360} \times \pi \times 5^2 = 21.816\ 6\ \text{cm}^2$$

Area 1 on its own forms a triangle.
Its area is:

$$\tfrac{1}{2} \times 5 \times 5 \times \sin 100° = 12.310\ 1\ \text{cm}^2$$

Thus the area of the segment, Area 2, is:

$$21.816\ 6 - 12.310\ 1 = 9.506\ 5\ \text{cm}^2$$
$$= \underline{9.51\ \text{cm}^2}\ (3\ \text{s.f.})$$

EXERCISE 25.3

1 In triangle ABC, AB = 15 cm and BC = 18 cm.
Angle ABC = 40°.

Calculate the area of triangle ABC.
Give your answer correct to 3 significant figures.

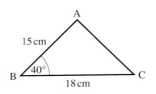

2 In triangle PQR, PQ = 9 cm and QR = 11.4 cm.
Angle PQR = 142°.

Calculate the area of triangle PQR.
Give your answer correct to 3 significant figures.

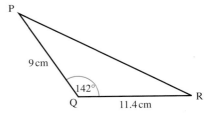

3 In triangle ABC, AB = 5 cm and BC = 8 cm.
The area of triangle ABC =15 m².
Angle ABC is acute.
  a) Calculate the size of angle ABC.
    Give your answer correct to the nearest 0.1°
  b) Calculate the perimeter of triangle ABC.
    Give your answer correct to 3 significant figures.

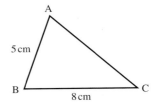

4 In triangle LMN, LM = 6 cm, MN = 8 cm and LN = 12 cm.
  a) Calculate the size of angle LMN.
    Give your answer correct to the nearest 0.1°.
  b) Calculate the area of triangle LMN.
    Give your answer correct to 3 significant figures.

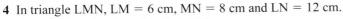

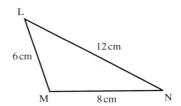

5 A chord AB is drawn across a circle of radius 10 cm.
The chord AB is of length 15 cm.
  a) Use the cosine rule to find angle AOB.
    Give your answer to the nearest 0.1°.
  b) Hence find the area of the segment shaded in the diagram.

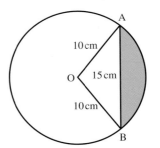

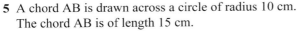

6 In triangle PQR, PQ = 6 cm and PR = 7 cm.
The area of triangle PQR is 11 cm².
  a) Calculate the size of angle QPR.
    Give your answer correct to the nearest 0.1°.
  b) Calculate the perimeter of triangle QPR.
    Give your answer correct to 3 significant figures.

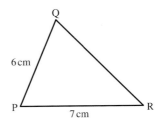

7 The diagram shows a regular hexagon inscribed inside a circle of radius 12 cm.

   a) Work out the area of the circle.

   b) Work out the area of the segment shaded on the diagram.

   c) Hence work out the area of the hexagon.

   Give all your answers correct to 3 significant figures.

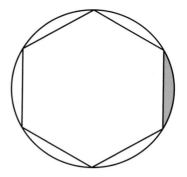

8 An isosceles triangle has sides of length 12 cm, 12 cm and 8 cm. Work out the area of this triangle.

25.4 Trigonometry in 3-D

You have already met the idea of using Pythagoras' theorem in 3-D. The approach is to break the problem down into two or more 2-D triangles, and solve them using the regular Pythagorean methods.

The same principle can be used for sines and cosines, to find unknown angles and lengths.

EXAMPLE

The diagram shows a wedge.
The base of the wedge is a horizontal rectangle measuring 60 cm by 80 cm.
The sloping face ABPQ is inclined at 25° to the horizontal.

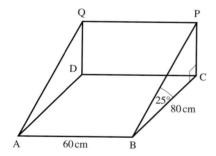

a) Calculate the lengths AC and PC.

b) Calculate the length AP.

c) Calculate the angle that AP makes with the horizontal plane ABCD.

SOLUTION

a) By Pythagoras' theorem:
$$x^2 = 60^2 + 80^2$$
$$= 3600 + 6400$$
$$= 10\ 000$$
$$x = \sqrt{10\ 000}$$
$$= 100 \text{ cm}$$

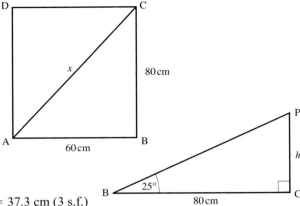

$$h = 80 \times \tan 25°$$
$$= 37.3 \text{ cm (3 s.f.)}$$

Thus AC = 100 cm and PC = 37.3 cm (3 s.f.)

b) From the prism, pick out triangle ACP.

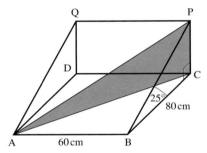

This triangle may be turned flat:

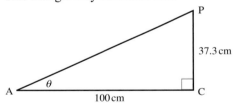

By Pythagoras' theorem:
$$AP^2 = 100^2 + 37.3^2$$
$$= 10\ 000 + 1391.6$$
$$= 113\ 91.6$$
$$AP = \sqrt{113\ 91.6}$$
$$= 106.731\ 5\ldots$$
$$= 106.7 \text{ cm (4 s.f.)}$$

c) Using the same triangle as above:
$$\tan \theta = \frac{37.3}{100}$$
$$= 0.373$$
$$\theta = \tan^{-1}(0.373)$$
$$= 20.5°$$

EXERCISE 25.4

1 The diagram shows a box in the shape of a cuboid ABCDEFGH. AB = 20 cm, BC = 30 cm, AE = 25 cm. A string runs diagonally across the box from C to E.

a) Calculate the length of the string CE.
Give your answer correct to 3 significant figures.
b) Work out the angle between the string CE and the horizontal plane ABCD.
Give your answer correct to the nearest 0.1°.

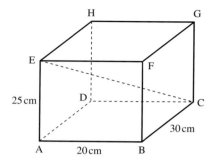

2 The diagram shows a square-based pyramid ABCDX. AB = BC = 20 cm. The point M is the centre of the square base ABCD. XM = 25 cm.

a) Calculate the length of AC. Give your answer correct to 3 significant figures.
b) Work out the length of the slanting edge AX.
Give your answer correct to 3 significant figures.
c) Work out the angle between the edge AX and the horizontal plane ABCD.
Give your answer correct to the nearest 0.1°.

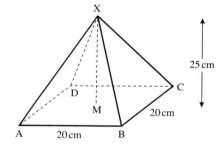

3 The diagram shows a wedge in the shape of a prism PQRSUV. PQ = 40 cm, QR = 90 cm, UR = 10 cm.

a) Calculate the angle UQR.
Give your answer correct to the nearest 0.1°.
b) Calculate the length PU.
Give your answer correct to 3 significant figures.
c) Work out the angle between PU and the horizontal plane PQRS.
Give your answer correct to the nearest 0.1°.

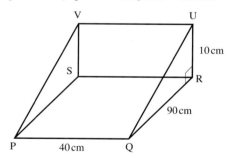

4 The diagram shows a cuboid ABCDEFGH. AB = 16 cm, BC = 18 cm, EC = 34 cm.

a) Calculate the length AE.
b) Work out the angle between CE and the horizontal plane ABCD.
Give your answer correct to the nearest 0.1°.

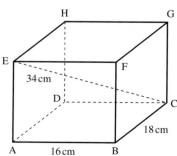

5 The cuboid ABCDEFGH has a square base ABCD.
AB = BC = 25 cm, EA = 8 cm.

  a) Calculate the length BD. Give your answer correct to
  3 significant figures.

  b) Calculate the length BH. Give your answer correct to
  3 significant figures.

  c) Work out the angle between BH and the horizontal plane EFGH.
  Give your answer correct to the nearest 0.1°.

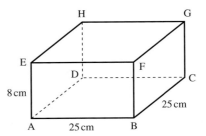

REVIEW EXERCISE 25

1 AB = 11.7 cm, BC = 28.3 cm, angle ABC = 67°.

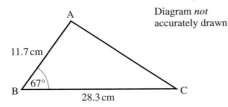

Diagram *not*
accurately drawn

Calculate the area of triangle ABC.

2 The diagram shows triangle ABC.
AC = 7.2 cm. BC = 8.35 cm.
Angle ACB = 74°.

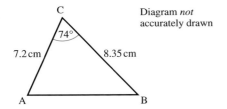

Diagram *not*
accurately drawn

  a) Calculate the area of triangle ABC.
  Give your answer correct to 3 significant figures. Give the units with your answer.

  b) Calculate the length of AB. Give your answer correct to 3 significant figures.   [Edexcel]

3

C
80°
8 cm   10 cm
A   B

Diagram *not*
accurately drawn

  a) Calculate the length of AB.
  Give your answer, in centimetres, correct to 3 significant figures.

  b) Calculate the size of angle ABC.
  Give your answer correct to 3 significant figures.   [Edexcel]

4 AB = 3.2 cm. BC = 8.4 cm.
The area of triangle ABC is 10 cm².

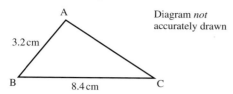

Diagram *not*
accurately drawn

Calculate the perimeter of triangle ABC.
Give your answer correct to 3 significant figures.

[Edexcel]

5 Angle ACB = 150°. BC = 60 m.
The area of triangle ABC is 450 m².

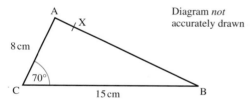

Diagram *not*
accurately drawn

Calculate the perimeter of triangle ABC.
Give your answer correct to 3 significant figures.

[Edexcel]

6 In triangle ABC, AC = 8 cm, CB = 15 cm, angle ACB = 70°.

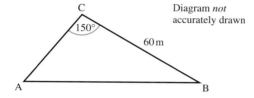

Diagram *not*
accurately drawn

a) Calculate the area of triangle ABC.
Give your answer correct to 3 significant figures.

X is the point on AB such that angle CXB = 90°.
b) Calculate the length of CX.
Give your answer correct to 3 significant figures.

[Edexcel]

7 This question relates to the Starter exercise.
The diagram shows a triangle ABX. AB = 25 m.
Angle XAB = 32°. Angle XBY = 65°.
The aim is to find the height of the church steeple, XY.
Follow these steps.

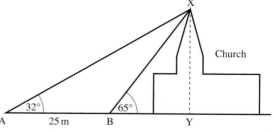

a) Make a copy of triangle ABX. Mark the
values of angles ABX and AXB.
b) Use the sine rule to work out the length AX.
c) Now look at triangle AXY. Use trigonometry
to calculate the height XY.
How closely does your answer agree with your scale drawing from the Starter?

8 The diagram represents a cuboid ABCDEFGH.
AB = 5 cm. BC = 7 cm. AE = 3 cm.

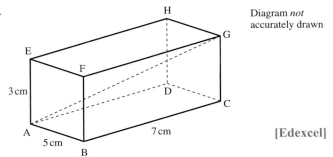

Diagram *not* accurately drawn

  a) Calculate the length of AG.
Give your answer correct to
3 significant figures.

  b) Calculate the size of the angle
between AG and the face ABCD.
Give your answer correct to 1 decimal
place.

[Edexcel]

9 The diagram represents a prism. AEFD is a
rectangle. ABCD is a square.
EB and FC are perpendicular to plane ABCD.
AB = 60 cm. AD = 60 cm. Angle ABE = 90°.
Angle BAE = 30°.
Calculate the size of the angle that the line DE
makes with the plane ABCD.
Give your answer correct to 1 decimal place.

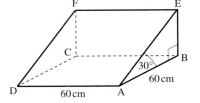

Diagram *not* accurately drawn

[Edexcel]

Key points

1 Unknown sides and angles in triangles may be found by using the sine rule or the cosine rule.

2 The sine rule is generally used to relate two sides and two angles.
The examination formula book gives you the sine rule in this form:

$$\frac{a}{\sin A} = \frac{b}{\sin B} = \frac{c}{\sin C}$$

3 The cosine rule is generally used to relate three sides and one angle.
The examination formula book gives you the cosine rule in this form:

$$a^2 = b^2 + c^2 - 2bc \cos A$$

4 If you know two sides of a triangle, and the included angle (i.e. the angle between the two given sides) then the area can be computed using this result, which is also in the examination formula book:

$$\text{Area of a triangle} = \tfrac{1}{2} ab \sin C$$

This is also a helpful formula for problems about finding the area of a segment of a circle.

5 For the IGCSE examination you will be expected to know how to solve trigonometry problems in three dimensions. The approach is to form right-angled triangles inside the given 3-D object. Such triangles should be drawn out in 2-D to help you see what calculations to do.

Internet Challenge 25

Heron's formula

Look at this triangle. How might you work out its area?

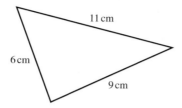

A rather tedious method is to use the cosine rule to find one of the angles. Then use $\frac{1}{2}ab \sin C$ to compute the area.

A much quicker way is to use Heron's formula. You start by working out the **semiperimeter** s of the triangle. For this example:

$$s = \tfrac{1}{2}(6 + 9 + 11) = 13$$

1   Use the internet to find Heron's formula.

2   Use Heron's formula to find the area of the triangle above.

Here are three more triangles. Their areas are almost the same.

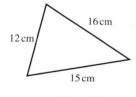

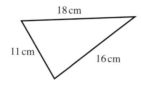

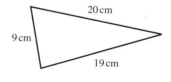

Triangle A   Triangle B   Triangle C

3   Use Heron's formula to work out the area of each triangle. Which of A, B or C has the largest area?

4   Use the internet to find out when Heron's formula was first used.

5   Try to find out how to prove Heron's formula.

CHAPTER 26

Graphs and transformations

In this chapter you will learn how to:

- recognise plot and draw graphs of curves
- draw graphs of $y = \cos$, $y = \sin x$ and $y = \tan x$
- apply and interpret transformations of functions.

You will also be challenged to:

- investigate famous curves.

Starter: **Making waves**

Make sure your calculator is in DEG (degree) mode. Then copy and complete this table of values for sin x.

x	0	30	45	60	90	120	135	150	180
$\sin x$	0	0.5			1			0.5	

x	210	225	240	270	300	315	330	360	390
$\sin x$								0	

Now plot your results on graph paper, or a copy of the grid below.

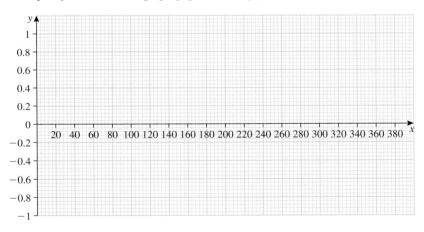

What do you notice?

Do you get a similar result if you use cosine instead of sine?

26.1 Plotting and using graphs of curves

In this chapter you will be working with functions like these:

$$y = x^2 - 3 \qquad \text{a } \textbf{quadratic} \text{ function}$$
$$y = x^3 + x \qquad \text{a } \textbf{cubic} \text{ function}$$
$$y = \frac{1}{x} \text{ and } y = \frac{1}{x^2} \quad \textbf{reciprocal} \text{ functions}$$

You will need to be able to draw up tables of values for such functions. You will be given the x values and be asked to work out corresponding y values.

A reciprocal function like $y = \frac{1}{x}$ can be tabulated in a similar way to the quadratic equations you met in Chapter 22.

However, there cannot be an entry for $x = 0$ because you cannot divide by 0. We say this function is **not defined** when $x = 0$.

EXAMPLE

Complete the table of values for the function $y = \frac{1}{x}$.

x	-4	-2	-1	-0.5	0	0.5	1	2	4
y					not defined				

SOLUTION

As before, begin with $\frac{1}{-4} = -0.25$, then $\frac{1}{-2} = -0.5$, and so on.

Here is the final table:

x	-4	-2	-1	-0.5	0	0.5	1	2	4
y	-0.25	-0.5	-1	-2	not defined	2	1	0.5	0.25

Once you have made a table of values, the points can be plotted on graph paper to make a graph of the function. You should join the points using a *smooth* curve. Use a pencil, so that you can easily change your graph if it is not quite right at the first attempt.

EXAMPLE

The table shows some values for the function $y = x^3 - 3x$.

x	-2	-1	0	1	2	3
y	-2					18

a) Complete the missing values of y in the table.
b) Plot the points on a graph and join them with a smooth curve.
c) Use your graph to find all the solutions to the equation $x^3 - 3x = 0$.
d) By drawing a suitable a horizontal line on the grid, use your graph to solve the equation $x^3 - 3x = 8$.

SOLUTION

a)

x	-2	-1	0	1	2	3
y	-2	2	0	-2	2	18

b)

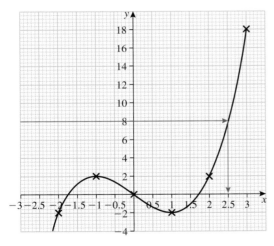

c) The solutions to the equation $x^3 - 3x = 0$ occur when $y = 0$, marked with arrows on the diagram above. Thus $\underline{x = -1.7}$, $\underline{x = 0}$ or $\underline{x = 1.7}$ (1 d.p.)

d) Rearrange $x^2 - 3x - 8 = 0$ to give $x^3 - 3x = 8$.

To solve the equation $x^3 - 3x = 8$, it is necessary to draw a horizontal line representing $y = 8$ on the graph, and see where this crosses $y = x^3 - 3x$:

So, the solution is $\underline{x = 2.5}$ (1 d.p.)

Note that whenever you plot the graph of a quadratic function, the curve will always have a distinctive bowl shape: it is a **parabola**.

The graph of a cubic function may be plotted in a similar way. The shape of a cubic function is not a parabola. The graph is now S-shaped, like these examples:

Graph of $y = x^3$

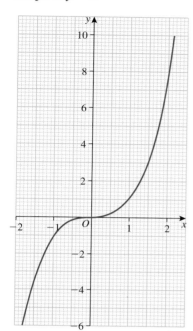

Graph of $y = x^3 - 3x$

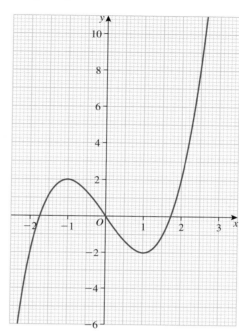

You can use graphs to find approximate solutions to equations. For example, the graph of $y = x^3 - 3x$ can be used to solve the equations $x^3 - 3x = 0$, $x^3 - 3x = 5$, and so on.

EXERCISE 26.1

1 Complete the table of values for the function $y = x^3 + x$.
Two values have been filled in for you.

x	-2	-1	0	1	2	3
y		-2			10	

2 Complete the table of values for the function $y = x + \dfrac{1}{x}$.

x	-2	-1	-0.5	0	0.5	1	2
y				not defined			

3 The table shows some values for the function $y = x^3 - 4x$.

x	-3	-2	-1	0	1	2	3
y	-15				-3		

    a) Copy and complete the table.

    b) Draw a set of coordinate axes on squared paper, so that x can run from -3 to 3 and y from -15 to 15. Plot these points on your graph, and join them with a smooth curve.

    c) Use your graph to solve the equation $x^3 - 4x = 0$.

    d) Use your graph to solve the equation $x^3 - 4x = 5$.

4 The table shows some values of the function $y = \dfrac{12}{x}$. The function is not defined when $x = 0$.

x	-3	-2	-1	0	1	2	3	4
y	4			not defined		6		

    a) Copy and complete the table.

    b) Draw a set of coordinate axes on squared paper, so that x can run from -3 to 4 and y from -15 to 15. Plot these points on your graph.

    c) Join the first three points with a smooth curve.

    d) Join the last four points with another smooth curve.

    e) Use your graph to solve the equation $\dfrac{12}{x} = 9$.

5   **a)** Complete the table of values for $y = \dfrac{4}{x^2}$.

x	-4	-2	-1	-0.5	0.5	1	2	4
y								

    b) Draw the graph of $y = \dfrac{4}{x^2}$.

    c) Use your graph to solve $\dfrac{4}{x^2} = 5$.

6   **a)** Complete the table of values for $y = 2 + \dfrac{1}{x^2}$.

x	-4	-2	-1	-0.5	-0.25	0.25	0.5	1	2	4
y										

    b) Draw the graph of $y = 2 + \dfrac{1}{x^2}$.

    c) Use your graph to solve $2 + \dfrac{1}{x^2} = x$.

7 The diagram shows part of the graph of the function
$y = x^2 - 7x + 9$.

 a) Use the graph to find the two solutions to the equation
 $x^2 - 7x + 9 = 0$.

 b) Use the graph to find the two solutions to the equation
 $x^2 - 7x + 9 = 5$.

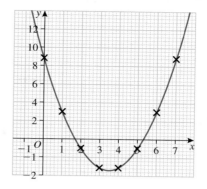

8 The diagram opposite shows part of the graph of $y = \dfrac{1}{x}$.

Use the graph to find a solution to the equation $\dfrac{1}{x} = 0.8$.

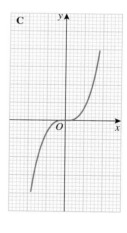

9 Here are six graphs.

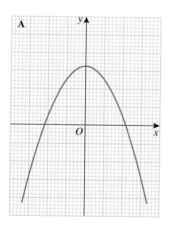

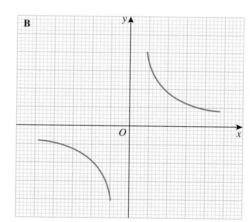

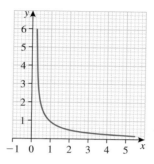

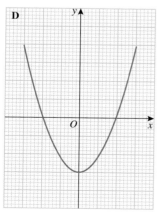

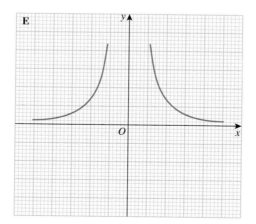

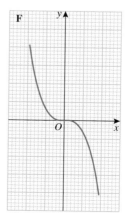

Copy and complete the table below with the letter of the graph that could represent each given equation.

Equation	Graph
$y = 3 - x^2$	
$y = x^3$	
$y = \dfrac{3}{x^2}$	
$y = \dfrac{3}{x}$	
$y = x^2 - 3$	
$y = -x^3$	

10 Ali draws the graph of $y = x^2 + 6x + 4$ on a grid.

He draws a straight line on his grid and uses his graph to solve $x^2 + 6x + 1 = 0$.

a) Find the equation of the straight line which Ali should use.

Find the equation of the straight line Ali should draw on his grid to solve:

b) $x^2 + 4x + 5 = 0$

c) $x^2 + 8x + 2 = 0.$

26.2 Graphs of sine, cosine and tangent functions

In this section you will learn about the behaviour of the graphs of the sine, cosine and tangent functions. The IGCSE examination does not explicitly require a detailed understanding of these, but you may find some of the exam questions on functions (Chapter 24) to be more accessible once you have completed this section.

The starter for this chapter, Making waves, shows you that you can work out the value of sin x for *any* angle, not just those between 0 and 90°. A graph helps you visualise what is happening.

Here is the graph of $y = \sin x$ for values of x between 0 and 90°.

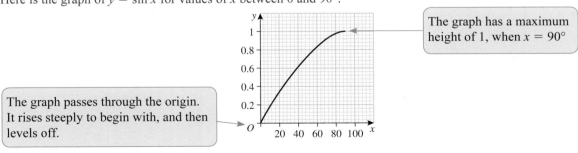

The graph has a maximum height of 1, when $x = 90°$

The graph passes through the origin. It rises steeply to begin with, and then levels off.

By extending the x axis up to 360° we obtain a **sine wave**:

Graph of $y = \sin x$ from 0° to 360°

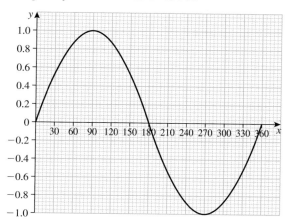

The wave repeats regularly every 360°; we say $\sin x$ is a **periodic** function.

Graph of $y = \sin x$ from −180° to 720°

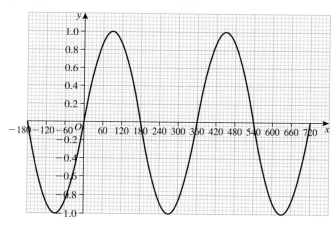

The graph of $\cos x$ behaves in a very similar way. The graph of $y = \cos x$ does not pass through the origin, however. Instead, it has a **maximum value** (of 1) when $x = 0°$:

Graph of $y = \cos x$ from −180° to 720°

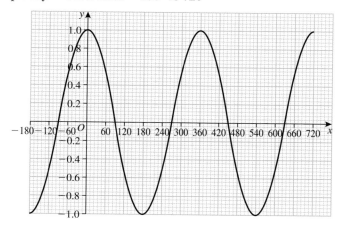

This table summarises some properties of the sine and cosine functions. Study it carefully – there are some obvious similarities, but also subtle differences between the two.

	$y = \sin x$	$y = \cos x$
Behaviour at $x = 0$	Passes through the origin	Has a maximum at $(0, 1)$
Type of symmetry	Rotation symmetry of order 2 about the origin	Reflection symmetry in the y axis
Range of values	-1 to $+1$	-1 to $+1$

The function $y = \tan x$ behaves in a completely different way from $y = \sin x$ and $y = \cos x$. Unlike sine and cosine, the tangent function does not have a maximum value of 1. You can use your calculator to verify that $\tan 45° = 1$, $\tan 50° = 1.1918$, $\tan 60° = 1.7321$, and so on. By the time you get to 89°, you will find that $\tan 89° = 57.2890$. As the value of x gets closer and closer to 90°, the value of $\tan x$ increases without limit!

Here is the graph of $y = \tan x$ for values of x between 0 and 90°.

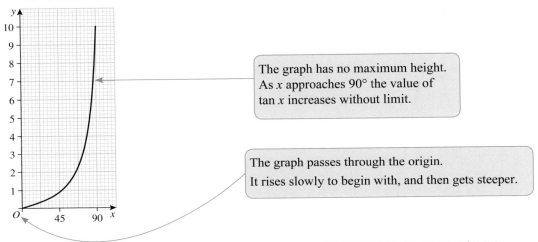

The graph has no maximum height. As x approaches 90° the value of $\tan x$ increases without limit.

The graph passes through the origin. It rises slowly to begin with, and then gets steeper.

Here is the graph of $y = \tan x$ over a wider range of x values:

Like the sine function, the graph of $y = \tan x$ passes through the origin, and has rotational symmetry of order 2 about the origin. The curve is disconnected at 90°, 270° and so on.

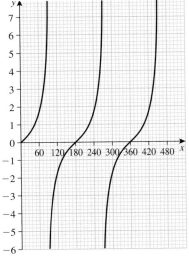

EXERCISE 26.2

1 Draw three sketch graphs to show:
   a) $y = \sin x$
   b) $y = \cos x$,
   c) $y = \tan x$,
   as x ranges from 0 to 450°.

2 Here are some statements about trigonometric functions. Decide whether each one is true or false. Try to deduce the answer from graphical considerations rather than using a calculator.
   a) $\sin 30°$ and $\sin 150°$ have the same value.
   b) $\cos 30°$ and $\cos 150°$ have the same value.
   c) $\tan x$ always lies between 0 and 1.
   d) $\cos x$ cannot be negative provided x lies between 0° and 180°.
   e) $\sin 10°$ and $\sin 370°$ have the same value.

3 Here are some clues about trigonometric functions. For each one, decide whether it is referring to $y = \sin x$, $y = \cos x$ or $y = \tan x$.
   a) This function has a maximum when $x = 0$.
   b) This function is not defined when $x = 90°$.
   c) The graph of this function passes through the point $(90°, 1)$.
   d) The graph of this function has reflection symmetry in the y axis.
   e) This function has a minimum when $x = 270°$.

26.3 Transformations of graphs – translations

You can transform a graph by translating (moving) it vertically.
Look at these curves. The red curve and the blue curve can be translated onto each other.

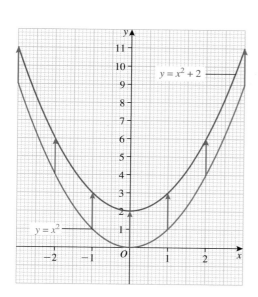

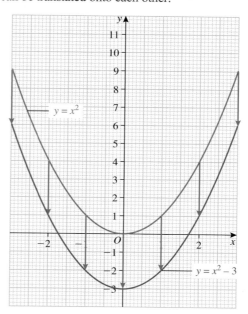

When the graph of $y = x^2$ is:

- translated by the vector $\begin{pmatrix} 0 \\ 2 \end{pmatrix}$ it maps onto the graph of $y = x^2 + 2$

- translated by the vector $\begin{pmatrix} 0 \\ -3 \end{pmatrix}$ it maps onto the graph of $y = x^2 - 3$

Usually function notation is used to describe transformations of curves.
So for the curve $y = f(x)$ you can write:

$f(x) \rightarrow f(x) + 2$ is a translation by the vector $\begin{pmatrix} 0 \\ 2 \end{pmatrix}$ ← 2 units UP

$f(x) \rightarrow f(x) - 3$ is a translation by the vector $\begin{pmatrix} 0 \\ -3 \end{pmatrix}$ ← 3 units DOWN

In general,

$f(x) \rightarrow f(x) + a$ is a translation by the vector $\begin{pmatrix} 0 \\ a \end{pmatrix}$ ← When a is negative, the graph moves DOWN.

You can translate curves horizontally in a similar way.

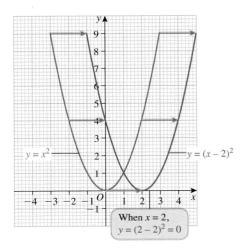

When $x = 2$,
$y = (2 - 2)^2 = 0$

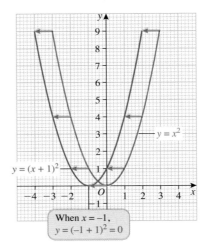

When $x = -1$,
$y = (-1 + 1)^2 = 0$

When the graph of $y = x^2$ is:

- translated by the vector $\begin{pmatrix} 2 \\ 0 \end{pmatrix}$ it maps onto the graph of $y = (x - 2)^2$

- translated by the vector $\begin{pmatrix} -1 \\ 0 \end{pmatrix}$ it maps onto the graph of $y = (x + 1)^2$

So for the curve $y = f(x)$ you can write:

$f(x) \rightarrow f(x - 2)$ is a translation by the vector $\begin{pmatrix} 2 \\ 0 \end{pmatrix}$ ← 2 units RIGHT

$f(x) \rightarrow f(x+1)$ is a translation by the vector $\begin{pmatrix} -1 \\ 0 \end{pmatrix}$ ← 1 unit LEFT

In general:

$f(x) \rightarrow f(x-a)$ is a translation by the vector $\begin{pmatrix} a \\ 0 \end{pmatrix}$ ← *a* units RIGHT

$f(x) \rightarrow f(x+a)$ is a translation by the vector $\begin{pmatrix} -a \\ 0 \end{pmatrix}$ ← *a* units LEFT

EXAMPLE

The diagram shows part of the line with equation $y = f(x)$.

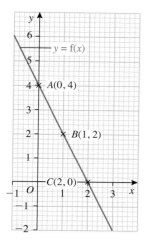

$y = f(x)$ is mapped onto each of the following:

a) $y = f(x) + 4$ **b)** $y = f(x+3)$ **c)** $y = f(x+3) + 4$

For each case, work out the coordinates of the points onto which *A*, *B* and *C* are mapped.

SOLUTION

a) $f(x) \rightarrow f(x) + 4$ is a translation by the vector $\begin{pmatrix} 0 \\ 4 \end{pmatrix}$ ← 4 units UP

$A(0, 4) \rightarrow \underline{A'(0, 8)}$ $B(1, 2) \rightarrow \underline{B'(1, 6)}$ $C(2, 0) \rightarrow \underline{C'(2, 4)}$ ← So all the *y* coordinates increase by 4.

b) $f(x) \rightarrow f(x+3)$ is a translation by the vector $\begin{pmatrix} -3 \\ 0 \end{pmatrix}$ ← 3 units LEFT

$A(0, 4) \rightarrow \underline{A'(-3, 4)}$ $B(1, 2) \rightarrow \underline{B'(-2, 2)}$ $C(2, 0) \rightarrow \underline{C'(-1, 0)}$ ← So all the *x* coordinates decrease by 3.

c) $f(x) \rightarrow f(x+3) + 4$ is a translation by the vector $\begin{pmatrix} -3 \\ 4 \end{pmatrix}$ ←

The order doesn't matter. So you can translate:

$A(0, 4) \rightarrow \underline{A'(-3, 8)}$ $B(1, 2) \rightarrow \underline{B'(-2, 6)}$ $C(2, 0) \rightarrow \underline{C'(-1, 4)}$

4 UP and then 3 LEFT
or 3 LEFT and then 4 UP.

EXAMPLE

The graph of $y = f(x)$ is shown below.

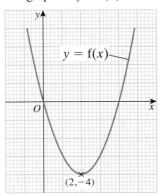

Sketch the curve with equation $y = f(x - 1) - 2$.

SOLUTION

$f(x) \to f(x - 1) - 2$ is a translation by the vector . $\begin{pmatrix} 1 \\ -2 \end{pmatrix}$

The minimum will move from:

$(2, -4) \to (3, -6)$ ◄──── 1 RIGHT and 2 DOWN

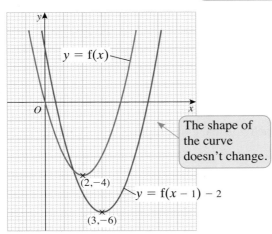

The shape of
the curve
doesn't change.

EXAMPLE

$f(x) = 3x + 2$

Find an equation in terms of x for:

a) $y = f(x) + 3$ **b)** $y = f(x - 2)$

SOLUTION

a) $f(x) = 3x + 2$

$$y = f(x) + 3$$
$$= 3x + 2 + 3 \longleftarrow \boxed{\text{Simplify.}}$$

So $y = 3x + 5$

b) To find $y = f(x - 2)$ substitute $(x - 2)$ for x in $f(x) = 3x + 2$

$$y = f(x - 2)$$
$$= 3(x - 2) + 2 \longleftarrow \boxed{\text{Expand brackets.}}$$
$$= 3x - 6 + 2 \longleftarrow \boxed{\text{Simplify.}}$$

So $y = 3x - 4$

EXERCISE 26.3

1 Describe fully the transformation that maps the graph of $y = f(x)$ onto the graph of:

    a) $y = f(x) + 3$ **b)** $y = f(x - 3)$ **c)** $y = f(x + 3)$

    d) $y = f(x) - 3$ **e)** $y = f(x - 3) - 3$

2 The diagram shows part of the graph of $y = f(x)$.

On a copy of the diagram, draw the graph of:

    a) $y = f(x) + 1$

    b) $y = f(x - 2)$

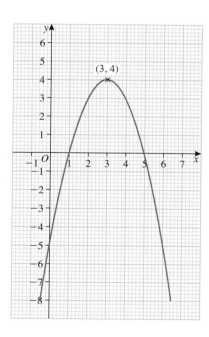

3 The diagram shows part of the graph of $y = f(x)$.

It has one minimum point at $(3, -1)$.

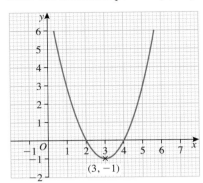

Write down the coordinates of the minimum point of the curve with equation

a) $y = f(x) + 1$

b) $y = f(x - 2)$

c) $y = f(x - 2) + 1$

4 a) The diagram shows part of the graph of $f(x) = a + \sin x°$.
Find the value of a.

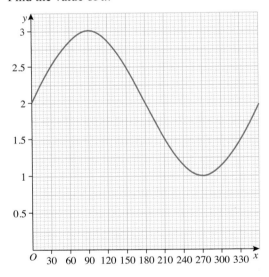

b) Sketch the graph of $y = \sin x° - 1$

5 The diagram shows part of the graph of $y = f(x)$ where $f(x) = \cos(x + 60)°$.

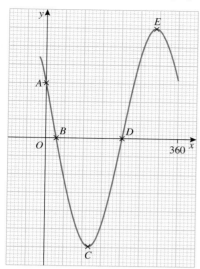

Write down the coordinates of the points A, B, C, D and E.

6 A line has equation $y = f(x)$ where $f(x) = 2x - 1$.
Find an equation in terms of x for the line:

a) $y = f(x) + 3$

b) $y = f(x - 3)$

c) $y = f(x + 2)$

d) $y = f(x + 2) - 1$

7 A curve has equation $y = f(x)$ where $f(x) = x^2$.
Find an equation in terms of x for the curve:

a) $y = f(x) - 1$

b) $y = f(x - 1)$

c) $y = f(x - 1) + 1$

8 a) Write $y = x^2 - 4x + 7$ in the form $y = (x + a)^2 + b$.

b) Describe fully the transformation which maps the graph of $y = x^2$ onto the graph of $y = x^2 - 4x + 7$.

c) Write down the coordinates of the turning point.

d) Hence, or otherwise, explain why the graph of $y = x^2 - 4x + 7$ does not intersect the x-axis.

9 a) Write $y = x^2 + 6x + 4$ in the form $y = (x + a)^2 + b$.

b) Describe fully the transformation which maps the graph of $y = x^2$ onto the graph of $y = x^2 + 6x + 4$.

c) Hence sketch the graph of $y = x^2 + 6x + 4$.

Label the turning point with its coordinates.

26.4 Further transformations

You can transform also transform a graph by reflecting it.

The table shows the coordinates of four points on the curve $y = f(x)$ and of four points on the curve $y = -f(x)$.

$y = f(x)$	$y = -f(x)$
(5, 2)	(5, −2)
(3, 0)	(3, 0)
(0, −2)	(0, 2)
(−4, −3)	(−4, 3)

The y coordinate of each point on $y = f(x)$ is multiplied by −1.

The diagram shows the graph of $y = f(x)$ and the graph of $y = -f(x)$.

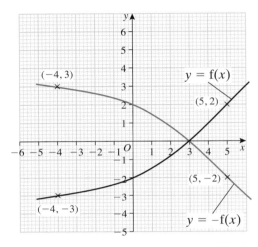

The transformation which maps the graph of $y = f(x)$ onto the graph of $y = - f(x)$ is **a reflection in the x axis**.

The table shows the coordinates of four points on the curve $y = f(x)$ and four points on the curve $y = f(-x)$

$y = f(x)$	$y = f(-x)$
(5, 2)	(−5, 2)
(3, 0)	(−3, 0)
(0, -2)	(0, − 2)
(−4, −3)	(4, −3)

The x coordinate of each point on $y = f(x)$ is multiplied by -1.

The diagram shows the graph of $y = f(x)$ and the graph of $y = f(-x)$.

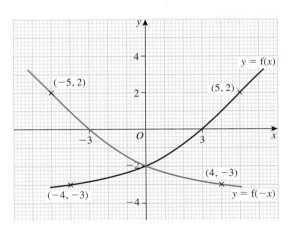

The transformation which maps the graph of $y = f(x)$ onto the graph of $y = f(-x)$ is **a reflection in the y axis**.

EXAMPLE

Here is a sketch of the graph of the curve with equation $y = f(x)$.

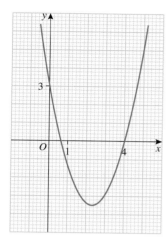

Sketch the graph of **a)** $y = -f(x)$ **b)** $y = f(-x)$

SOLUTION

a) The curve with equation $y = -f(x)$ is a reflection in the x-axis of the curve with equation $y = f(x)$.

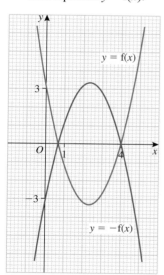

b) The curve with equation $y = f(-x)$ is a reflection in the y axis of the curve with equation $y = f(x)$.

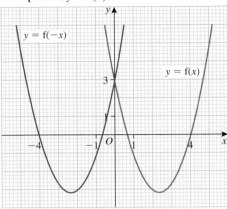

EXAMPLE

$f(x) = 2x + 1$

Find the equation of the graph when the graph of $y = f(x)$ is:

a) reflected in the x axis

b) reflected in the y axis.

SOLUTION

a) When the graph is reflected in the x axis, $f(x) \rightarrow -f(x)$

$$y = -f(x) = -(2x+1)$$
$$\underline{y = -2x - 1}$$

b) When the graph is reflected in the y axis, $f(x) \rightarrow f(-x)$

$$y = f(-x) = 2(-x) + 1$$
$$\underline{y = -2x + 1}$$

The graphs of $y = f(x)$, $y = 2f(x)$ and $y = \frac{1}{2}f(x)$ have been drawn on the grid for the function $f(x) = x(8 - x)$.

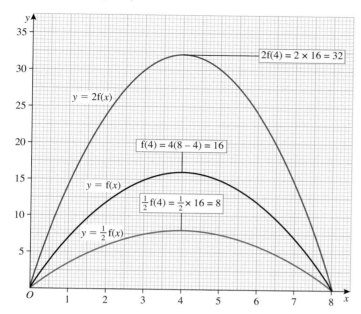

Notice that the shape of the curve has changed – the curves can be 'stretched' vertically onto each other.

- The graph of $y = 2f(x)$ is obtained from the graph of $y = f(x)$ by multiplying the y coordinate of each point by 2

- The graph of $y = \frac{1}{2}f(x)$ is obtained from the graph of $y = f(x)$ by multiplying the y coordinate of each point by ½

- In general, the graph of $y = af(x)$ is obtained from the graph of $y = f(x)$ by multiplying the x coordinate of each point by a.

The graphs of $y = f(x)$, $y = f(2x)$ and $y = f\left(\frac{1}{2}x\right)$ have been drawn on the grid for the function $f(x) = x(8 - x)$.

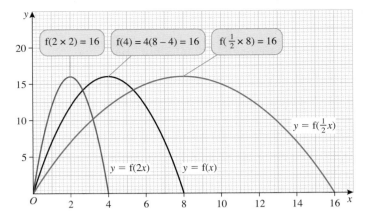

Notice that the shape of the curve has changed – the curves can be 'stretched' horizontally onto each other.

- The graph of $y = f(2x)$ is obtained from the graph of $y = f(x)$ by multiplying the x coordinate of each point by ½

- The graph of $y = f\left(\frac{1}{2}x\right)$ is obtained from the graph of $y = f(x)$ by multiplying the x coordinate of each point by 2

- In general, the graph of $y = f(ax)$ is obtained from the graph of $y = f(x)$ by multiplying the x coordinate of each point by $\frac{1}{a}$

EXAMPLE

Here is the graph of the curve $y = f(x)$

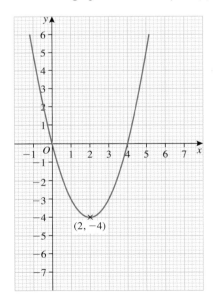

Sketch the curve with equation

a) $y = \frac{1}{2}f(x)$

b) $y = f(2x)$

c) $y = \frac{1}{2}f(2x)$

SOLUTION

a) $f(x) \rightarrow \frac{1}{2}f(x)$

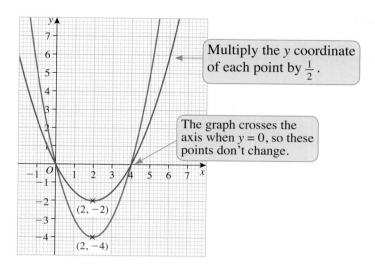

Multiply the y coordinate of each point by $\frac{1}{2}$.

The graph crosses the axis when $y = 0$, so these points don't change.

b) $f(x) \rightarrow f(2x)$

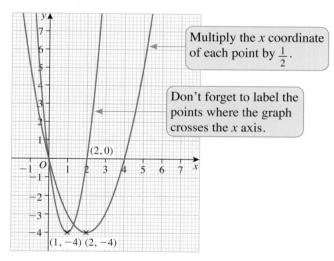

Multiply the x coordinate of each point by $\frac{1}{2}$.

Don't forget to label the points where the graph crosses the x axis.

c) $f(x) \rightarrow \frac{1}{2}f(2x)$

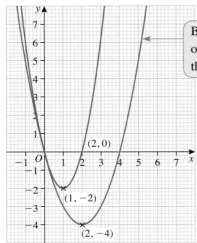

Because both the x and the y coordinate of each point have been multiplied by $\frac{1}{2}$, the graph is the same shape – only smaller!

In fact, it is an enlargement, centre the origin, scale factor $\frac{1}{2}$.

(2, 0)

(1, −2)

(2, −4)

EXERCISE 26.4

1 The diagram shows part of the graph of $y = f(x)$.

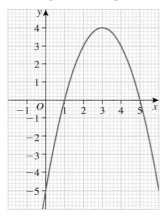

On a copy of the diagram, draw the graphs of:

a) $y = f(-x)$ **b)** $y = -f(x)$

2 $y = f(x)$ has a turning point at $(-2, 7)$.
Write down the coordinates of the turning points of the curve with equation:

a) $y = f\left(\frac{1}{2}x\right)$ **b)** $y = 3f(x)$

c) $y = f(-x)$ **d)** $y = -f(x)$

3 $f(x) = 3x - 2$.
Find in terms of x:

a) $-f(x)$ **b)** $f(-x)$

c) $3f(x)$ **d)** $f(2x)$

4 $f(x) = x^2 + x$.

Find in terms of x:

a) $f(-x)$ **b)** $-f(x)$

c) $2f(x)$ **d)** $f(2x)$

5 a) The graph of $y = 4x - 1$ is reflected in the x axis.
Find the equation of the new graph.

b) The graph of $y = 4x - 1$ is reflected in the y-axis.
Find the equation of the new graph.

6 The diagram shows part of the curve $y = a\cos bx°$.

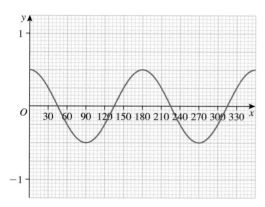

Use the graph to find the value of:

a) a **b)** b.

REVIEW EXERCISE 26

1 Here are six graphs.

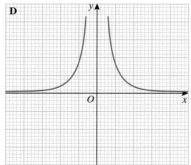
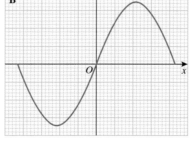
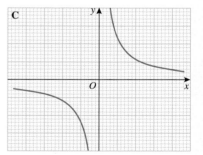
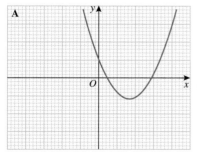
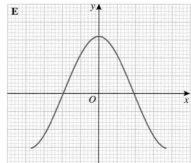
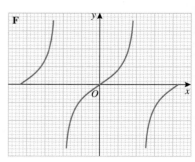

Complete the table here with the letter of the graph that could represent each given equation.

Equation	Graph
$y = \sin x$	
$y = \cos x$	
$y = \dfrac{1}{x^2}$	
$y = \dfrac{2}{x}$	
$y = \tan x$	
$y = x^2 - 3x + 1$	

2 a) Copy and complete the table of values for $y = 2x^2$.

x	-3	-2	-1	0	1	2	3
y	18				2	8	

b) On a copy of the grid below, draw the graph of $y = 2x^2$.

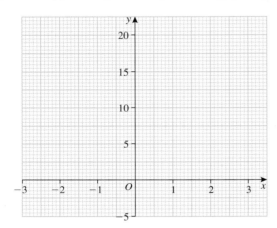

c) Use your graph to find:
  (i) the value of y when $x = 2.5$
  (ii) the values of x when $y = 12$.   [Edexcel]

3 a) Copy and complete the table of values for the graph $y = x^3 + 2$.

x	-3	-2	-1	0	1	2
$y = x^3 + 2$	-25					10

b) On a copy of the grid, draw the graph of $y = x^3 + 2$.

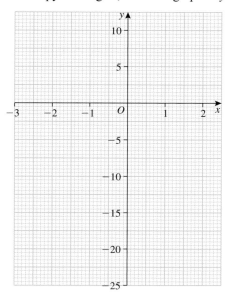

c) Use your graph to find:
   (i) an estimate of the solution of the equation $x^3 + 2 = 0$
   (ii) an estimate of the solution of the equation $x^3 + 2 = 8$.

[Edexcel]

4 The diagram shows part of the graph of $y = x^3 - 5x + 1$.

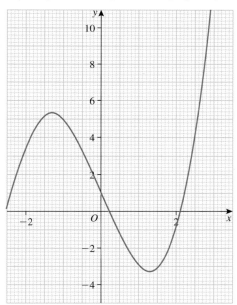

a) Copy the graph and use it to estimate the values of the two positive solutions to the equation $x^3 - 5x + 1 = 0$. Give your answers to 1 decimal place.

b) By drawing a suitable horizontal line, estimate the value of the solution to the equation $x^3 - 5x - 1 = 0$.

5 The table shows some values of the function $y = x^3 - 16x + 16$.

x	-4	-3	-2	-1	0	1	2	3	4
y	16		40		16		-8		

a) Copy and complete the table.

b) On a copy of the grid below, draw the graph of $y = x^3 - 16x + 16$.

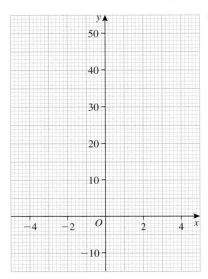

c) By drawing the graph of $y = 5x$ on your graph, find a value of x for which $x^3 - 16x + 16 = 5x$.

d) Rewrite the equation $x^3 - 16x + 16 = 5x$ so it is in the form $x^3 + ax + b = 0$.

6 The diagram below shows part of the graph of $y = x + \dfrac{1}{x}$.

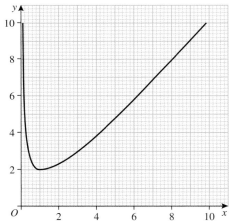

a) Copy the graph and use it to estimate the two values of x for which $x + \dfrac{1}{x} = 3$.

b) Copy the graph and use it to estimate the two values of x for which $x + \dfrac{1}{x} = 10 - x$.

7 The red curve is the graph of $y = \sin x°$.
   Find the equation of the **a)** blue curve   **b)** green curve.

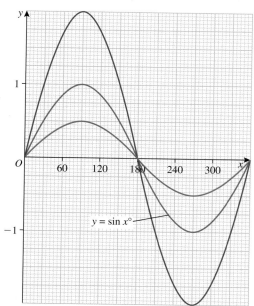

8 The red curve is the graph of $y = \cos x°$.
   Find the equation of the **a)** blue curve   **b)** green curve.

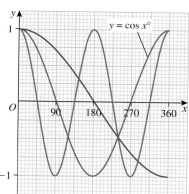

9 Describe fully the transformation that maps the graph of $y = f(x)$ onto the graph of:

a) $y = f(x) + 1$

b) $y = f(x - 1)$

c) $y = f(x + 1)$

d) $y = f(x - 1) + 1$

e) $y = f(-x)$

f) $y = -f(x)$

10 The diagram shows part of the graph of $y = f(x)$.
It has one turning point, which is at $(2, 4)$.

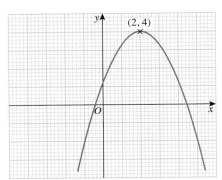

Work out the coordinates of the turning point of the curve with equation:

a) $y = f(x) - 1$

b) $y = 2f(x)$

c) $y = -f(x)$

d) $y = f(x - 2) + 1$

e) $y = f(-x)$

f) $y = 3f(2x)$

11 A line has equation $y = f(x)$ where $f(x) = 2x - 5$.
Find in terms of x:

a) $y = f(x) + 3$

b) $y = f(x - 3)$

c) $y = f(-x)$

d) $y = 3f(x)$

Key points

1 Graphs of quadratics and cubics may be plotted by drawing up a table of values.

2 Reciprocal graphs such as $y = \dfrac{1}{x}$ may be constructed in a similar way, but will have

one point where the curve is not defined (you cannot divide by zero) and this causes the curve to break into two disconnected parts.

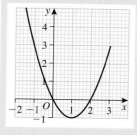

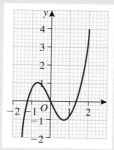

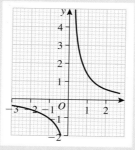

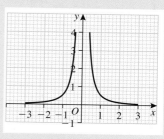

A typical quadratic curve $y = x^2 - 2x$

A typical cubic curve $y = x^3 - 2x$

A typical reciprocal curve $y = \dfrac{1}{x}$

A typical reciprocal curve $y = \dfrac{1}{x^2}$

Make sure you can recognise the graphs of $y = \sin x$, $y = \cos x$ and $y = \tan x$.

3 You can solve equations graphically by finding out where a curve crosses the x axis, or another appropriate straight line.

4 You can apply the following transformations to the graph of $y = f(x)$:

$f(x) \rightarrow f(x) + a$	translation by the vector $\begin{pmatrix} 0 \\ a \end{pmatrix}$	Add a to each y coordinate
$f(x) \rightarrow f(x - a)$	translation by the vector $\begin{pmatrix} a \\ 0 \end{pmatrix}$	Add a to each x coordinate
$f(x) \rightarrow -f(x)$	reflection in the x axis	
$f(x) \rightarrow f(-x)$	reflection in the y axis	
$f(x) \rightarrow af(x)$	vertical stretch	Multiply each y coordinate by a
$f(x) \rightarrow f(ax)$	horizontal stretch	Multiply each x coordinate by $\dfrac{1}{a}$

Internet Challenge 26

Famous curves

Here are nine famous curves in mathematics.

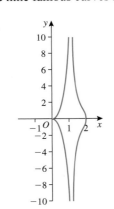
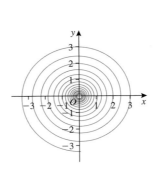
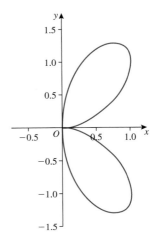

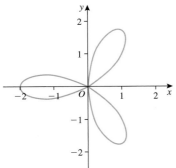
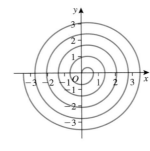
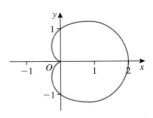

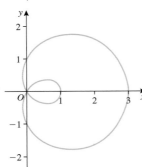
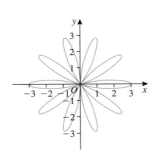
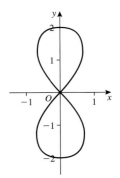

In dictionary order, these curves are:

Archimedean spiral	cardioid	conchoid	double folium	equiangular spiral
lemniscate	limacon of Pascal	rose curve	trifolium	

Use the internet to help you match the names to the right curves. Try to find one interesting fact about each curve.

CHAPTER 27

Vectors

In this chapter you will learn how to:

- write vectors as column vectors, and find their magnitudes
- add and subtract vectors, and multiply a vector by a number
- use vectors to prove geometric theorems.

You will also be challenged to:

- investigate Queens on a chessboard.

Starter: **Knight's tours**

When a knight moves on a chessboard, it can move two squares in a straight line and one square at right angles, like this:

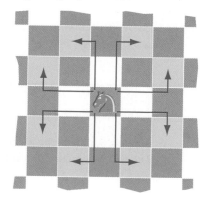

A knight can also move one square in a straight line and two squares sideways.

See if you can work out how to move a knight around a chessboard so that it visits all 64 squares. The first three moves have been done to start you off.

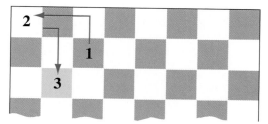

If possible, try to find a route so that the 64th square is a knight's hop away from the first square; this will close the tour so that the knight can get back to its starting position.

27.1 Introducing vectors

A **vector** is a quantity that has a magnitude (length) and a direction. Vectors are often described using **column vector** notation such as $\begin{pmatrix} 3 \\ 5 \end{pmatrix}$. You have already met column vector notation in Chapter 14. Here are some diagrams to remind you how the notation works:

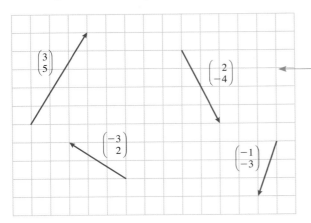

The column vector $\begin{pmatrix} x \\ y \end{pmatrix}$ means that the vector can be drawn by going x units to the *right* and y units *up*.
(*Negative* values indicate *left* or *down*.)

The magnitude, or modulus, of a vector is simply its length, regardless of direction. In simple cases the magnitude of a vector may be seen by inspection, but often Pythagoras' theorem is needed.

EXAMPLE

a) Illustrate the vector $\mathbf{a} = \begin{pmatrix} 6 \\ 8 \end{pmatrix}$ on a square grid.

b) Work out the magnitude of the vector $\mathbf{a}$.

The vector $\mathbf{a} = \begin{pmatrix} 6 \\ 8 \end{pmatrix}$ means 6 units to the *right* (in the x direction) and 8 units *up* (in the y direction).

SOLUTION

a)

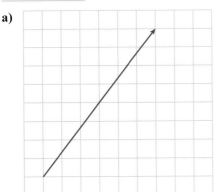

b) Magnitude of $\mathbf{a}$ is $\sqrt{6^2 + 8^2}$
$= \sqrt{36 + 64}$
$= \sqrt{100}$
$= 10$

The magnitude of $\mathbf{a}$ is its length, so we can use Pythagoras' theorem to find it.

EXERCISE 27.1

1 The diagram below shows some vectors drawn on a grid of unit squares.
Write down column vectors to describe each one.

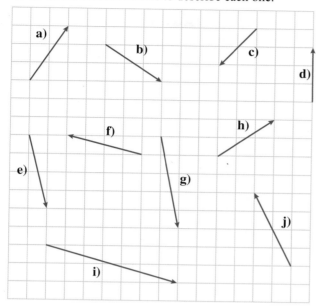

2 Draw a sketch of each of these vectors on squared paper.

a) $\begin{pmatrix} 7 \\ 1 \end{pmatrix}$ b) $\begin{pmatrix} -5 \\ 5 \end{pmatrix}$ c) $\begin{pmatrix} 4 \\ -6 \end{pmatrix}$ d) $\begin{pmatrix} -3 \\ -2 \end{pmatrix}$

3 Calculate the magnitude of each of the following vectors:

a) $\begin{pmatrix} 4 \\ 3 \end{pmatrix}$ b) $\begin{pmatrix} 5 \\ 12 \end{pmatrix}$ c) $\begin{pmatrix} 9 \\ 12 \end{pmatrix}$ d) $\begin{pmatrix} 24 \\ 7 \end{pmatrix}$ e) $\begin{pmatrix} -4 \\ 7.5 \end{pmatrix}$

27.2 Adding and subtracting vectors

You can add two vectors using simple arithmetic. For example:

$$\begin{pmatrix} 7 \\ 1 \end{pmatrix} + \begin{pmatrix} 5 \\ 3 \end{pmatrix} = \begin{pmatrix} 7 + 5 \\ 1 + 3 \end{pmatrix} = \begin{pmatrix} 12 \\ 4 \end{pmatrix}$$

Subtraction is done in a similar way. For example:

$$\begin{pmatrix} 8 \\ 4 \end{pmatrix} - \begin{pmatrix} 2 \\ -3 \end{pmatrix} = \begin{pmatrix} 8 - 2 \\ 4 - -3 \end{pmatrix} = \begin{pmatrix} 6 \\ 7 \end{pmatrix}$$

Geometrically, addition corresponds to placing the two vectors head to tail like this:

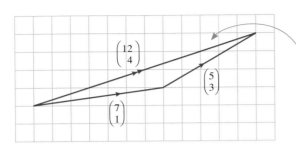

The sum of the two vectors is called the **resultant**.

Vectors are often named using letter **a**, **b**, **c**, etc. The letters are usually <u>underlined</u> if written by hand, but they are in **bold type** in examination papers and textbooks.

EXAMPLE

The vectors **a**, **b** and **c** are given by $\mathbf{a} = \begin{pmatrix} 6 \\ 1 \end{pmatrix}$, $\mathbf{b} = \begin{pmatrix} 2 \\ -3 \end{pmatrix}$ and $\mathbf{c} = \begin{pmatrix} 4 \\ 2 \end{pmatrix}$

Work out: **a) a + b**   **b) a − c**   **c) a − b + c**

SOLUTION

a) $\mathbf{a} + \mathbf{b} = \begin{pmatrix} 6 \\ 1 \end{pmatrix} + \begin{pmatrix} 2 \\ -3 \end{pmatrix} = \begin{pmatrix} 6+2 \\ 1+-3 \end{pmatrix} = \begin{pmatrix} 8 \\ -2 \end{pmatrix}$

b) $\mathbf{a} - \mathbf{c} = \begin{pmatrix} 6 \\ 1 \end{pmatrix} - \begin{pmatrix} 4 \\ 2 \end{pmatrix} = \begin{pmatrix} 6-4 \\ 1-2 \end{pmatrix} = \begin{pmatrix} 2 \\ -1 \end{pmatrix}$

c) $\mathbf{a} - \mathbf{b} + \mathbf{c} = \begin{pmatrix} 6 \\ 1 \end{pmatrix} - \begin{pmatrix} 2 \\ -3 \end{pmatrix} + \begin{pmatrix} 4 \\ 2 \end{pmatrix} = \begin{pmatrix} 6-2+4 \\ 1--3+2 \end{pmatrix} = \begin{pmatrix} 8 \\ 6 \end{pmatrix}$

To find the negative of a vector, just reverse the signs of the numbers.

For example, if $\mathbf{a} = \begin{pmatrix} 5 \\ 2 \end{pmatrix}$ then $-\mathbf{a} = \begin{pmatrix} -5 \\ -2 \end{pmatrix}$

When drawn on a grid, the vector $-\mathbf{a}$ will be *parallel* to the vector **a**, but will point in the *opposite direction*.

EXAMPLE

Given that $\mathbf{p} = \begin{pmatrix} 6 \\ 1 \end{pmatrix}$ and $\mathbf{q} = \begin{pmatrix} 3 \\ -1 \end{pmatrix}$, work out:

a) p + q   **b) p − q**
Illustrate your answers graphically.

SOLUTION

a) $p + q = \begin{pmatrix} 6 \\ 1 \end{pmatrix} + \begin{pmatrix} 3 \\ -1 \end{pmatrix} = \begin{pmatrix} 9 \\ 0 \end{pmatrix}$

b) $p - q = \begin{pmatrix} 6 \\ 1 \end{pmatrix} - \begin{pmatrix} 3 \\ -1 \end{pmatrix} = \begin{pmatrix} 3 \\ 2 \end{pmatrix}$

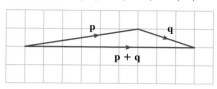

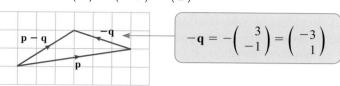

$-q = -\begin{pmatrix} 3 \\ -1 \end{pmatrix} = \begin{pmatrix} -3 \\ 1 \end{pmatrix}$

EXERCISE 27.2

The vectors **a**, **b** and **c** are given by $a = \begin{pmatrix} 2 \\ 5 \end{pmatrix}$, $b = \begin{pmatrix} -3 \\ 8 \end{pmatrix}$ and $c = \begin{pmatrix} 2 \\ -2 \end{pmatrix}$.

Work out each of these as a column vector. Illustrate your answer with a diagram.

1 $a + b$ **2** $b - c$ **3** $a + c$ **4** $c - b$

5 Work out $a - c + b$ **6** Work out $b - a + c$

7 The vectors **p**, **q** and **r** are given by $p = \begin{pmatrix} 2 \\ -1 \end{pmatrix}$, $q = \begin{pmatrix} 5 \\ 4 \end{pmatrix}$ and $r = \begin{pmatrix} 8 \\ 5 \end{pmatrix}$.

Work out each of these as a column vector. Illustrate your answer with a diagram.

8 $p - q$ **9** $q + r$ **10** $r - p$ **11** $r - q$

12 Work out $p + q + r$ **13** Work out $p - r + q$

14 You are given that $\begin{pmatrix} 5 \\ 6 \end{pmatrix} + \begin{pmatrix} 1 \\ x \end{pmatrix} = \begin{pmatrix} 6 \\ 9 \end{pmatrix}$. Find the value of x.

15 You are given that $\begin{pmatrix} x \\ 6 \end{pmatrix} - \begin{pmatrix} 7 \\ -1 \end{pmatrix} = \begin{pmatrix} 2 \\ y \end{pmatrix}$. Find the values of x and y.

16 You are given that $\begin{pmatrix} 5 \\ y \end{pmatrix} + \begin{pmatrix} x \\ -3 \end{pmatrix} = \begin{pmatrix} 2 \\ 12 \end{pmatrix}$. Find the values of x and y.

27.3 Multiplying a vector by a number (scalar multiplication)

You can **multiply** a vector by an ordinary number, say k. The direction of the vector remains unaltered, but the magnitude is changed by factor k.

For example, $3\begin{pmatrix} 2 \\ 1 \end{pmatrix} = \begin{pmatrix} 3 \times 2 \\ 3 \times 1 \end{pmatrix} = \begin{pmatrix} 6 \\ 3 \end{pmatrix}$

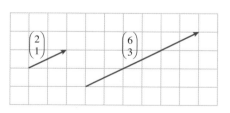

Questions about multiplication are often combined with addition and subtraction.

EXAMPLE

The vectors **a**, **b** and **c** are given by $\mathbf{a} = \begin{pmatrix} 3 \\ -1 \end{pmatrix}$, $\mathbf{b} = \begin{pmatrix} 4 \\ 2 \end{pmatrix}$ and $\mathbf{c} = \begin{pmatrix} 1 \\ 5 \end{pmatrix}$

a) Work out 3**a**. Give your answer as a column vector.
b) Work out 2**a** − **c**. Give your answer as a column vector.
   Illustrate with a diagram.
c) Work out 4**a** − 3**b** + 2**c**.

SOLUTION

a) $3\mathbf{a} = 3\begin{pmatrix} 3 \\ -1 \end{pmatrix} = \begin{pmatrix} 3 \times 3 \\ 3 \times -1 \end{pmatrix} = \begin{pmatrix} 9 \\ -3 \end{pmatrix}$

b) $2\mathbf{a} - \mathbf{c} = 2\begin{pmatrix} 3 \\ -1 \end{pmatrix} - \begin{pmatrix} 1 \\ 5 \end{pmatrix} = \begin{pmatrix} 6 \\ -2 \end{pmatrix} - \begin{pmatrix} 1 \\ 5 \end{pmatrix} = \begin{pmatrix} 5 \\ -7 \end{pmatrix}$

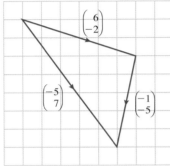

c) $4\mathbf{a} - 3\mathbf{b} + 2\mathbf{c} = 4\begin{pmatrix} 3 \\ -1 \end{pmatrix} - 3\begin{pmatrix} 4 \\ 2 \end{pmatrix} + 2\begin{pmatrix} 1 \\ 5 \end{pmatrix} = \begin{pmatrix} 12 \\ -4 \end{pmatrix} - \begin{pmatrix} 12 \\ 6 \end{pmatrix} + \begin{pmatrix} 2 \\ 10 \end{pmatrix} = \begin{pmatrix} 2 \\ 0 \end{pmatrix}$

EXERCISE 27.3

The vectors **a**, **b** and **c** are given by $\mathbf{a} = \begin{pmatrix} 4 \\ -2 \end{pmatrix}$, $\mathbf{b} = \begin{pmatrix} 1 \\ 5 \end{pmatrix}$ and $\mathbf{c} = \begin{pmatrix} -1 \\ 1 \end{pmatrix}$

Work out:

1 3**a**   2 2**b** + **c**

3 **a** + 3**c**   4 3**c** − 5**b**

5 4**a** + 5**b**   6 2**a** − 4**c**

The vectors **p**, **q** and **r** are given by $\mathbf{p} = \begin{pmatrix} -2 \\ 5 \end{pmatrix}$, $\mathbf{q} = \begin{pmatrix} 0 \\ 4 \end{pmatrix}$ and $\mathbf{r} = \begin{pmatrix} 3 \\ -1 \end{pmatrix}$

Work out:

7 5**p**   8 −3**r**

9 2**r** − 3**p**   10 4**p** + 2**q** + **r**

11 5**r** − 3**q**   12 2**p** − 3**r** + **q**

13 You are given that $3\begin{pmatrix} 2 \\ 3 \end{pmatrix} + \begin{pmatrix} 1 \\ x \end{pmatrix} = \begin{pmatrix} 7 \\ 6 \end{pmatrix}$. Find the value of x.

14 You are given that $3\begin{pmatrix} 1 \\ 4 \end{pmatrix} - 2\begin{pmatrix} 1 \\ -5 \end{pmatrix} = \begin{pmatrix} x \\ y \end{pmatrix}$. Find the values of x and y.

15 You are given that $4\begin{pmatrix} x \\ 5 \end{pmatrix} + 2\begin{pmatrix} 4 \\ y \end{pmatrix} = \begin{pmatrix} 20 \\ 18 \end{pmatrix}$. Find the values of x and y.

27.4 Using vectors

You can use vectors to solve geometric problems, and to prove some theorems about parallel lines. If one vector is a (scalar) multiple of another, then the two vectors must be parallel. The size of the multiple will tell you the scale factor.

In these problems it is often helpful to use $\overrightarrow{AB}$, for example, to represent the vector that would translate you from A to B. You can always rewrite the vector if you need to travel via an intermediate point P:

$$\overrightarrow{AB} = \overrightarrow{AP} + \overrightarrow{PB}$$

EXAMPLE

ABCD is a parallelogram. $\overrightarrow{AB} = \mathbf{p}$, $\overrightarrow{BC} = \mathbf{q}$.

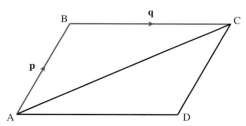

Find, in terms of $\mathbf{p}$ and $\mathbf{q}$, expressions for:

a) $\overrightarrow{BA}$ **b)** $\overrightarrow{AC}$ **c)** $\overrightarrow{BD}$

SOLUTION

a) $\overrightarrow{BA} = -\overrightarrow{AB}$
$\quad = -\mathbf{p}$

b) $\overrightarrow{AC} = \overrightarrow{AB} + \overrightarrow{BC}$
$\quad = \mathbf{p} + \mathbf{q}$

c) $\overrightarrow{BD} = \overrightarrow{BA} + \overrightarrow{AD}$
$\quad = -\mathbf{p} + \mathbf{q}$

EXAMPLE

The diagram shows a triangle ABC.
M is the midpoint of AB and
N is the midpoint of AC.

$\overrightarrow{AM} = \mathbf{p}$ and $\overrightarrow{AN} = \mathbf{q}$.

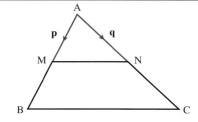

a) Find an expression for $\overrightarrow{MN}$ in terms of **p** and **q**.

b) Find an expression for $\overrightarrow{BC}$ in terms of **p** and **q**.

c) Use your results from **a)** and **b)** to prove that MN is parallel to BC.

SOLUTION

a) $\overrightarrow{MN} = \overrightarrow{MA} + \overrightarrow{AM}$

$= -\overrightarrow{AM} + \overrightarrow{AN}$

$= (-\mathbf{p}) + \mathbf{q}$

$= -\mathbf{p} + \mathbf{q}$

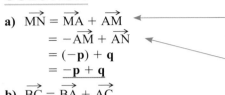

To get $\overrightarrow{MN}$ in terms of **p** and **q**, go from M to N via the point A.

$\overrightarrow{MA}$ has the same length as $\overrightarrow{AM}$ but points in the opposite direction, so $\overrightarrow{MA} = -\overrightarrow{AM}$

b) $\overrightarrow{BC} = \overrightarrow{BA} + \overrightarrow{AC}$

$= (-2\mathbf{p}) + 2\mathbf{q}$

$= -2\mathbf{p} + 2\mathbf{q}$

c) $\overrightarrow{BC} = -2\mathbf{p} + 2\mathbf{q}$

$= 2(-\mathbf{p} + \mathbf{q})$

$= 2 \times \overrightarrow{MN}$

Therefore BC is parallel to MN.

Some exam questions might refer to a line being divided in a certain *ratio*.
For example, you might be told that X is the point on AB for which
AX : XB = 2 : 1. This simply means that AX is twice as long as XB, so that
X is two-thirds of the way along AB.

EXAMPLE

The diagram shows a parallelogram ABCD. $\overrightarrow{AB} = 6\mathbf{p}$ and $\overrightarrow{BC} = 6\mathbf{q}$.

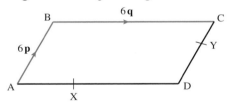

X is the point on AD for which AX : XD = 1 : 2
Y is the point on DC for which DY : YC = 2 : 1

Find, in terms of **p** and **q**, expressions for:

a) $\overrightarrow{AC}$ **b)** $\overrightarrow{AD}$ **c)** $\overrightarrow{DC}$

d) $\overrightarrow{XD}$ **e)** $\overrightarrow{DY}$ **f)** $\overrightarrow{XY}$

Hence prove that AC is parallel to XY.

SOLUTION

a) $\overrightarrow{AC} = \overrightarrow{AB} + \overrightarrow{BC}$

$= 6\mathbf{p} + 6\mathbf{q}$

b) $\overrightarrow{AD} = \overrightarrow{BC}$ since they are opposite sides of the parallelogram

$= 6\mathbf{q}$

c) $\vec{DC} = \vec{AB}$ since they are opposite sides of the parallelogram

$\qquad = \underline{6\mathbf{p}}$

d) $\vec{XD} = \frac{2}{3} \times \vec{AD}$

$\qquad = \frac{2}{3} \times 6\mathbf{q}$

$\qquad = \underline{4\mathbf{q}}$

e) $\vec{DY} = \frac{2}{3} \times \vec{DC}$

$\qquad = \frac{2}{3} \times 6\mathbf{p}$

$\qquad = \underline{4\mathbf{p}}$

f) $\vec{XY} = \vec{XD} + \vec{DY}$

$\qquad = 4\mathbf{q} + 4\mathbf{p}$

$\qquad = \underline{4\mathbf{p} + 4\mathbf{q}}$

Now $\vec{AC} = 6\mathbf{p} + 6\mathbf{q} = 6(\mathbf{p} + \mathbf{q})$

and $\vec{XY} = 4\mathbf{p} + 4\mathbf{q} = 4(\mathbf{p} + \mathbf{q})$

Thus $\vec{AC} = 1.5 \times \vec{XY}$, and therefore <u>AC is parallel to XY</u>.

EXERCISE 27.4

1 The diagram shows two squares ABXY and CDYX.
$\vec{AB} = \mathbf{p}$ and $\vec{AY} = \mathbf{q}$.

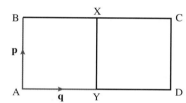

Find, in terms of **p** and **q**, expressions for:

a) $\vec{BX}$ **b)** $\vec{AX}$ **c)** $\vec{AD}$ **d)** $\vec{AC}$

2 The diagram shows a trapezium PQRS.
$\vec{PQ} = \mathbf{a}$ and $\vec{QR} = \mathbf{b}$.
PS is twice the length of QR.

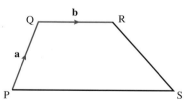

Find, in terms of **a** and **b**, expressions for:

a) $\vec{QP}$ **b)** $\vec{PR}$ **c)** $\vec{PS}$ **d)** $\vec{QS}$

3 The diagram shows a triangle ABC. $AP = \frac{1}{3}AB$, and $AQ = \frac{1}{3}AC$.
$\overrightarrow{AP} = \mathbf{p}$ and $\overrightarrow{AQ} = \mathbf{q}$.

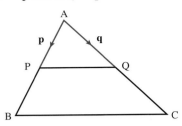

a) Find, in terms of **p** and **q**, expressions for:

   (i) $\overrightarrow{PQ}$ **(ii)** $\overrightarrow{AB}$ **(iii)** $\overrightarrow{AC}$ **(iv)** $\overrightarrow{BC}$

b) Use your results from **a)** to prove that PQ is parallel to BC.

4 A quadrilateral ABCD is made by joining points A (1, 1), B (5, 8), C (11, 11) and D (7, 4).

a) Write column vectors for:

   (i) $\overrightarrow{AB}$ **(ii)** $\overrightarrow{DC}$

b) What do your answers to part **a)** tell you about AB and DC?

c) Write column vectors for:

   (i) $\overrightarrow{BC}$ **(ii)** $\overrightarrow{AD}$

d) What kind of quadrilateral is ABCD?

5 The diagram shows a parallelogram PQRS.
$\overrightarrow{PQ} = \mathbf{a}$ and $\overrightarrow{PS} = \mathbf{b}$.
E is the mid-point of QS.

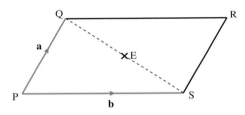

a) Find, in terms of **a** and **b**:

   (i) $\overrightarrow{QS}$ **(ii)** $\overrightarrow{QE}$ **(iii)** $\overrightarrow{PE}$

b) Explain why $\overrightarrow{SR} = \mathbf{a}$.

c) Find $\overrightarrow{PR}$ in terms of **a** and **b**.

d) What can you deduce about the diagonals of a parallelogram?

6 A quadrilateral ABCD is made by joining A $(-3, -3)$, B (9, 3), C (3, 7) and D $(-1, 5)$.

a) Write column vectors for:

   (i) $\overrightarrow{AB}$ **(ii)** $\overrightarrow{DC}$

b) What do your answers to part **a)** tell you about AB and DC?

c) What kind of quadrilateral is ABCD?

7 The diagram shows a quadrilateral PQRS.
$\overrightarrow{PQ} = 2\mathbf{a}$, $\overrightarrow{PS} = 2\mathbf{b}$ and $\overrightarrow{SR} = 2\mathbf{c}$.
E, F, G and H are the mid-points of PQ, PS, SR and QR respectively.

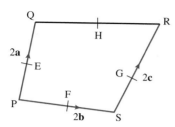

a) Explain why $\overrightarrow{QR} = -2\mathbf{a} + 2\mathbf{b} + 2\mathbf{c}$.
b) Find $\overrightarrow{EH}$ in terms of $\mathbf{a}$, $\mathbf{b}$ and $\mathbf{c}$.
c) Find $\overrightarrow{FG}$ in terms of $\mathbf{a}$, $\mathbf{b}$ and $\mathbf{c}$.
d) What can you deduce about the line segments EH and FG?
e) What type of quadrilateral is EFGH?

REVIEW EXERCISE 27

1 Given that $3\begin{pmatrix} x \\ 5 \end{pmatrix} - \begin{pmatrix} 2 \\ 4 \end{pmatrix} = \begin{pmatrix} 16 \\ y \end{pmatrix}$, find the values of x and y.

2 P is the point $(5, 4)$ and Q is the point $(-1, 12)$.
a) Write $\overrightarrow{PQ}$ and $\overrightarrow{QP}$ as column vectors.
b) Work out the length of the vector $\overrightarrow{PQ}$.

3 A is the point $(2, 3)$ and B is the point $(-2, 0)$.
a) (i) Write $\overrightarrow{AB}$ as a column vector.
   (ii) Find the length of the vector $\overrightarrow{AB}$.

D is the point such that $\overrightarrow{BD}$ is parallel to $\begin{pmatrix} 0 \\ 1 \end{pmatrix}$ and the length of $\overrightarrow{AD}$ = the length of $\overrightarrow{AB}$.

O is the point $(0, 0)$.

b) Find $\overrightarrow{OD}$ as a column vector.

C is a point such that ABCD is a rhombus. AC is a diagonal of the rhombus.
c) Find the coordinates of C.

[Edexcel]

4 OPQ is a triangle.
R is the midpoint of OP.
S is the midpoint of PQ.
$\overrightarrow{OP} = \mathbf{p}$ and $\overrightarrow{OQ} = \mathbf{q}$.

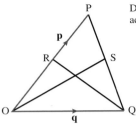

Diagram *not*
accurately drawn

a) Find $\overrightarrow{OS}$ in terms of $\mathbf{p}$ and $\mathbf{q}$.

b) Show that RS is parallel to OQ. [Edexcel]

5 OPQ is a triangle.
T is the point on PQ for which PT : TQ = 2 : 1.
$\overrightarrow{OP} = \mathbf{a}$ and $\overrightarrow{OQ} = \mathbf{b}$.

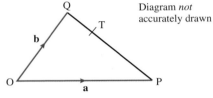

Diagram *not*
accurately drawn

a) Write down, in terms of $\mathbf{a}$ and $\mathbf{b}$, an expression for $\overrightarrow{PQ}$.

b) Express $\overrightarrow{OT}$ in terms of $\mathbf{a}$ and $\mathbf{b}$. Give your answer in its simplest form. [Edexcel]

6 OABC is a parallelogram.
P is the point on AC such that $AP = \frac{2}{3}AC$.
$\overrightarrow{OA} = 6\mathbf{a}$ and $\overrightarrow{OC} = 6\mathbf{c}$.

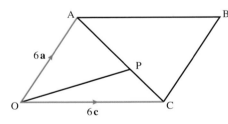

Diagram *not*
accurately drawn

a) Find the vector $\overrightarrow{OP}$. Give your answer in terms of $\mathbf{a}$ and $\mathbf{c}$.

The midpoint of CB is M.

b) Prove that OPM is a straight line. [Edexcel]

7 PQRS is a parallelogram.
T is the midpoint of QR.
U is the point on SR for which SU : UR = 1 : 2.
$\overrightarrow{PQ} = \mathbf{a}$ and $\overrightarrow{PS} = \mathbf{b}$.

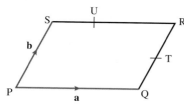

Diagram *not*
accurately drawn

Write down, in terms of **a** and **b**, expressions for:
a) $\overrightarrow{PT}$
b) $\overrightarrow{TU}$

[Edexcel]

8 ABCD is a quadrilateral.

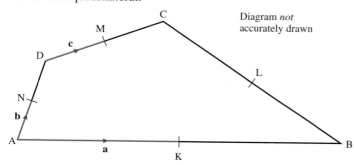

Diagram *not*
accurately drawn

K is the midpoint of AB. L is the midpoint of BC.
M is the midpoint of CD. N is the midpoint of AD.
$\overrightarrow{AK} = \mathbf{a}$, $\overrightarrow{AN} = \mathbf{b}$ and $\overrightarrow{DM} = \mathbf{c}$

a) Find, in terms of **a**, **b** and **c**, the vectors:
    (i) $\overrightarrow{KN}$ **(ii)** $\overrightarrow{AC}$ **(iii)** $\overrightarrow{BC}$ **(iv)** $\overrightarrow{LM}$

b) Write down two geometrical facts about the lines KN and LM which could be deduced
from your answers to part **a)**.

[Edexcel]

9 The diagram shows a regular hexagon ABCDEF with centre O.

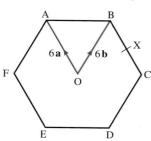

Diagram *not*
accurately drawn

$\overrightarrow{OA} = 6\mathbf{a}$ and $\overrightarrow{OB} = 6\mathbf{b}$.

a) Express in terms of **a** and/or **b**.

    (i) $\overrightarrow{AB}$ **(ii)** $\overrightarrow{EF}$

X is the midpoint of BC.

b) Express $\overrightarrow{EX}$ in terms of **a** and/or **b**.

Y is the point on AB extended, such that AB : BY = 3 : 2.

c) Prove that E, X and Y lie on the same straight line. [Edexcel]

10 OPQR is a trapezium. PQ is parallel to OR. $\overrightarrow{OP} = \mathbf{b}$, $\overrightarrow{PQ} = 2\mathbf{a}$, $\overrightarrow{OR} = 6\mathbf{a}$.
    M is the midpoint of PQ. N is the midpoint of OR.

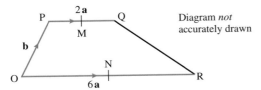

Diagram *not* accurately drawn

a) Find, in terms of **a** and **b**, the vectors:

    (i) $\overrightarrow{OM}$ **(ii)** $\overrightarrow{MN}$

X is the midpoint of MN.

b) Find, in terms of **a** and **b**, the vector $\overrightarrow{OX}$.

The lines OX and PQ are extended to meet at the point Y.

c) Find, in terms of **a** and **b**, the vector $\overrightarrow{NY}$. [Edexcel]

11 The vector $\mathbf{a} = \begin{pmatrix} 1.5 \\ 2 \end{pmatrix}$ and the vector $\mathbf{b} = 4\mathbf{a}$.

    a) Work out the magnitude of vector **a**.

    b) Hence write down the magnitude of vector **b**.

12 $\overrightarrow{PQ}$ has magnitude 6 cm, and $\overrightarrow{PR} = 3 \times \overrightarrow{PQ}$.

    a) What can you deduce about the directions of the vectors $\overrightarrow{PQ}$ and $\overrightarrow{PR}$?

    b) Find the magnitude of the vector $\overrightarrow{PR}$.

Key points

1 A vector has a direction and a length, or magnitude. Vectors are usually written in column form, such as $\begin{pmatrix} 4 \\ 6 \end{pmatrix}$, which represents a translation of 4 units in the x direction and 6 in the y direction.

2 Vectors are often used in examination questions to prove geometric theorems. The method is to use given base vectors **a**, **b**, **c**, etc. and then express other lines in terms of these, for example 2**a** + **b**.

3 Two vectors will be parallel if one is a scalar multiple of the other. For example, 6**a** + 3**b** is parallel to 2**a** + **b**, since 6**a** + 3**b** = 3 × (2**a** + **b**)

Internet Challenge 27

Queens on a chessboard

Here is another chessboard problem.

The Queen is the most powerful piece on a chessboard. A Queen can attack any squares in a straight line from it, forwards, backwards, left, right or diagonal. The diagram below shows this in green for one position of the Queen:

Place eight Queens on a chessboard so that no two Queens attack each other.
You may want to use squared paper to record your attempts. This problem does have more than one solution. Once you have solved it, you might want to use the internet to help answer the following questions.

1   How many different distinct solutions does this problem have?

2   How many solutions are there in which no three Queens lie on an oblique line?

3   What is a Latin square? Is this a Latin square problem?

4   How many knights can be placed on a chessboard so that no knight attacks any other?

5   How about bishops?

Obviously it is not possible to place nine Queens on a board without at least two Queens attacking each other. (Why not?) There is, however, a 'nine Queens' problem:

Place nine Queens and one pawn on a chessboard so that no two Queens attack each other.

6   Try to solve the nine Queens problem. Use the internet if you get stuck.

CHAPTER 28

Calculus

In this chapter you will learn how to:

- estimate the gradient of the tangent to a curve by measurement
- calculate the gradient of the tangent to a curve by differentiation
- use calculus to find maximum and minimum points on curves
- use turning points to help sketch parabolas
- solve simple kinematics problems about distance, speed and acceleration.

You will also be challenged to:

- complete a calculus crossword.

Starter: **Steeper and steeper**

Here are some sketches of straight line segments. Work out the gradient of each one, and arrange them in order of steepness, lowest to highest.
Remember that gradient = rise/run.

A
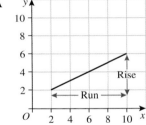

B
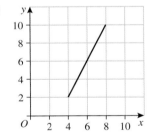

C
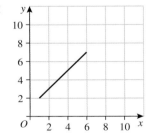

D
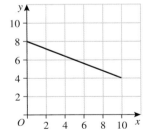

28.1 Gradient of a curve

In Chapter 6 you measured the gradient of a straight line graph by dividing the height ('rise') by the horizontal distance ('run'). The gradient of a curve can be measured in a similar way, but you have to draw a **tangent** first; this is a straight line touching the curve at the point of interest. Then measure the gradient of the tangent, and take this as the gradient of the curve at that point.

EXAMPLE

The diagram shows part of the curve of $y = x^2 + 1$. By drawing suitable tangents, estimate the gradient of the curve:

a) at the point $(1, 2)$
b) at the point $(2, 5)$.

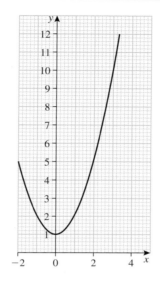

SOLUTION

a)

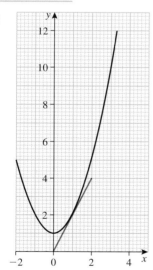

b)

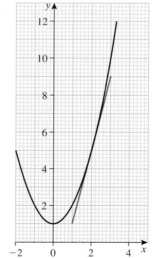

At the point $(1, 2)$ the gradient of the tangent is $\dfrac{4 - 0}{2 - 0} = \dfrac{4}{2} = 2$.

Hence the gradient of the curve at $(1, 2)$ is estimated as 2.

At the point $(2, 5)$ the gradient of the tangent is $\dfrac{9 - 1}{3 - 1} = \dfrac{8}{2} = 4$.

Hence the gradient of the curve at $(2, 5)$ is estimated as 4.

You can also use gradients to measure rate of change in real-life graphs, for example, distance–time graphs. The gradient of a distance–time graph gives velocity ('speed'), and the gradient of a velocity–time graph gives acceleration.

EXAMPLE

The distance–time graph shows the distance travelled by a cyclist during the first 12 seconds of a journey. The distance s metres travelled is plotted against the time t seconds.

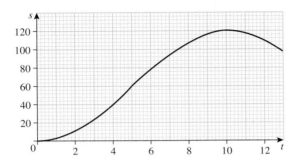

a) At what times is the cyclist stationary?
b) Estimate the cyclist's speed at time $t = 7$ seconds.

SOLUTION

a) The cyclist is stationary when the graph is horizontal.
This occurs when $t = 0$ and again when $t = 10$.

b)

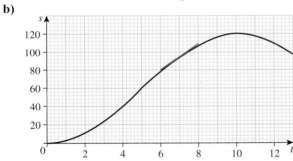

At time $t = 7$ the gradient of the tangent is $\dfrac{110 - 80}{8 - 6} = \dfrac{30}{2} = 15$.

Hence the speed at time $t = 7$ is estimated as 15 m/s.

EXERCISE 28.1

1 The diagram shows part of the graph of $y = x^2 + 2$.
   a) Write down the gradient of the graph when $x = 0$.
   b) By drawing a suitable tangent on a copy of the graph, estimate the gradient of the graph when $x = 1$.

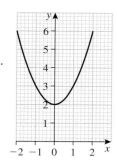

2 The diagram shows part of the graph of $y = x^2 - 3x$.
   By drawing suitable tangents on a copy of the graph, estimate the gradient of $y = x^2 - 3x$ when:
   a) $x = 2$
   b) $x = 0$

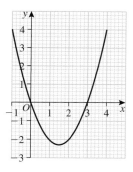

3 The diagram shows part of the graph of $y = x^2 - 1$.
   a) Draw a tangent at the point (2, 3) on a copy of the graph.
   b) Hence estimate the gradient of $y = x^2 - 1$ at the point (2, 3).

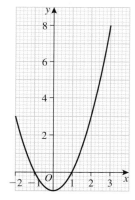

4 The distance–time graph shows the distance travelled by an athlete during the first 12 seconds of a training run.
   The athlete runs away from his start point for a time, then turns around and runs back.
   The distance s metres travelled is plotted against the time t seconds.
   a) At what time does the athlete turn around?
   b) Write down the distance travelled when $t = 10$.
   c) By constructing a suitable tangent on a copy of the graph, estimate the velocity of the athlete when $t = 6$.

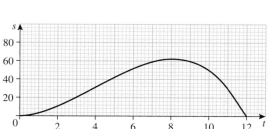

28.2 Gradient of a curve – differentiation

If you were to draw the graph of $y = x^2$ you would obtain a distinctive curve, called a parabola:

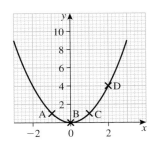

The gradient of the curve may be found at various places, such as A, B, C, D above, by constructing and measuring tangents. The table shows the results:

Point	A	B	C	D
x coordinate	-1	0	1	2
Gradient	-2	0	2	4

Note that in every case the gradient is twice the x value. It turns out that this relationship holds at every point on the graph of $y = x^2$ (though the IGCSE does not require you to be able to prove this). We say that

$y = x^2$ has a gradient function $\dfrac{dy}{dx} = 2x$

The process of finding a gradient function is called differentiation. Here is a general rule you can use for differentiating any power of x:

if $y = ax^n$ then $\dfrac{dy}{dx} = anx^{n-1}$

EXAMPLE

Differentiate:
a) $y = 9x^4$ b) $y = 8x^3 + 4x^2$

SOLUTION

a) If $y = 9x^4$ then $\dfrac{dy}{dx} = (9 \times 4)\, x^{(4-1)}$

$= 36x^3$

b) If $y = 8x^3 + 4x^2$ then $\dfrac{dy}{dx} = (8 \times 3)\, x^{(3-1)} + (4 \times 2)\, x^{(2-1)}$

$= 24x^2 + 8x$

If you need to differentiate a constant, for example $y = 4$, think of it as $y = 4x^0$.
Then $\dfrac{dy}{dx} = (4 \times 0)\,x^{-1} = 0$.
It follows that any constant will differentiate to 0.

EXAMPLE

Differentiate:
a) $y = x^3 + 5x^2 - 2x + 4$

b) $y = x^8 - 5x^4 - 7$

SOLUTION

a) $y = x^3 + 5x^2 - 2x + 4$
   Then $\dfrac{dy}{dx} = 3x^2 + 10x - 2$

b) $y = x^8 - 5x^4 - 7$
   Then $\dfrac{dy}{dx} = 8x^7 - 20x^3$

EXERCISE 28.2

Find $\dfrac{dy}{dx}$ in each case:

1 $y = x^3 + 6x^2$

2 $y = x^4 - 2x^3 + x^2$

3 $y = 2x^4 + 3x^3 + 10x$

4 $y = x^5 - 5x^2$

5 $y = 3x^4 - 4x^3$

6 $y = x^2 - 6x + 8$

7 $y = x^{10} + 3x^3$

8 $y = 6x^3 + 12x$

9 $y = 9x^3 + 4x + 11$

10 $y = x^3 - x^4 + 2x^5$

28.3 Harder differentiation

The rule that if $y = ax^n$ then $\dfrac{dy}{dx} = nax^{n-1}$ may be extended to negative powers of x as well.

EXAMPLE

Differentiate:
a) $y = \dfrac{10}{x}$

b) $y = \dfrac{8}{x^2}$

SOLUTION

a) $y = \dfrac{10}{x}$

   $= 10x^{-1}$

   $\dfrac{dy}{dx} = (10 \times -1)x^{-1-1}$

   $= -10x^{-2}$

   $= \dfrac{-10}{x^2}$

b) $y = \dfrac{8}{x^2}$

   $= 8x^{-2}$

   $\dfrac{dy}{dx} = (8 \times -2)x^{-2-1}$

   $= -16x^{-3}$

   $= \dfrac{-16}{x^3}$

Special care must be taken when expressions are written in brackets. If you need to differentiate the product of two brackets you should multiply out and simplify the brackets first.

EXAMPLE

a) Multiply out and simplify $(x + 5)(2x - 1)$.
b) Differentiate $y = (x + 5)(2x - 1)$.

SOLUTION

a) $(x + 5)(2x - 1) = 2x^2 + 10x - x - 5$
$$= 2x^2 + 9x - 5$$

b) $y = (x + 5)(2x - 1)$
$$= 2x^2 + 9x - 5$$

Thus $\dfrac{dy}{dx} = 4x + 9$

Some questions may give you a string of algebraic terms as the top of an algebraic fraction, with a constant denominator. In such cases, simply differentiate the whole of the top, and divide this by the same denominator at the end.

EXAMPLE

Differentiate $\dfrac{x^3 + 5x^2 - 4x}{5}$.

SOLUTION

If $y = \dfrac{x^3 + 5x^2 - 4x}{5}$

then $\dfrac{dy}{dx} = \dfrac{3x^3 + 10x - 4}{5}$

EXERCISE 28.3

1 Find $\dfrac{dy}{dx}$ if $y = 4x^2 + 5x + \dfrac{1}{x}$.

2 Differentiate $8x + \dfrac{4}{x^2}$.

3 Find $\dfrac{dy}{dx}$ if $y = x^2 - \dfrac{3}{x}$.

4 Differentiate $2x + 3 + \dfrac{4}{x} + \dfrac{5}{x^2}$.

5 a) Multiply out and simplify $(x + 3)(2x + 1)$.

  b) Find $\dfrac{dy}{dx}$ if $y = (x + 3)(2x + 1)$.

6 a) Multiply out and simplify $(2x - 5)(x + 1)$.
  b) Differentiate $(2x - 5)(x + 1)$.

7 Find $\dfrac{dy}{dx}$ if $y = \dfrac{x^2 + 5x + 4}{3}$.

8 Differentiate $\dfrac{2x^3 + 3x^2 - 5 + 1}{6}$.

9 $y = \dfrac{3x^2 - 4x + 7}{4}$. Find the value of $\dfrac{dy}{dx}$ when $x = 2$.

10 $y = \dfrac{x^3 - 5x + 4}{2}$. Find the value of $\dfrac{dy}{dx}$ when $x = -1$.

28.4 Maximum and minimum points on curves

When a curve passes through a **maximum** or **minimum point**, the gradient of the tangent becomes zero. This means that such points can be found by solving the equation $\dfrac{dy}{dx} = 0$.

Any point on a curve where the gradient of the tangent is zero is called a stationary point. Maximum and minimum points are, therefore, examples of stationary points.

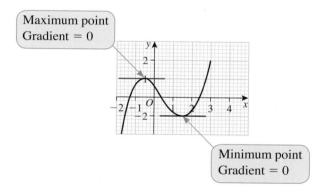

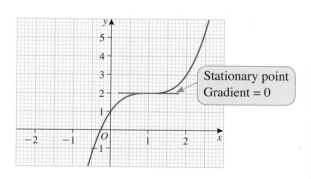

Maximum and minimum points are known collectively as **turning points**, since the curve changes direction there.

A quadratic will have only a single turning point. This will be a minimum if the x^2 coefficient is positive, and a maximum if the x^2 coefficient is negative.

x^2 coefficient *positive*
e.g. $y = 2x^2 + 3x - 2$

x^2 coefficient *negative*
e.g. $y = -2x^2 + 3x - 2$

547

EXAMPLE

A curve has equation $y = x^3 - 3x^2 + 3x + 1$

a) Find an expression for $\dfrac{dy}{dx}$.

b) Find the coordinates of the stationary points of this curve.

SOLUTION

a) $\dfrac{dy}{dx} = 3x^2 - 6x + 3$

b) At a stationary point $\dfrac{dy}{dx} = 0$

When $\dfrac{dy}{dx} = 0$

then $\quad 3x^2 - 6x + 3 = 0$

> Divide each term by 3.

So $\quad x^2 - 2x + 1 = 0$

> Factorise.

$(x - 1)(x - 1) = 0$

So $x = 1$

When, $x = 1$, $y = 1^3 - 3 \times 1^2 + 3 \times 1 + 1 = 2$

> Don't forget to substitute x back into the equation of the curve to find the y coordinate.

So the coordinates of the stationary point are $(1, 2)$

The curve $y = x^3 - 3x^2 + 3x + 1$ is shown in the diagram on page 547.

EXAMPLE

A curve has equation $y = x^2 - 8x + 3$.

a) Find an expression for $\dfrac{dy}{dx}$.

b) Work out the value of x for which $\dfrac{dy}{dx} = 0$.

c) Find the coordinates of the turning point of this curve.

d) Sketch the curve.

SOLUTION

a) $\dfrac{dy}{dx} = \underline{2x - 8}$

b) When $\dfrac{dy}{dx} = 0$,

$2x - 8 = 0$

$2x = 8$

$\underline{x = 4}$

c) At the turning point, $x = 4$, so

$$y = x^2 - 8x + 3$$
$$= 16 - 32 + 3$$
$$= -13$$

The turning point is at $(4, -13)$.

d) Since the curve has a positive x^2 coefficient, the turning point will be a minimum.

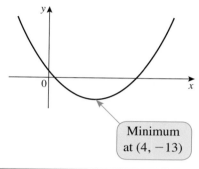

Minimum at $(4, -13)$

In the exam, you may be asked to distinguish between maximum and minimum points.

EXAMPLE

A curve has equation $y = 2x^3 + 3x^2 - 36x + 1$.

a) Find an expression for $\dfrac{dy}{dx}$.

b) Work out the values of x for which $\dfrac{dy}{dx} = 0$.

c) Find the coordinates of the turning points of this curve.
d) Sketch the curve.

SOLUTION

a) $\dfrac{dy}{dx} = 6x^2 + 6x - 36$

b) When $\dfrac{dy}{dx} = 0$,

$$6x^2 + 6x - 36 = 0$$
$$x^2 + x - 6 = 0$$
$$(x + 3)(x - 2) = 0$$
$$x = -3 \text{ or } 2$$

c) At the turning point given by $x = -3$:

$$y = 2x^3 + 3x^2 - 36x + 1$$
$$= -54 + 27 + 108 + 1$$
$$= 82$$

549

At the turning point given by $x = 2$:
$$y = 2x^3 + 3x^2 - 36x + 1$$
$$= 16 + 12 - 72 + 1$$
$$= -43$$

The turning points are at $(-3, 82)$ and at $(2, -43)$.

d) One of these turning points is a maximum and the other a minimum. The greater y coordinate shows that $(-3, 82)$ is the maximum, and $(2, -43)$ the minimum, so the curve looks like this.

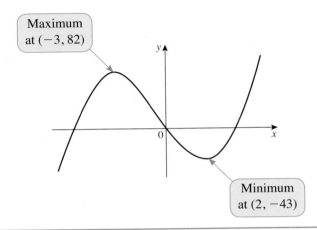

Maximum at $(-3, 82)$

Minimum at $(2, -43)$

EXERCISE 28.4

1 Find the coordinates of the minimum point on the curve $y = x^2 - 6x + 11$.

2 Find the coordinates of the minimum point on the curve $y = 2x^2 - 8x + 7$.

3 Find the coordinates of the maximum point on the curve $y = -x^2 + 10x - 23$.

4 Find the coordinates of the maximum point on the curve $y = 3 + 2x - x^2$.

5 A curve is given by $y = 3x^2 + 6x + 5$.
  a) Find $\dfrac{dy}{dx}$.
  b) Find the coordinates of the turning point on this curve.
  c) State whether the turning point is a maximum or a minimum.

6 a) Find the coordinates of the turning point of the curve $y = -2x^2 + 16x - 28$.
  b) State whether the turning point is a maximum or a minimum.

7 A curve is given by $y = x^2 + 12x + 45$.
  a) Find $\dfrac{dy}{dx}$.
  b) Find the coordinates of the turning point on this curve.
  c) Sketch the curve.

8 A curve is given by $y = -4x^2 + 8x + 1$.

   a) Find $\dfrac{dy}{dx}$.

   b) Find the coordinates of the turning point on this curve.

   c) Sketch the curve.

9 A curve is given by $y = 1 - 12x + 9x^2 - 2x^3$.

   a) Find $\dfrac{dy}{dx}$.

   b) Find the coordinates of the turning points on this curve.

   c) Sketch the curve.

10 A curve is given by $y = 2x^3 + 15x^2 + 24x$.

   a) Find $\dfrac{dy}{dx}$.

   b) Find the coordinates of the turning points on this curve.

   c) Sketch the curve.

11 The curve with equation $y = 16x + \dfrac{1}{x^2}$ has one stationary point.

Find the coordinates of this stationary point.
Show your working clearly.

12 A curve has equation $y = 4x + \dfrac{1}{x}$

   a) Find the coordinates of the stationary points of the curve.

   b) Identify the nature of each stationary point.

13 a) Expand and simplify $(x - 2)^3$.
A curve has equation $x = (x - 2)^3 - 4$

   b) Find $\dfrac{dy}{dx}$.

   c) The curve has one stationary point.

Find the coordinates of the stationary point.

28.5 Further problems on maximum and minimum

Calculus problems about graphs normally use x and y as the variables, but other symbols might be encountered in different situations, especially practical (or 'applied') problems.

EXAMPLE

An economist reckons that the total revenue £R generated by selling goods at a price of £m each may be modelled by the equation $R = 2000 + 500m - 200m^2$.
a) Find the value of m for which the revenue £R is a maximum.
b) Hence find the maximum revenue.

SOLUTION

a) $R = 2000 + 500m - 200m^2$

Thus $\dfrac{dR}{dm} = 500 - 400m$

$= 0$ for a maximum

Then $400m = 500$

$m = \dfrac{500}{400}$

$= 1.25$

b) Then R is $2000 + 500 \times (1.25) - 200 \times (1.25)^2 = 2312.5$

So the maximum revenue is £2312.50.

In some questions you might not be given the expressions, but would need to obtain them first from the context of a particular problem.

EXAMPLE

A rectangle has perimeter 40 cm. The length along one side is x cm.

a) Obtain an expression for the area, A cm^2, in terms of x.

b) Find $\dfrac{dA}{dx}$, and hence determine the value of x for which the area is a maximum.

SOLUTION

a) Since the perimeter is 40 cm, the total length of two adjacent sides is 20 cm.
Thus if the length is x cm then the width is $(20 - x)$ cm:

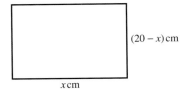

$(20-x)\,\text{cm}$

$x\,\text{cm}$

To obtain the area we multiply these two dimensions together, so

$A = x(20 - x)$

b) Since $A = x(20 - x) = 20x - x^2$

then $\dfrac{dA}{dx} = 20 - 2x$

$= 0$ for a maximum

Thus $2x = 20$

$x = 10$

EXERCISE 28.5

1 The fuel economy E of a car travelling at v miles per hour is modelled by the equation $E = -v^2 + 100v + 60$.

    a) Find an expression for $\dfrac{dE}{dv}$ in terms of v.

    b) Hence find the value of v for which the fuel economy takes its maximum value.

2 A colony of bacteria is allowed to grow over a period of 100 hours. During this time, the number of bacteria N in the colony at time t hours after the start is modelled by the equation $N = 2000 + 80t - t^2$.

    a) Write down the number of bacteria at time $t = 0$.

    b) Find an expression for $\dfrac{dN}{dt}$ in terms of t.

    c) Find the maximum number of bacteria in the colony during the 100 hour period.

3 Two quantities P and n are related by the equation $P = n^2 - 40n + 450$.

    a) Find the value of P:

        (i) when $n = 10$ **(ii)** when $n = 25$.

    b) Find an expression for $\dfrac{dP}{dn}$.

    c) Find the minimum value of P, and the value of n for which it occurs.

4 A mathematician is investigating the quality of a particular type of wine as it ages. He notices that the quality of the wine seems to improve for a few years, then decline. He suggests that the quality Q of the wine at age t years can be modelled by the equation $Q = 36 + 16t - t^2$.

    a) Work out the value of Q when:

        (i) $t = 5$ **(ii)** $t = 10$

    b) Find an expression for $\dfrac{dQ}{dt}$.

    c) Work out the maximum value of Q.

5 Two quantities Q and z are related by the equation $Q = (z + 8)(12 - z)$.

    a) Find an expression for $\dfrac{dQ}{dz}$.

    b) Find the maximum value of Q, and the value of z for which it occurs.

6 A square card of side 8 cm has four small squares of side x cm removed from the corners as shown opposite. The card is folded along the dashed lines to make an open box of volume V cm^3.

    a) Show that $V = x(8 - 2x)^2$.

    b) Expand this expression for V, and hence find $\dfrac{dV}{dx}$.

    c) Find the maximum volume of the box.

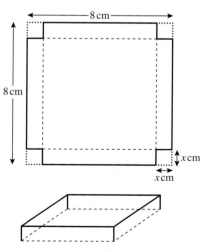

7 The diagram shows a rectangular field ABCD.

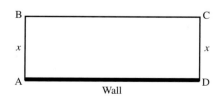

The sides AB, BC and CD are to be enclosed using fencing. The side AD does not need to be fenced, as there is an existing wall along this side. The total length of fencing to be used is 60 metres.

a) Find an expression for the length BC, in terms of x.

b) Show that the area A of the field is given by $A = x(60 - 2x)$.

c) Multiply out and simplify this expression. Use your result to find an expression for $\dfrac{dA}{dx}$.

d) Hence find the value of x for which the area of the field is a maximum.

e) State the dimensions of the field when the area is a maximum.

8 The demand M for electricity t hours after midnight is modelled by the equation $M = 1200t + \dfrac{19\,200}{t}$.

a) Find an expression for $\dfrac{dM}{dt}$.

b) Find the minimum value of M, and the value of t for which it occurs.

28.6 Distance, velocity and acceleration

Suppose a particle is travelling in a straight line — the x axis — and that its **distance** s from the origin at any time is given in terms of t.
For example, $s = t^3 + 5t^2$.

The **velocity** is the rate at which distance changes, in other words, $\dfrac{ds}{dt}$.

So, for this example, the velocity v of the particle is $v = \dfrac{ds}{dt} = 3t^2 + 10t$.

Similarly, the **acceleration** is the rate at which velocity changes, in other words, $\dfrac{dv}{dt}$.

So, for this example, the acceleration a of the particle is $a = \dfrac{dv}{dt} = 6t + 10$.

You might be asked to find the distance, velocity and acceleration of a particle at a specified value of t. The method is to use differentiation first, then substitute the given value of t into the appropriate formula at the end.

EXAMPLE

A particle travels such that its distance s metres from the origin at time t seconds is given by the equation $s = t^3 + 10t^2 - 3t$.

a) Find expressions for:
   (i) the velocity
   (ii) the acceleration of the particle at time t.

b) Find the velocity when $t = 3$.

c) Find the acceleration when $t = 4$.

SOLUTION

a) (i) Velocity $v = \dfrac{ds}{dt}$

$\qquad\qquad = \underline{3t^2 + 20t - 3}$

(ii) Acceleration $a = \dfrac{dv}{dt}$

$\qquad\qquad = \underline{6t + 20}$

b) When $t = 3$,

velocity $v = 3 \times 9 + 20 \times 3 - 3$

$= 27 + 60 - 3$

$= 84$ m/s

c) When $t = 4$,

acceleration $a = 6 \times 4 + 20$

$= 24 + 20$

$= 44$ m/s^2

EXERCISE 28.6

1 A particle moves so that its distance s metres from the origin at time t seconds is given by
$s = t^3 + 6t^2 + 5t - 1$.
   a) Find expressions for:
      (i) the velocity   **(ii)** the acceleration of the particle at time t.
   b) Find the velocity when $t = 2$.
   c) Find the acceleration when $t = 3$.

2 A particle moves so that its distance s metres from the origin at time t seconds is given by the equation
$s = t^3 + 16t + 2$.
   a) Find an expression for the velocity v in terms of t.
   b) Find the value of the velocity when $t = 3$, stating your units clearly.
   c) Find the value of the acceleration when $t = 4$, stating your units clearly.

3 The dispacement s metres of a particle at time t seconds is given by the equation $s = t^3 + 14t^2$.
   a) Find the velocity of the particle after 3 seconds.
   b) Find the acceleration of the particle after 2 seconds.

4 The displacement s metres of a particle at time t seconds is given by the equation $s = t^2 - 8t - 10$.
   a) Find the velocity of the particle after 5 seconds.
   b) At what time is the velocity equal to zero?

5 A stone is thrown from a cliff. Its distance s metres from the point of projection at time t seconds is given
by $s = 4.9t^2 + 2t$.
   a) Find the velocity when $t = 0$.
   b) Prove that the acceleration is constant, and state its value.

REVIEW EXERCISE 28

1 The diagram shows part of the graph of a function $y = f(x)$.
   a) Copy the graph. Draw a tangent to the curve at the point $(3, 0)$.
   b) Hence estimate the gradient of the curve at
      the point $(3, 0)$.

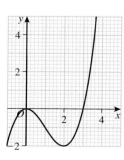

2 Differentiate $x^3 + 5x^2$.

3 Find $\dfrac{dy}{dx}$ if $y = x^2 + 8x - 3$.

4 Differentiate $\dfrac{x^3}{9} + \dfrac{3x^2}{4} - \dfrac{2x}{5}$.

5 Find $\dfrac{dy}{dx}$ if $y = 4x^2 + 10x + \dfrac{1}{x}$.

6 Differentiate $\dfrac{2}{x} + \dfrac{1}{x^2}$.

7 Find an expression for the gradient of $y = \dfrac{x^3 - 5x^2 + 4 + 7}{10}$.

8 Find the gradient of $y = 5x^3 - 12x + 3$ at the point where $x = 2$.

9 Expand and differentiate $(x + 3)(2x + 5)$.

10 Find the gradient of $y = 8x^4 - 12x^3 + 7x$ at the point $(1, 3)$.

11 Expand and differentiate $(2x - 3)(3x - 1)$.

12 A curve has equation $y = x^2 - 10x + 1$.
   a) Find an expression for $\dfrac{dy}{dx}$.
   b) Work out the x coordinate of the minimum point on this curve.
   c) Work out the y coordinate of the minimum point on the curve.

13 A curve has equation $y = x^2 - 10x + 4$.
   a) Find an expression for $\dfrac{dy}{dx}$.
   b) Find the value of x for which $\dfrac{dy}{dx} = 0$.
   c) Obtain the coordinates of the stationary point of the curve $y = x^2 - 10x + 4$.

14 A curve has equation $y = 5 + 4x - x^2$.
   a) Find an expression for $\dfrac{dy}{dx}$.
   b) Find the coordinates of the stationary point of the curve $y = 5 + 4x - x^2$.
   c) Say whether this stationary point is a maximum or a minimum.
       Give a reason for your answer.

15 A curve has the equation $y = (x + 3)(2x - 5)$.
   a) Expand the brackets, and hence find an expression for $\dfrac{dy}{dx}$.
   b) Find the value of x for which $\dfrac{dy}{dx} = 0$.
   c) When the graph of $y = (x + 3)(2x - 5)$ is plotted,
       the result is a parabola.
       Explain briefly whether the graph looks like
       parabola A or parabola B.

Parabola A   Parabola B

16 Two quantities P and z are related by the equation $P = z^2(z - 6)$.
   a) Find an expression for $\dfrac{dP}{dz}$.
   b) Find the two values of z for which $\dfrac{dP}{dz} = 0$.
   c) Find the two corresponding values of P.

17 Two quantities E and r are related by the equation $E = 3r^2 - 24r + 50$.
   Find the minimum possible value of E.

18 The displacement s metres of a particle at time t seconds is given by the equation $s = t^3 + 5t + 3$.
   a) Find the velocity of the particle after 2 seconds.
   b) Find the acceleration of the particle after 5 seconds.
   State the units of your answers clearly.

19 The displacement s metres of a particle at time t seconds is given by the equation $s = 4t^2 - 4t + 1$.
   a) Find the velocity of the particle after 5 seconds.
   b) At what time is the velocity equal to zero?

20 The velocity v metres per second of a particle at time t seconds is given by the equation
   $v = 2t^2 - 12t + 20$.
   a) Find the velocity of the particle at time 4 seconds.
   b) Find the acceleration of the particle after 2 seconds.
   c) Find the minimum value of the velocity, giving the time at which this occurs.

21 A curve has equation $y = x^2 - 4x + 1$.
   a) For this curve find:
      (i) $\dfrac{dy}{dx}$ (ii) the coordinates of the turning point.
   b) State, with a reason, whether the turning point is a maximum or a minimum.
   c) Find the equation of the line of symmetry of the curve $y = x^2 - 4x + 1$.   [Edexcel]

22 A body is moving in a straight line which passes through a fixed point O.
   The displacement, s metres, of the body from O at time t seconds is given by
   $s = t^3 + 4t^2 - 5t$.

   a) Find an expression for the velocity, v m/s, at time t seconds.
   b) Find the acceleration after 2 seconds.   [Edexcel]

Key points

1 The gradient of a curve at a point may be estimated by drawing a tangent at that point; this is a straight line that touches the curve. The gradient of the tangent (rise ÷ run) is used as an estimate of the gradient of the curve.

2 If the equation of a curve is known, then the gradient may be found using differentiation. After differentiating the equation, substitute the x coordinate of the point into the result.

3 The general rule for differentiating powers of x is

$$y = ax^n \text{ differentiates to } \frac{dy}{dx} = anx^{n-1}$$

Some examples:

y	$\dfrac{dy}{dx}$
$4x^2 + 10x$	$8x + 10$
$x^3 + 10x^2 + 25x + 3$	$3x^2 + 20x + 25$
$\dfrac{1}{x}$	$\dfrac{1}{x^2}$
$\dfrac{1}{x^2}$	$\dfrac{2}{x^3}$
$\dfrac{2x^3 - 5x^2 - 4x + 1}{5}$	$\dfrac{6x^2 - 10x - 4}{5}$

4 Differentiation is used for finding stationary points of graphs, which occur when $\frac{dy}{dx} = 0$.

Maximum points and minimum points, which are known as turning points, are, therefore, types of stationary point.

5 If the displacement s of a particle at time t is given algebraically, then differentiation may be used to find the velocity v and the acceleration a:

$$v = \frac{ds}{dt} \quad \text{and} \quad a = \frac{dv}{dt}$$

For example, if $s = t^3 + 6t^2 - 5t$

then $\quad v = \dfrac{ds}{dt} = 3t^2 + 12t - 5$

and $\quad a = \dfrac{dv}{dt} = 6t + 12$

6 Differentiation is part of a branch of mathematics called calculus. It is a central idea in the first year of an A-level or IB mathematics course.

Internet Challenge 28

Calculus crossword

The answers to this crossword are key words and people in basic calculus; you have met some of these in this chapter, but you may need the internet to help with others. When you have completed the ten across clues, the shaded squares will spell another name for the result of carrying out a differentiation.

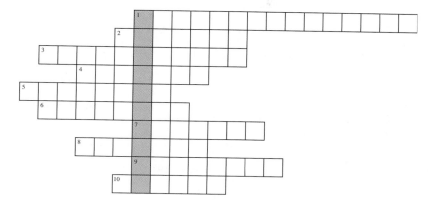

1 The process of finding the gradient function.

2 Gottfried Wilhelm von _____, German mathematician and co-discoverer of calculus.

3 Reverse process of 1 across, often used to find areas or volumes.

4 Turning point at which a curve has its greatest value.

5 If the function $y = ax^2 + bx + c$ has a *minimum* value, then the value of a will be this kind of number.

6 This kind of quantity has a gradient function of zero.

7 A straight line that touches a curve without crossing it.

8 Turning point at which a curve has its least value.

9 The gradient function of distance, with respect to time.

10 English mathematician, co-discoverer of calculus along with 2 across.

Index